Sustainable Development Goals Series

The Sustainable Development Goals Series is Springer Nature's inaugural cross-imprint book series that addresses and supports the United Nations' seventeen Sustainable Development Goals. The series fosters comprehensive research focused on these global targets and endeavors to address some of society's grand challenges. The SDGs are inherently multidisciplinary, and they bring people working across different fields together toward a common goal. In this spirit, the Sustainable Development Goals series is the first at Springer Nature to publish books under both the Springer and Palgrave Macmillan imprints, bringing the strengths of our imprints together.

The Sustainable Development Goals Series is organized into eighteen subseries: one subseries based around each of the seventeen respective Sustainable Development Goals, and an eighteenth subseries, "Connecting the Goals," which serves as a home for volumes addressing multiple goals or studying the SDGs as a whole. Each subseries is guided by an expert Subseries Advisor with years or decades of experience studying and addressing core components of their respective SDG.

The SDG Series has a remit as broad as the SDGs themselves, and contributions are welcome from scientists, academics, policymakers, and researchers working in fields related to any of the seventeen goals. If you are interested in contributing a monograph or curated volume to the series, please contact the Publishers: Zachary Romano [Springer; zachary.romano@springer.com] and Rachael Ballard [Palgrave Macmillan; rachael.ballard@palgrave.com].

More information about this series at http://www.springer.com/series/15486

Godwell Nhamo
David Chikodzi • Kaitano Dube
Editors

Sustainable Development Goals for Society Vol. 2

Food security, energy, climate action and biodiversity

Springer

Editors
Godwell Nhamo
Chief Researcher & Exxaro Chair in Business and Climate Change
Institute for Corporate Citizenship, University of South Africa
Pretoria, South Africa

David Chikodzi
Postdoctoral Fellow: Exxaro Chair in Business and Climate Change
Institute for Corporate Citizenship, University of South Africa
Pretoria, South Africa

Kaitano Dube
Department of Ecotourism Management
Vaal University of Technology
Vanderbijlpark, South Africa

ISSN 2523-3084 ISSN 2523-3092 (electronic)
Sustainable Development Goals Series
ISBN 978-3-030-70951-8 ISBN 978-3-030-70952-5 (eBook)
https://doi.org/10.1007/978-3-030-70952-5

This Springer imprint is published by the registered company Springer Nature Switzerland AG
The registered company address is: Gewerbestrasse 11, 6330 Cham, Switzerland

Preface

This book is part of the Sustainable Development Goals (SDGs) for Society book series, which comes in two volumes. This particular volume (Volume II) addresses current issues on four various themes: food security, energy, climate action and biodiversity. This volume is organised into five major parts with 19 chapters. The five parts are: (1) introduction and background; (2) food security and sustainable energy; (3) climate action for SDGs; (4) health, water and biodiversity engagements; and (5) conclusion and recommendations. The 2030 Agenda for Sustainable Development that enshrines the 17 interwoven SDGs, their 169 targets and many more indicators is broad. It covers a wide range of goals, including poverty eradication, economic growth, social inclusion, environmental sustainability, peace and partnerships for all by the year 2030. Post 2015, the world is working towards meeting the SDGs. Across the globe, the challenge of domesticating and localising the SDGs in terms of national and local development priorities requires a combination of technical, scientific, administrative and political input. It is clear that a collaborative research approach is needed to stay true to the SDGs' inclusive and bottom-up approach. Of particular interest is the notion that the SDGs represent a development agenda that should be realised by both the developed and developing countries. This provides researchers across and within disciplines with endless novel opportunities to engage with the SDGs, especially at societal levels. Given the foregone, the SDGs remain, therefore, an agenda for society. The world has gone past four years of SDGs implementation, and 2030 is fast approaching. To this end, it is high time to take stock and report what is on the ground to inform further implementation and to scale up implementation activities going forward. The 2030 Agenda for Sustainable Development is also facing severe funding challenges given that it has been disrupted by the Covid-19 pandemic that took centre stage in 2020. Hence, this book also assists in reminding key stakeholders that the SDGs agenda should remain the main agenda. To this end, the cases documented remain valuable as a contribution in highlighting what had been happening in societies relative to the implementation of the SDGs.

Pretoria, South Africa — Godwell Nhamo
Pretoria, South Africa — David Chikodzi
Vanderbijlpark, South Africa — Kaitano Dube

Acknowledgements

The editors, Prof Godwell Nhamo, Dr David Chikodzi and Dr Kaitano Dube wish to thank all the blind peer reviewers for their invaluable inputs during the writing and publishing process of the book. We thank Springer for taking this book project on board and for a product of high standard and quality. We also wish to extend gratitude to our families for their ongoing support of our work. The book project was coordinated through the Exxaro Chair in Business and Climate Change, a research chair that is funded by the Exxaro Resources (Pty) Ltd Chairman's Fund and hosted by the Institute for Corporate Citizenship at the University of South Africa. The Exxaro Chair was established in 2008, and is currently in its fourth term ending in 2022.

This book is double-blind peer-reviewed. Apart from this being the international best practice norm, this double-blind peer-review process is mandatory for South Africa–based authors to fulfil the requirements of the Department of Higher Education and Training's (DHET) policy for recognised research outputs for subsidy purposes. The authors invested their time to incorporate observations from the blind peer-review process, an aspect that enhanced the quality of the product.

Contents

Part I Introduction and Background

1 Making SDGs Work to End Hunger, Sustain Energy, Resolve Climate Change, and Reverse Biodiversity Loss 3
Kaitano Dube, David Chikodzi, and Godwell Nhamo

Part II Food Security and Sustainable Energy

2 The Contribution of Responsible Leadership in Raising Funding to Support Organisational Mandate and the SDGs: Case of the Land Bank of South Africa 19
Carolien Samson and Dawie (D.A.J.) Bornman

3 Confronting Poverty, Hunger, and Food Insecurity: Lessons from Malawi and Zimbabwe 33
Mavis Thokozile Macheka and Gift Wasambo Kayira

4 Preventing Fall Armyworm (*Spodoptera frugiperda* JE Smith) Damage in Maize by Altering Planting Time and Using Varied Genotypes 47
Leonard Nyabanga, Ronald Mandumbu, Joyful T. Rugare, Never Mafuse, Emmanuel Zivenge, Handsen Tibugari, George Nyamadzawo, and Christopher T. Gadzirayi

5 Enhancing Urban Farming for Sustainable Development Through Sustainable Development Goals 63
Nyasha Chaminuka, Ernest Dube, Itai Kabonga, and Smart Mhembwe

6 Water and Sanitation Access in the Shamva District for Sustainability and Development of the Zimbabwean Smallholder Farming Sector 79
Theresa Tendai Rubhara and Olawaseun Samuel Oduniyi

7 **Responsible Leadership and the Implementation of SDG 7: The Case of the UNDP Botswana Biogas Project** 91
Newton Tawanda Runyowa and Willem Fourie

8 **Elements of Responsible Leadership in Driving Climate Action (SDG 13)** 107
Karien Erasmus and Yolande Steenkamp

9 **Leadership Capabilities for Successful Implementation of SDG 7 Targets at Energy Company X** 123
Tatenda Mangondo and Andani Thakhathi

10 **Designing Effective Social Protection for Food and Nutrition Security Among Farm Workers: Lessons from Masvingo, Zimbabwe** 139
Joseph Tinarwo

Part III Climate Action for SDGs

11 **Mitigating Climate Change Through Carbon Sequestration for Sustainable Development: Empirical Evidence from Cameroon's Forest Economy** 155
Ernest L. Molua

12 **Private Sector Sustainable Development Goals' Localisation: Case of Kruger Mpumalanga International Airport, South Africa** 177
Kaitano Dube and Godwell Nhamo

13 **Scaling up University Engagement with the Water SDG for General Environmental Stewardship and Climate Change Resilience** 191
Godwell Nhamo and David Chikodzi

14 **Climate Change in Zimbabwe's Vulnerable Communities: A Case Study of Supporting Enhanced Climate Action Project (SECA Project) in Bulilima District** 211
Veronica Nonhlanhla Gundu-Jakarasi and Justice Nhidza

15 **Climate Resilience Strategies and Livelihood Development in Dry Regions of Zimbabwe** 225
Leonard Chitongo

16 **Climate Action at International Airports: An Analysis of the Airport Carbon Accreditation Programme** 237
Kaitano Dube

Part IV Health, Water and Biodiversity Engagements

17 Protected Areas Interventions and SDGs: The Case of Bolsa Floresta Programme in the Brazilian Amazon 255
Anne-Elisabeth Laques, Ana I. R. Cabral, Romero Gomes Pereira da Silva, and Carlos Hiroo Saito

18 Implementation of the SDGs Through Greening Household Responses to Water, Energy and Food Shortages in Newlands West, Durban 271
Muchaiteyi Togo and Hirshwyn B. Arulappan

Part V Conclusion and policy recommendations

19 Summary of Findings, Conclusions and Policy Recommendations 287
David Chikodzi, Kaitano Dube, and Godwell Nhamo

Correction to: Preventing Fall Armyworm (Spodoptera Frugiperda JE Smith) Damage in Maize by Altering Planting Time and Using Varied Genotypes C1

Index 301

About the Editors

Godwell Nhamo is Full Professor and Exxaro Chair in Business and Climate Change at the University of South Africa (UNISA), South Africa. He is a National Research Foundation (NRF) C-Rated researcher in the fields of Climate Change and Governance, Green Economy and Sustainable Development. He holds a PhD from Rhodes University (South Africa), an MSc from the University of Botswana (Botswana) and a BSc (Honours) from the University of Zimbabwe (Zimbabwe).

David Chikodzi is a post-doctoral fellow with the Exxaro Chair in Business and Climate Change at the University of South Africa. He holds a PhD in Geography and Environment Science from the University of the Western Cape (South Africa). He also holds a master's degree in Environmental Policy and Planning, as well as a Bachelor of Arts (Honours) Degree in Geography from the University of Zimbabwe (Zimbabwe).

Kaitano Dube is an Ecotourism Management Lecturer at Vaal University of Technology (South Africa). He is a tourism geographer researching in the area of tourism climate change and sustainability. He holds a PhD and an MSc from the University of South Africa (South Africa). He graduated with a BSc (Honours) from Midlands State University in Gweru (Zimbabwe).

Contributing Authors

Godwell Nhamo, PhD (Editor-in-Chief and Accounting Officer), is Full Professor and Exxaro Chair in Business and Climate Change at the University of South Africa (UNISA). He is a National Research Foundation (NRF) C-Rated researcher undertaking research in the fields of climate change and governance, sustainable tourism, green economy and sustainable development. Prof Nhamo has conceptualised and completed 11 book projects. The most recent being: *Counting the Cost of COVID-19 on the Global Tourism Industry*, by Springer (2020); *Scaling Up SDGs Implementation: Emerging Cases from the State, Development and Private Sectors*, published by Springer (2020); *SDGs and Institutions of Higher Education*, published by Springer (2020); and *SDG 7 – Ensure Access to Affordable, Reliable, Sustainable and Morden Energy*, by Emerald (2020). Prof Nhamo has also published over 90 journal articles. Since 2013, Prof Nhamo has graduated 11 PhDs students and hosted 10 postdoctoral fellows. Currently, Prof Nhamo is leading a mega research project on cyclones, tornados and floods in the era of SDGs in Southern Africa. Prof Nhamo also sits in a number of both international and national boards and has received several awards and recognitions for his outstanding work both locally and internationally. Finally, Prof Nhamo was one of the four-member African Union High-Level Panel drafting the Green Innovation Framework for the continent. Email: nhamog@unisa.ac.za

David Chikodzi, PhD (Co-Editor), is currently a postdoctoral fellow with the Exxaro Chair in Business and Climate Change at the University of South Africa. His research interests are in climate change, water resources management, tourism, sustainable development and application of Earth Observation technologies for societal benefit. He has worked for over 10 years in academia at Great Zimbabwe University (Zimbabwe) and has published over 30 journal articles and book chapters. Dr Chikodzi has also taken part in several local and internationally funded research projects across Southern Africa and has previously worked as a Research Scientist at the Scientific and Industrial Research and Development Centre in Zimbabwe. Dr Chikodzi is a former member of the ISIbalo Africa Young Statisticians and the Zimbabwe Young Academy of Sciences (ZIMYAS). Email: chikod@unisa.ac.za

Kaitano Dube, PhD (Co-Editor), is an Ecotourism Management Lecturer at the Vaal University of Technology, South Africa. He is one of Africa's leading tourism geographers researching in the area of tourism, climate change, sustainability and green aviation. He has published in high-impact, international, peer-reviewed journals, with his work receiving global attention. His work has received extensive media coverage, including in National Geographic, Wunderground, Atlasobscura and Agence France-Presse (AFP), to mention but a few. He has granted several international television and radio interviews. Dr Dube holds a PhD and an MSc from the University of South Africa. He graduated with a BSc (Honours) from Midlands State University in Gweru, Zimbabwe. He holds several other qualifications from UNISA, Vaal University of Technology and University of the Witwatersrand Business School in Johannesburg, South Africa. Dr Dube is an executive member of the Tourism Educators Association of South Africa and Tourism Sector Human Resource Development Governance and Institutional Coordination Forum hosted by South Africa's National Department of Tourism. Email: kaitanod@vut.ac.za

Ana I. R. Cabral, PhD is a senior researcher at Forest Research Centre, School of Agriculture, University of Lisbon, Portugal. She is a Remote Sensing and Geographic Information System expert, working in tropical regions, mainly in African countries and Brazil. Her main areas of research are deforestation and forest degradation, mapping and modelling of land cover/land-use change scenarios, quantification of carbon emissions and landscape fragmentation. Her most recent articles have been published in *Applied Geography*, *Human Ecology*, *ISPSR Journal of Photogrammetry and Remote Sensing* and *Swarm and Evolutionary Computation*. anaicabral70@gmail.com

Anne-Elisabeth Laques, PhD is a geographer and an expert in landscape analysis. She uses satellite images at different scales to define landscape indicators, markers of socio-environmental dynamics. In Brazil, she coordinated a multidisciplinary knowledge-sharing project to help control the effects of spatial events on human health (dam, urbanisation, deforestation, health policies, etc.). Currently, her research focuses on landscape indicators capable of assessing the impact of public policies in the field of environmental protection. She is a senior research fellow at the IRD – Institut de Recherche pour le Développement, France, and is linked to UMR ESPACE-DEV. She is also coordinator of the Centre of Scientific Competence on Landscape CES/Centre d'Expertise Scientifique "Paysage". Email: anne-elisabeth.laques@ird.fr

Andani Thakhathi, PhD is currently an SSAUF nGAP Lecturer in Business, Strategic and Responsible Management at the University of Pretoria, South Africa, Department of Business Management in the Faculty of Economic and Management Sciences. He holds a PhD in Ethics and Responsible Leadership in Business obtained through the Wittenberg Center for Global Ethics and the Martin-Luther-Universität of Halle-Wittenberg in Germany. Outside of academia, Andani has worked as a leadership development consultant for the African Leadership Group, where he led interventions on leadership development for corporate management and employees in the state-owned and private sector. He is an inspired researcher who believes that all knowledge should help uplift humanity. Dr Thakhathi's research interests lie in what he refers to as "Transcendent Development"; a novel holistic approach to realising well-being in the world's developing, underdeveloped and least developed in the twenty-first century. Email: andani.thakhathi@up.ac.za

Carlos Hiroo Saito, PhD is a biologist working in interdisciplinary research involving environmental education, spatial analysis and water security. He works with conceptual modelling for science literacy in a systemic approach, and how the understanding of social and environmental processes can strengthen social participation. He is a research fellow of CNPq, and executive coordinator of the National Institute of Science and Technology (INCT-Observatory of Socio-Environmental Dynamics). He is a full professor at the University of Brasilia, Brazil, linked to the Department of Ecology/Institute of Biological Sciences and the Centre for Sustainable Development. He is a researcher associated with UMR Espace-DEV, Montpellier, France. Email: carlos.h.saito@hotmail.com

Carolien Samson started her career in policy-making roles in the South African government. A stint working on land reform redistribution policy resulted in a long-term move to focus on agricultural finance both in development finance and commercial banking environments. She is a banking generalist with experience in product development, management reporting, risk management and business strategy. Her current focus is on the development and implementation of social and environmental policies and practices within financial institutions. She has degrees in Economics and Law as well as Town and Regional Planning and formed part of the inaugural cohort of students who completed the Master's in Development Practice offered by the University of Pretoria in South Africa.

Christopher T. Gadzirayi is Associate Professor in the Department of Agricultural Economics and Extension at Bindura University of Science Education. He has more than 55 research and review articles in peer-reviewed journals.

Emmanuel Zivenge holds a PhD in Agricultural Economics and is a lecturer in the Department of Agricultural Economics, Extension and Education at Bindura University of Science Education. He has 7 publications in peer-reviewed journals

Ernest Dube, PhD is a senior lecturer and researcher for Development Sciences at Marondera University of Agricultural Sciences and Technology (MUAST), Zimbabwe. Dr Dube has an interest in the fields of disaster risk reduction and climate change, poverty and livelihoods, as well as food security and sustainable development. Dr Dube published extensively in his areas of interest and has done review work for more than 15 journal publishers. He has been in the academic field for more than 8 years. Amongst Dr Dube's research output are over 15 publications in internationally accredited journals. Dr Dube is also a consultant in disaster risk reduction and development related issues. He has also carried out funded research projects to completion. Email: edube@muast.ac.zw

Ernest L. Molua, PhD is a professor of Agricultural Economics in the Faculty of Agriculture and Veterinary Medicine, University of Buea, Cameroon. He obtained academic degrees from the Georg-August University, Göttingen, Germany and the Royal Veterinary and Agricultural University, Copenhagen, Denmark. He is a member of the College of Research Associates of the United Nations University Institute for Natural Resources in Africa, a continent-wide network of senior research scientists for natural resources management. A Fulbright Research Fellow at Yale University, USA, he also serves as a visiting professor to the United Nations Institute for Economic Development and Planning, Dakar, Senegal. emolua@yahoo.com

George Nyamadzawo is a full professor in the department of Environmental Sciences at Bindura University of Science education and holds a PhD in Environmental Science. He has more than 50 research articles in peer-reviewed journals.

Gift Wasambo Kayira, PhD lectures at the History Department of the University of Malawi, Chancellor College. His research focuses on histories of development and poverty in the twentieth and twenty-first centuries in Malawi. He has published in the *Journal of Eastern African Studies*, *Journal of Public Administration and Development Alternatives* and the *African Studies Quarterly*, among others. He also has some book chapters to his credit. Email: gkayira@cc.ac.mw

Handsen Tibugari holds a PhD in Crop Science and is a lecturer at Gwanda state University. He has more than 25 research articles in peer-reviewed journals.

Hirshwyn B. Arulappan holds a Bachelor of Science (Honours) degree in Environmental Management from the University of South Africa. At the time of his studies his research focused on household responses to basic resource shortages as a result of climate changes. He currently works in the environmental monitoring and management sector which focuses on industries operating listed activities and assisting them to meet environmental compliance obligations. He is also a specialist in Air Quality Management and is currently a member of NACA and SIAOH. He holds Legal Knowledge Certification for Occupational Hygiene obtained from North-West University.

Itai Kabonga holds academic qualifications in Development Studies, Sociology, Project Management and Monitoring and Evaluation (M and E). He has also taught several courses in Development Studies that include Non-

Governmental Organisations (NGOs) and Civil Society in Africa, Gender Studies and International Development Corporations at Zimbabwe Ezekiel Guti University (ZEGU) and ARRUPE Jesuit University in Zimbabwe. Itai's research interests include NGOs and development, civil society-state relations, volunteerism in Africa and Zimbabwe's development crisis since 1980. Focusing on these areas, Itai has published 10 articles in reputable journals. Email: vakabonga@gmail.com

Joseph Tinarwo, PhD is an experienced Social Scientist and Researcher with more than 10 years of demonstrated experience in public policy analysis, food and nutrition security, and governance. He is currently the coordinator and lecturer of the Politics and Public Management Programme at Great Zimbabwe University in Zimbabwe and is finalising a PhD in Public Management and Governance with the University of Johannesburg in South Africa, researching on food policy system in Zimbabwe. Joseph has presented research and policy-relevant papers in both national and international conferences and has published in peer-reviewed journals. Over the years, Joseph has been actively involved in various assignments as a consultant to the Government of Zimbabwe and international development partners in the fields of agriculture and food security, health and development sector, with key result areas including research, knowledge management institutional capacity assessments, training, designing and managing special studies, as well as data management.

Joyful T. Rugare holds a PhD in Agronomy and is a senior lecturer in the Department of Crop Science at the University of Zimbabwe. He has more than 30 publications in peer-reviewed journals.

Justice Nhidza has a Bachelor of Science Education in Natural Resources Management from Bindura University of Science and Technology, Zimbabwe. He is a United Nations Volunteer in Climate Change Issues as well as the National researcher at Young Volunteers for the Environment organisation. His research interests are in the fields of climate change, disaster risk management, poverty and food security and sustainable development. Email: juniornhidzah@gmail.com

Karien Erasmus is a principal climate change advisor and holds a master's degree in Development Practice from the University of Pretoria. She has 15 years' professional experience in the development, sustainability and climate change fields. Karien started her career as a development planner working widely in Africa and South Africa on strategic development projects. Her additional postgraduate qualifications include diplomas in community development and mine closure and ecological rehabilitation. Karien joined Promethium Carbon in 2015, and leads climate change risk and vulnerability assessments, climate-change-related policy development and adaptation and mitigation plans for various sectors. Her interests include localised climate change adaptation, community resilience, the socio-economic context of climate change vulnerability and renewable energy. Karien has worked passionately on the "land, community and energy" nexus concept over the past three years. This nexus concept integrates mine land rehabilitation with agriculture, biomass production and local communities through renewable energy generation. Email: karien@promethium.co.za

Leonard Chitongo, PhD is a hardworking and self-motivated individual and always excited to face new challenges in his academic and professional career. He is a postdoctoral research fellow under SARChI Chair in Sustainable Local (Rural) Livelihoods in the School of Management, IT and Governance at the University of Kwazulu-Natal, South Africa. He has a strong interest in researching on issues that affect people's livelihoods. To date he has published several articles on rural and urban resilience, housing, livelihoods and public policy. Email: vachitongo@gmail.com

Leonard Nyabanga holds an MSc in Food Security and Sustainable Agriculture at Bindura University of Science Education.

Mavis Thokozile Macheka, PhD is a lecturer for Development Studies in the Department of Archaeology, History and Development Studies at Great Zimbabwe University. She is also a part-time lecturer at Zimbabwe Open University. Dr Macheka's teaching and research straddle the fields of rights of vulnerable groups of the society, political ecology, community development, livelihoods and sustainability, agriculture and development, human rights and development. To date, she has published with *Journal of Cultural Heritage Management and Sustainable Development* and *South African Geographical Journal*, among others. Email: mavythoko@gmail.com

Muchaiteyi Togo holds a PhD from Rhodes University (South Africa). She works for UNISA in the Environmental Science Department. Dr Togo's current research focusses on university responses to SDGs, household greening and climate change coping strategies (with a special focus on the water-energy-food nexus). Dr Togo's publications include a resource book, journal articles, peer-reviewed book chapters and conference proceedings on the same topics. She currently lectures a green economy module at the honours level and supervises honours, master's and PhD students in Environmental Sciences. Dr Togo has done research for international organisations like UNESCO, UNEP, the AAU, SARUA and the SADC-REEP and several South African organisations as a consultant. Since 2010 to early 2020, Dr Togo has been part of the *Southern African Journal for Environmental Education* editorial collective. She was a council member of the Environmental Education Association of Southern Africa holding the publications portfolio from 2016 to 2019. togom@unisa.ac.za

Never Mafuse holds a PhD in Agricultural Economics and is a lecturer at Bindura University of Science Education. He has more than 15 publications in peer-reviewed journals.

Newton Tawanda Runyowa is a development practitioner keenly interested in the Sustainable Development Agenda and Responsible Leadership Theory. A cyclist, health system strengthening professional and avid reader, Newton is a lifelong student of history, politics and philosophy. Having graduated recently with an MPhil in Development Practice from the University of Pretoria, South Africa, he is now dedicating his life to consciously and positively contributing to the achievement of an inclusive, socially just and sustainably developed Africa and the world. Email: ntrunyowa@gmail.com

Nyasha Chaminuka, Mr is a lecturer and researcher at Zimbabwe Open University (ZOU). His research interests are in urban agriculture, food security, disaster risk reduction and gender-based violence. Mr Chaminuka has published in the field of urban agriculture and has more than three publications in internationally accredited journals. He has been in the academic field for over seven years. Email: chaminukan2015@gmail.com

Olawaseun Samuel Oduniyi, PhD is a distinguished scholar and results-oriented researcher who is currently working at the University of South Africa. Samuel uses applied econometrics, spatial data, interdisciplinary approaches, and insights from behavioural economics to study issues related to agriculture and environmental and resource economics. His research focuses on agricultural sustainability, impact assessment, developmental economics, agricultural economics, food security and household welfare, environmental management, rural livelihood, climate change economic, adaptation and resilience. He has published more than 20 peer-reviewed papers, served as an editor for peer-reviewed journals and supervised many postgraduate students. Samuel has collaborated with researchers in about three countries to date.

Romero Gomes Pereira da Silva, PhD is a forest engineer and researcher at the Centre for Sustainable Development at the University of Brasília, Brazil. He is interested in geographic information systems and remote sensing applied to the planning and management of protected areas (conservation units and indigenous lands) and territorial planning and ordering with a focus on urban green spaces. Email: romerogomes1@hotmail.com

Ronald Mandumbu is Associate professor in the Department of Crop Science at Bindura University of Science Education. He holds a PhD in Crop Science and has more than 50 research and review articles in peer-reviewed journals.

Smart Mhembwe is a lecturer in the Gender Institute at Midlands State University (MSU). Mhembwe has an interest in rural development, gender, poverty and livelihoods as well as food security. He has been in the academic field for more than 6 years now and has done review work for more than 5 journal publishers. He has also carried out funded research projects to completion. Email: mhembwes@staff.msu.ac.zw

Tatenda Mangondo, Ms is currently a project administrator in the project preparation unit at the Independent Power Producers Office (IPPO). She holds a Master of Philosophy (MPhil) in Development Practice obtained from the Albert Luthuli Centre for Responsible Leadership at the University of Pretoria. Tatenda's career interests are in international development with a focus and passionate interest in sustainable development, her centre of attention being on quality education, gender equality and affordable and clean energy. Email: mangondotg@gmail.com

Theresa Tendai Rubhara, PhD is a post-doctoral researcher with the University of South Africa. She holds a PhD in Food Security from the University of KwaZulu Natal, South Africa. Emanating from the sub-Saharan region, Theresa's research interests are inspired by the need for sustainable livelihoods, improved food security and rural development in the region. Her research focus is agriculture as the main mechanism for food production. She is specifically interested in the economics of food production, distribution, policy framework influencing food security, agrarian change, climate change, environmental management and rural development. She has recently published several journal articles under the mentioned themes. Theresa is also interested in other developmental aspects of education, health, water and sanitation.

Veronica Nonhlanhla Gundu-Jakarasi, Mrs is the manager for climate finance and sustainability at the Infrastructure Development Bank of Zimbabwe (IDBZ). She is a seasoned climate change negotiator for Zimbabwe under the United Nations Convention on Climate Change (UNFCCC) and a reviewer for Annex 1 countries greenhouse gas inventories in the waste sector. Her areas of research interest are climate change, sustainability, development finance and green growth. She has published in the study fields of climate finance, climate change adaptation and disaster risk management and resilience. She is a seasoned trainer in climate-change-related issues, including climate diplomacy, climate finance, green procurement, and gender and climate change. She has over 12 years of experience in climate change, environment and natural resources management. Jakarasi is also a USA Techwomen 2017 Fellow and an Africa Climate Leadership Programme Fellow. Email: verogundu@gmail.com

Willem Fourie, PhD established the South African SDG Hub at the University of Pretoria, as well as the SDG-focussed Master's in Development Practice. He is an Alexander von Humboldt fellow and has published numerous peer-reviewed articles and chapters. Email: willemfourie@gmail.com

Yolande Steenkamp, PhD is a senior researcher at the Albert Luthuli Leadership Institute at the University of Pretoria, South Africa. Her research approach is largely interdisciplinary and addresses contemporary challenges. She contributes from a theological/philosophical perspective to questions of ethics, meaning and value in leadership, and explores the ethical dimensions of the imagination, narrative and hospitality as response to diversity in transcultural engagement. She is interested in how narrative and storytelling impacts on encounters with racial, economic, sexual, cultural and religious otherness, and investigates how this may address inequality, justice and development in post-colonial countries. She collaborates with Zeppelin University in Germany on two projects. The first studies leadership in multi-sector partnerships and the SDGs through the project Relational Leadership for the SDGs: Implications for a developed and a developing economy, while the second studies transcultural leadership and transcultural competence. Email: yolande.steenkamp@up.ac.za

Part I

Introduction and Background

1 Making SDGs Work to End Hunger, Sustain Energy, Resolve Climate Change, and Reverse Biodiversity Loss

Kaitano Dube, David Chikodzi, and Godwell Nhamo

Abstract

The adoption of the 17 ambitious global sustainable development goals (SDGs) and the 169 targets in September 2015 aimed to better the global society. The year 2020 marks 5 years post-adoption of these SDGs and signals the last decade in which the world is set to take stock of the progress made. This chapter sets the background and tone for the book that examines how society is interacting with some of the SDGs. A healthy society is one that is free of hunger (SDG 2), depends on clean and sustainable energy (SDG 7), and allows society to drive its socioeconomic desires and aspirations. Given the interconnectedness of society to the environment, there is a need to ensure healthy biodiversity (SDG 15) to support the thriving society through a variety of ecosystem services, including the ocean economy (SDG 14). However, anthropogenic climate change remains one of the biggest threats to human civilization and society's aspirations. Society, in many ways, drive climate change through activities that release greenhouse gas emissions and drive global warming, which is the biggest driver of climate change. This chapter tackles some of the crucial talking points and debates that will be further developed in various chapters of the book. It highlights significant challenges and discusses some initiatives on ending hunger, energy, biodiversity, water, and sanitation (SDG 6) and climate change action (SDG 13).

Keywords

SDGs · Climate change · Energy security · Renewable energy · Food security · Biodiversity

1 Introduction

The world has been suffering a myriad of environmental challenges attributed to global environmental change that has intensified over the past decade. The environmental challenges continue to raise concerns with fears that the biophysical environment is facing an imminent collapse (Sato and Lindenmayer 2018). There are also fears that such challenges will undermine the global society's quest for development and

K. Dube (✉)
Department of Hospitality, Tourism and PR, Vaal University of Technology, Vanderbijlpark, South Africa
e-mail: kaitanod@vut.ac.za

D. Chikodzi · G. Nhamo
Exxaro Chair in Business and Climate Change, Institute for Corporate Citizenship, University of South Africa, Pretoria, South Africa
e-mail: chikod@unisa.ac.za; nhamog@unisa.ac.za

G. Nhamo et al. (eds.), *Sustainable Development Goals for Society Vol. 2*, Sustainable Development Goals Series, https://doi.org/10.1007/978-3-030-70952-5_1

spark social unrest. The increase in global population has been partly to blame as it has led to increased demand for consumption (Infante-Amate et al. 2018; Kjaer et al. 2019). The growth in human population has led to a demand in more resources such as agricultural, industrial, and residential land, and other critical resources such as water and energy, whose extraction and exploitation result in environmental degradation and pollution.

The increase in global consumption due to increased human affluence has also been blamed for the environmental challenges facing the world today (Iizuka and Zanlungo 2016; Aydin et al. 2019; Wiedmann et al. 2020). Hence, the envisaged threat and the need to ensure sustainability forced global leaders to meet and adopt the 17 ambitious SDGs aimed at finding solutions to socioeconomic, environmental, and political challenges facing the world. According to the United Nations (2015), the 2030 Agenda for Sustainable Development that encompasses the SDGs and 169 targets aim to develop a plan of action for people, planet, and prosperity.

The boom in global population has been associated with an almost equally matched general deterioration in the quality of the environment (Weber and Sciubba 2019). Of key concern has been the disruption to the biophysical environment because of land pollution, water pollution both inland and ocean, and air pollution, which have threatened human civilization as already known. The degradation of the biosphere threatens ecological biodiversity and agricultural production, which have severe implications on the society. A critical concern across the world has been the impacts of the increasing anthropogenic air pollution, particularly that of greenhouse gases (GHG) that cause global warming, a key driver of climate change (Dube and Nhamo 2020a). Among the most significant contributors to global GHG emissions are the energy, agricultural, industrial, and transportation sectors. Figure 1.1 shows some of the sectors of interest.

Until the advent of COVID-19, the global economic growth was matched by an increase in GHG emissions with the amount of global carbon emissions rising from 9Gt in 1960 to about 35.3 Gt in 2018 (World Resources Institute 2019). Figure 1.2 profiles the top ten GHG polluting countries and shows that there has been an exponential growth in GHG emissions with the most significant increases being witnessed in the 1990s reaching an all-time high in 2018. In as

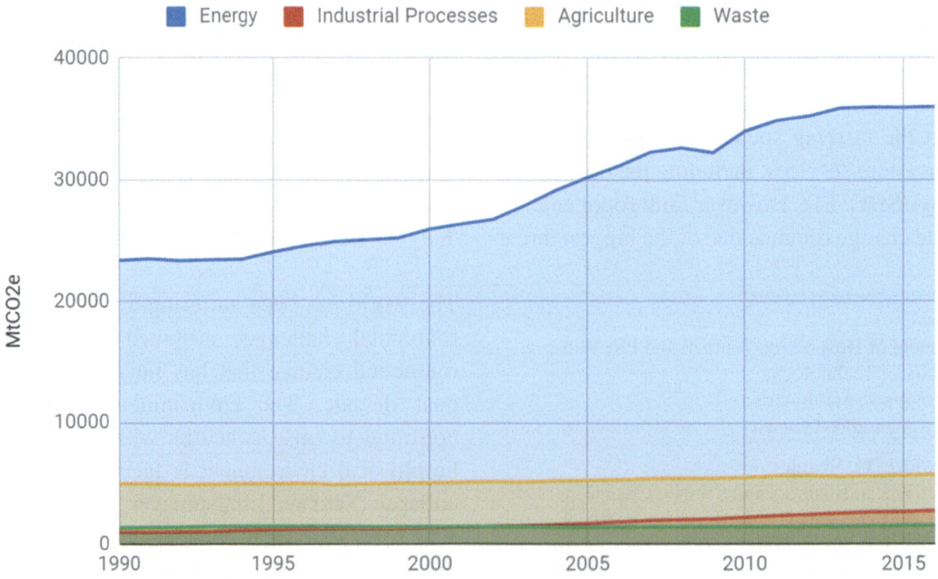

Fig. 1.1 Global historical emissions 1990–2016. (Source: Authors, data from the World Resources Institute (WRI) (2020)

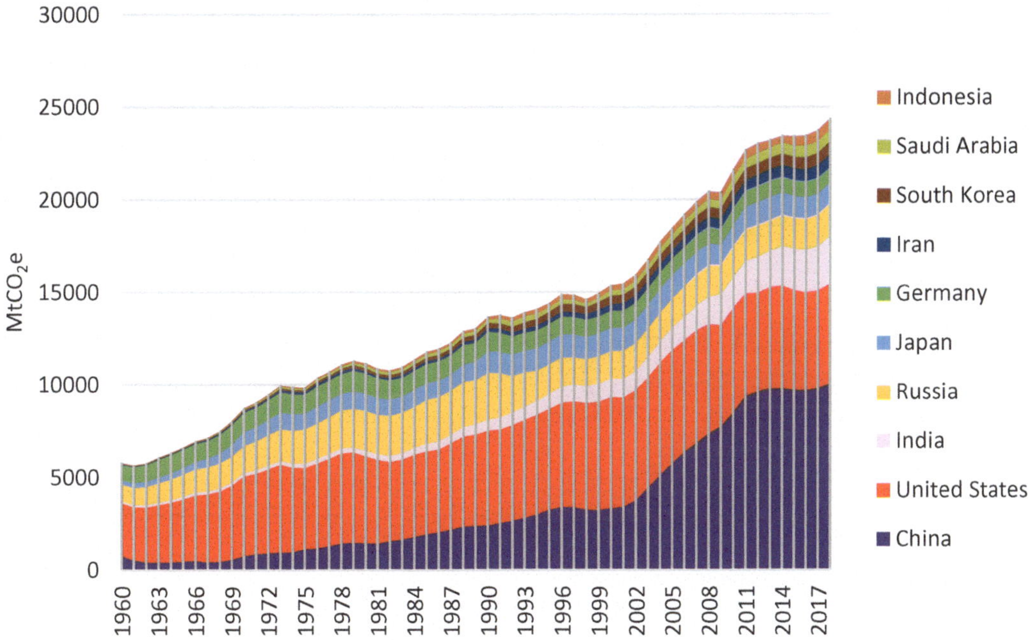

Fig. 1.2 Greenhouse emission trends in the top ten GHG polluting countries 1960–2018. (Source: Authors, data from the World Resources Institute (2020))

much as there might be a slowdown in carbon emissions globally, there seems to be a renewed growth run up to 2017 and 2018, which is quite a concerning development. In China and the USA, there seems to be some flattening of the curve, although a downward trajectory will be anticipated as there is a demand under the Paris Agreement to reduce GHG emissions by at least 40% by 2030 compared to 1990, under its broader 2030 climate and energy framework (European Union 2020).

Due to the increase in GHG emissions, the world has seen an increase in global average temperature. Figure 1.3a–d shows that global and regional temperatures have increased, with the most significant increases witnessed in the last two decades. However, the warming seems to be varied across the northern and southern hemispheres and the tropics. Evidence shows that the northern hemisphere has experienced a higher increase, followed by the tropics and the southern hemisphere coming in last. While there have been fluctuations on an annual basis, the general trend has been that of an increase across regions over time. Yu and Ruggieri (2019) made similar observations regarding global average temperature increase and also the trends. According to Yu and Ruggieri (2019), climatic patterns tend to shift multiple times over relatively short periods at change points. Over the period under their study, 2016 emerges as the warmest year on record confirming earlier reports by the World Meteorological Organization (2019), and all this has severe negative impacts on the attainment of the SDGs, especially SDG 13 focusing in climate action. However, the observed increase in global average temperature is of concern as the temperature is an important climatic driver. As such, any changes in temperature trigger a wide array of changes within the global, regional, and local climate systems. The Intergovernmental Panel on Climate Change (IPCC 2018) noted that an increase in temperature that surpasses the 1.5 °C mark pre-industrial levels would trigger and worsen the current global climate crisis characterized by the fury of extreme weather events.

Climate change has consequently added to the burden of global environmental challenges.

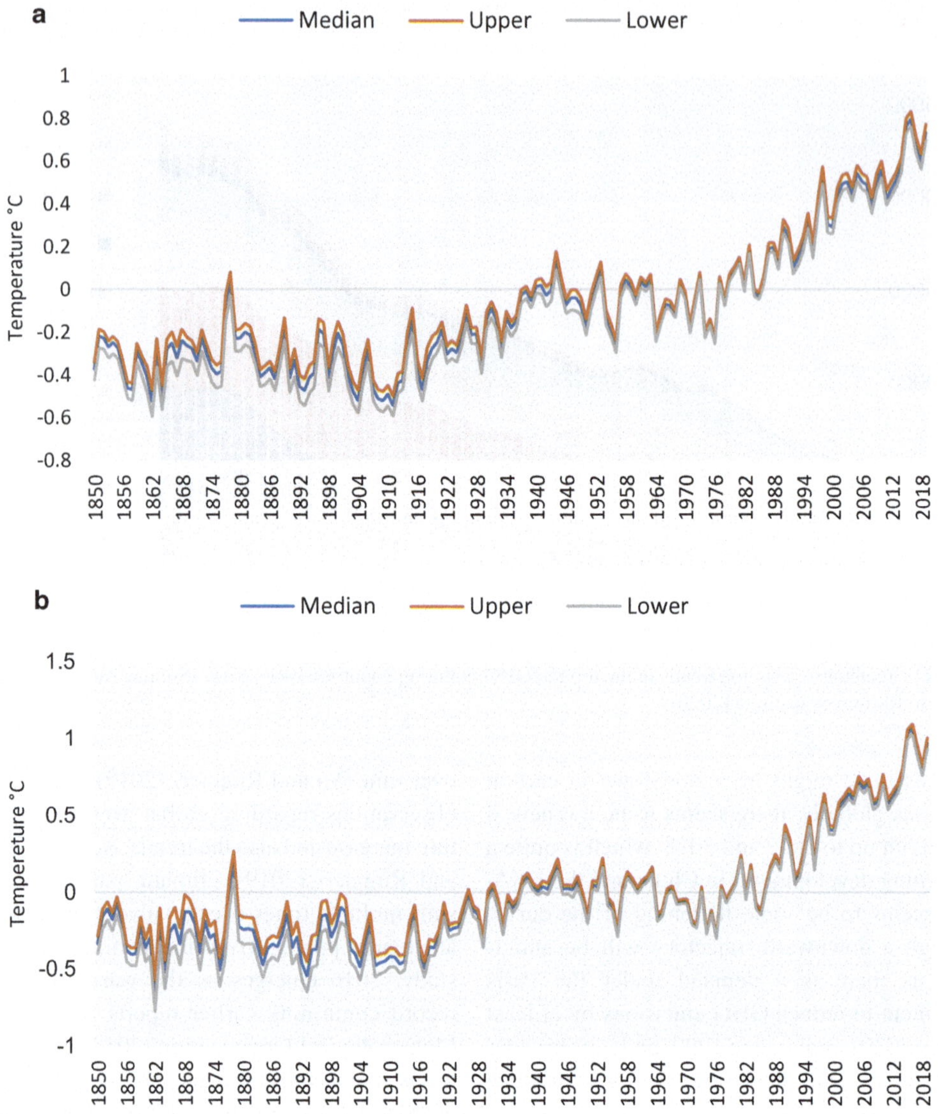

Fig. 1.3 Global and regional average land-sea temperature anomaly relative to the 1961–1990 average temperature. (**a**) Global average temperature annommally 1850–2019. (**b**) Northern Hemishphere average temperature annommally 1850–2019. (**c**) Southern Hemisphere average temperature annommally 1850–2019. (**d**) Tropical average temperature annommally 1850–2019. *The red line represents the median average temperature change, and gray lines represent the upper and lower 95% confidence intervals. (Source: Authors, data from Our World in Data – HadCRUT4 dataset published by Met Office Hadley Centre (2020)

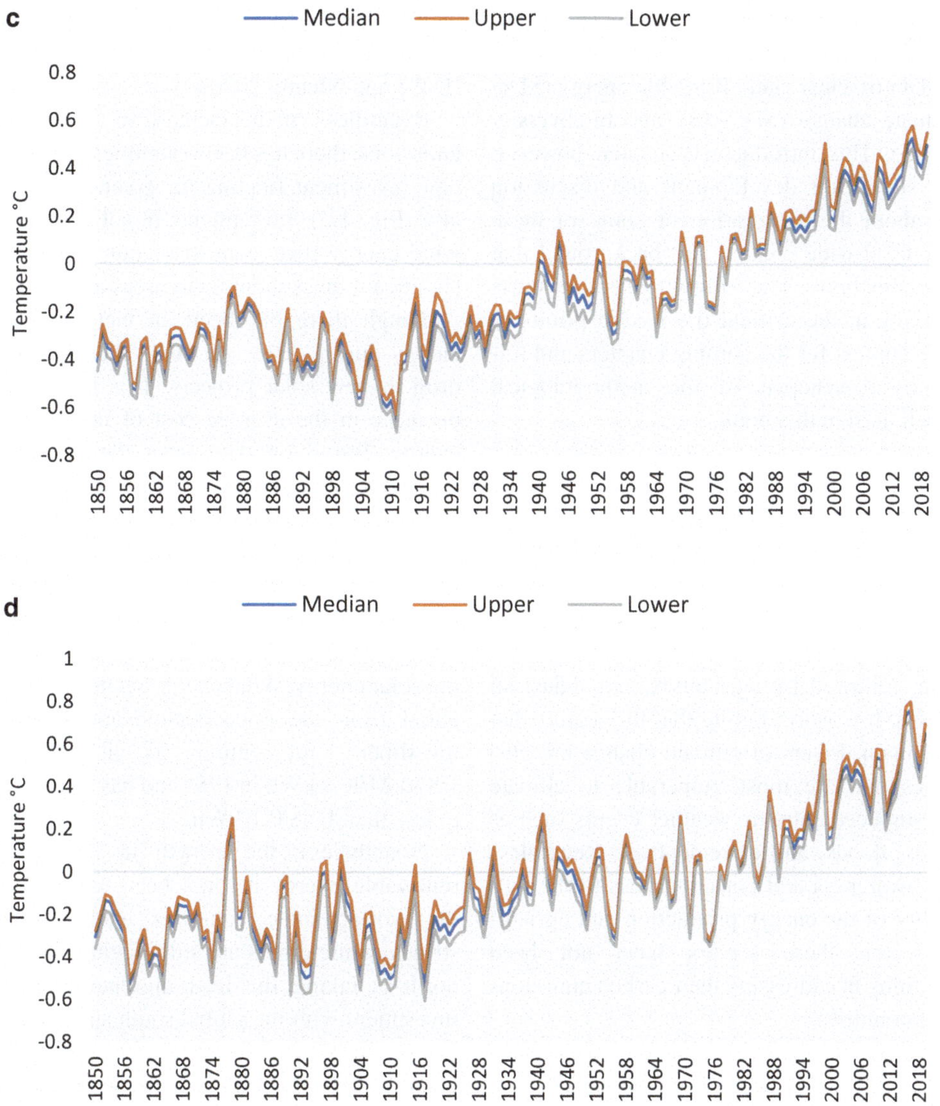

Fig. 1.3 (continued)

Climate change is a challenge because it has a dual relationship with almost every economic sector, social activities, and the natural environment. Climate change has also emerged as a threat to biodiversity (Kappelle et al. 1999), water security (Jaramillo and Nazemi 2018), energy production (Boadi and Owusu 2019; Dube and Nhamo 2020b), and agriculture threatening to derail the fight against hunger and food insecurity (Mason-D'Croz et al. 2019; Janssens et al. 2020). Given the centrality of these issues to society and its development, the 2030 Agenda for Sustainable Development has dedicated goals and targets aimed at addressing these challenges. SDG 2, for example, is aimed at addressing hunger, SDG 6 on water and sanitation, SDG7, affordable and clean energy, SDG13 on climate action, and SDG 15 focusing on life on land and biodiversity.

Given the foregoing, this book explores how society has interacted and sought to address issues related to several SDGs, among such, hunger and food security (SDG 2), water and sanitation (SDG 6), clean and affordable energy (SDG 7), climate change (SDG 13), and biodiversity (SDG 15). This introductory chapter, however, makes significant development and discussion points about the exploration of some of these aspects to provide context and background that informs this book. The following section starts by looking at the critical discussion points to provide context for the coming chapters and followed by a synopsis of the methodological approach used in this book.

2 Climate Change and the Clean Energy Debate

As discussed earlier, the major sectors contributing to the largest share of GHGs are energy production, followed by agriculture and industrial processes. It is worth noting that the sectors that are the main drivers of climate change are also the ones that are most vulnerable to climate change-induced extreme weather events such as droughts, floods, and extreme frost. Regardless of this awareness and consciousness to the vulnerability of the energy production and agricultural sector, these sectors have not been forthcoming in addressing their carbon emissions in most countries.

Consequently, the bulk of the energy produced globally still heavily depends on the use of fossil fuels such as coal, as can be seen in Fig. 1.4. The production of energy using green technology remains subdued globally. Green energy project seems to be slowly coming, while carbon-based energy demand and usage is increasing. It has been incredibly clear that the demand for coal and oil has increased in the recent past. At a time, there is an expectation that the demand declines and a robust uptick in the usage of solar, wind, gas, hydroelectricity, and biofuels. Notably, the production of hydroelectricity has been challenged by the impact of extreme weather events such as drought, which have resulted in some hydroelectricity power station failing to operate at their installed capacity. Power stations that have been worst affected in the recent past include the Kariba North Power station in Zambia and the Kariba South Power Station in Zimbabwe (Dube and Nhamo 2020).

Regardless of the increase in energy-related emissions, there has been a significant increase in new investment funding for green energy globally (Fig. 1.5). Investments in solar energy and wind energy that seem to account for the most significant investment focus areas are significant. Although there has been an increase in solar energy during 2018 and 2019, there has been a drop in new solar projects. This is despite the decrease in the average cost of installing solar energy. Based on data from the International Renewable Energy Agency (2020a), the cost of the solar installation fell from an average of between US$ 0.3/kWh and US$0.5/kWh in 2010 to low figures of between US$ 0.14kWh and US$0.04 in 2019. There has been a general decline on an annual basis for the cost of installing solar energy. Wind energy seems to have benefited from the price decline and has gained investment momentum as it was over US$0.24/K = kWh in 1984 and has since dropped to less than US$ 0,1/kWh.

Nonetheless, the growth in investment in renewable energy has not been uniform across the world; some countries made significant strides, while others are still lagging. The global north is taking the lead in renewable energy investment with the global south still a long way to go. No African countries are among the top countries that invested in renewables. For example, as of 2019, China and the USA had the largest investments of US$ 83.4 and the US$55.5 billion, respectively. Japan is in the third position, having invested $16.5 billion as of 2019 (Frankfurt School-UNEP 2020). A study conducted by the Frankfurt School and United Nations Environmental Programme (2020) found that globally in 2020, a commitment to invest about 826 gigawatts of new non-hydro renewable energy between 2020 and 2030 was made globally at a total cost of US$ 1.7 trillion which marks a falling decline from an investment of US$ 2.7 trillion made between 2010 and 2020. Consequently, the target is inadequate to allow the world to meet the target needed to cap global temperature at 2 °C by 2020 under SDGs 13 and

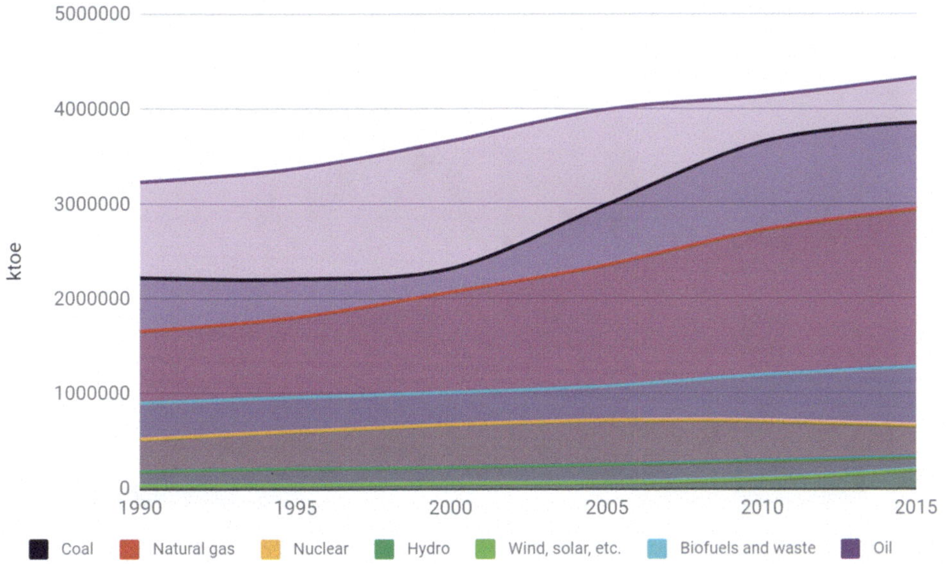

Fig. 1.4 Total primary energy supply (TPES) by source, World 1990–2017. (Source: Authors, data from the International Energy Agency (2020))

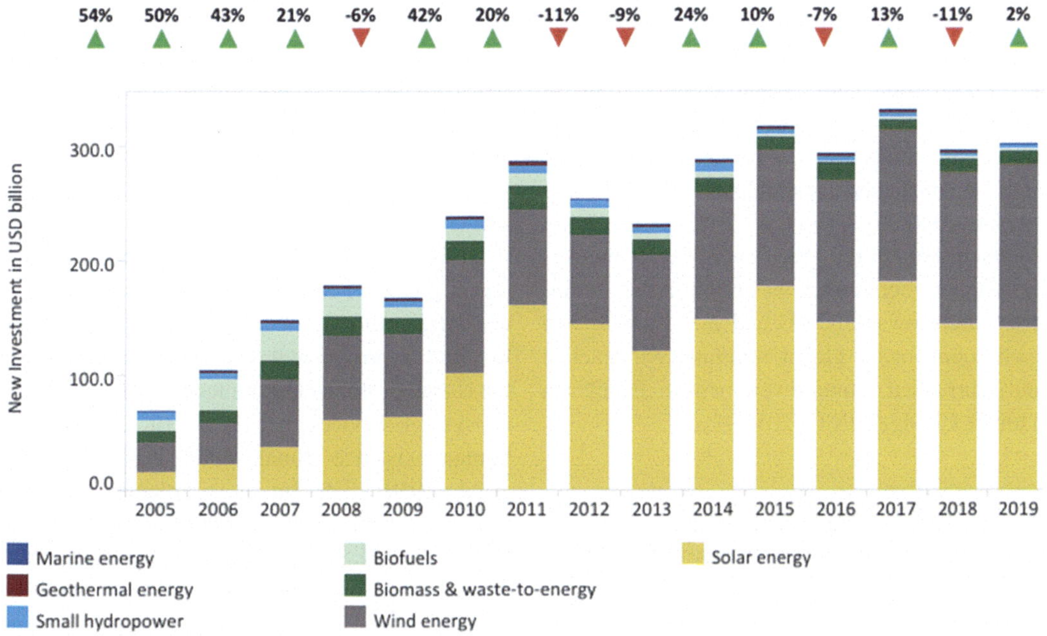

Fig. 1.5 New investment funding for green energy globally. (Source: International Renewable Energy Agency (2020))

7. The consequences for society and energy production will be dire.

The energy sector is equally vulnerable to extreme weather events. Hydro energy is believed to be severely threatened by climate change impacts across the world. The challenge associated with hydro energy is that it is widely viewed to be vulnerable to climate change (Schaeffer et al. 2012; Voisin et al. 2016; Klimenko et al. 2018). While current efforts by various economic sectors are commendable, there is need to rump up efforts to make the necessary and meaningful

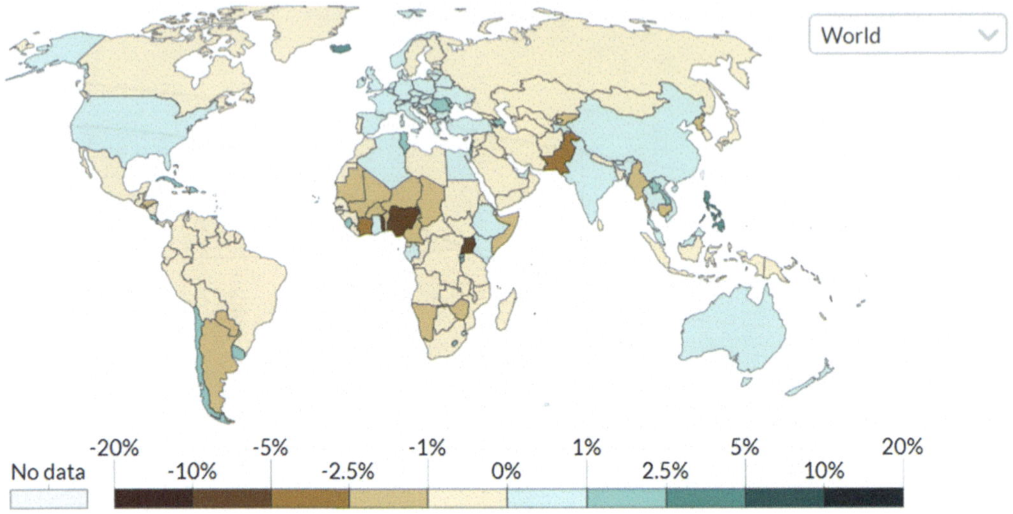

Fig. 1.6 Forest area net change rate between 2005 and 2015. (Source: SDG Tracker (2018))

impact in line with the commitments made by global leaders when they crafted the SDGs. There have been some efforts lately to increase investment in alternative energy by various private individuals at their homes, industry players such as airport companies, manufacturers, tourism establishments, and industry in general in line with the demands from the United Nations under SDG7 on clean and affordable energy (Dube and Nhamo 2020; Nhamo et al. 2020b). The main argument behind the green energy movement is to ensure carbon reductions and limit the continued march toward 1.5 °C and 2 °C, which will spell doom for the global community if reached and surpassed (Intergovernmental Panel on Climate Change (IPCC) 2018).

3 Crisis of Biodiversity and Marine Degradation

Climate change-induced extreme weather events ranging from droughts, floods, rising sea level, heat waves, extreme rainfall events, and warming oceans have had a devastating impacts on global communities threatening human existence as we know it. Climate change, agricultural development, and population growth have had serious implications on both maritime and terrestrial biodiversity (Nunez et al. 2019; Marques et al. 2019), which directly impact SDGs 14 and 15. Apart from these factors, urbanization and the demand for the built-up environment have also been highlighted as one of the major threats to biodiversity as forest and other vegetation are replaced by a concrete jungle (Opoku 2019). Terrestrial forests are crucial as they are home to about 80% of biodiversity; as such, their clearance has implications for habitat and biodiversity loss (Aerts and Honnay 2011).

Serious concerns have been raised in the recent past over the implications for the continued deterioration of the environment. Biodiversity loss and environmental degradation compromise the capacity of the environment to deliver crucial ecosystem services that are a crucial aspect of human existence (Smale et al. 2019; Juutinen et al. 2019). The Intergovernmental Science-Policy Platform on Biodiversity and Ecosystem Services (IPBES) rang a death bell of global natural stock in 2019, indicating that the world is fast losing several species due to increased human activity impact, desertification, climate change, and over-harvesting in some settings among others (2019).

Forests are considered to be some of the richest diversity areas in the world (Barlow et al. 2016; Liang et al. 2016). The concern is, however, being raised over the loss of forests across the world as it leads to biodiversity losses (Betts et al. 2017). Figure 1.6 shows that there are vast areas where forested areas declined with several hotspots in sub-Saharan Africa where forests are declining by as much as more than 5% per annum in countries such as Nigeria and Uganda. The highest declines globally have been recorded in Togo, where the forest annual loss rate is 8.11%. The loss of forest is particularly problematic in arid and semi-arid areas as such ecosystems are fragile, sensitive, and vulnerable, often triggering deforestation. The death of the forest aids and abates the release of carbon emissions that precipitates climate change (Lingbeek et al. 2017; Nwilo et al. 2020). Climate change is one of the biggest threats to global forests.

There is a worry that the declines are taking place in the least developed countries. Marques et al. (2019) observe that most of the forest losses occurring in central and southern America, Africa, and Asia were extractive in nature and buoyed by international trade. This means that there could be little value for member states in terms of economic and social benefits which could undermine the area's capacity to contribute to development in the future. Such a development could propagate increased poverty and inequality. This is likely to have severe implications on future socioeconomic development, as many developing countries depend on direct ecosystem services. This notion is supported by Alova et al. (2018) who observe that the development of Peru is hinged on its abundant natural capital for economic growth, development, and human well-being, and as such changes in biodiversity would have ramifications for development going forward.

For continued benefit from biodiversity, there have been calls for the protection of both terrestrial (SDG 15) and maritime (SDG 14) ecosystem. Biodiversity loss has been cited as a threat to society. Addison et al. (2020) argue that in light of the biodiversity threat to society, there is need to develop a model to ensure that businesses can develop biodiversity indicators for various business operations to enhance biodiversity protection. The United Nations makes a clarion call for the protection of biodiversity in the 2030 Agenda for Sustainable Development in a number of SDG targets (Box 1.1).

Box 1.1: SDGs and Targets on Biodiversity

- SDG 14a: Increase scientific knowledge, develop research capacity, and transfer marine technology, taking into account the Intergovernmental Oceanographic Commission Criteria and Guidelines on the Transfer of Marine Technology, to improve ocean health and to enhance the contribution of marine biodiversity to the development of developing countries, in particular, small island developing States and least developed countries
- SDG 15: Protect, restore, and promote sustainable use of terrestrial ecosystems, sustainably manage forests, combat desertification, halt and reverse land degradation, and halt biodiversity loss
- SDG 15.4: By 2030, ensure the conservation of mountain ecosystems, including their biodiversity, to enhance their capacity to provide benefits, which are essential for sustainable development
- SDG 15.5: Take urgent and significant action to reduce the degradation of natural habitats, halt the loss of biodiversity, and, by 2020, protect and prevent the extinction of threatened species
- SDG 15.9: By 2020, integrate ecosystems and biodiversity values into national and local planning, development processes, and poverty reduction strategies and accounts
- SDG 15a: Mobilize and significantly increase financial resources from all sources to conserve and sustainably use biodiversity and ecosystems

Source: Authors, based on United Nations (2015: 24–25).

4 Trends in Global Food Security and Implications for Achieving SDG 2

As could be seen in earlier sections, the global food supply chain is replete with sources of carbon emissions and contribute significantly to the total global carbon emissions. The agriculture sector, which is responsible for the global food supply chain, is one of the main emitters of GHGs that lead to global warming and, ultimately, climate change. According to Poore and Nemecek (2018), food production is responsible for 26% of global GHG emissions, with only 5% of that portion being attributed to non-food agriculture and other drivers of deforestation. The global agriculture industry is also responsible for other hosts of environmental challenges, such as acidification and eutrophication, which also contribute to global biodiversity losses. Agriculture has, however, a dual relationship with climate change. It is affected by climate change, which demands serious attention to be paid in the context of SDG2 to eradicate hunger, and its activities also emit harmful GHGs. Hunger cannot be addressed without paying attention to the threat of the agricultural sector to climate and ensuring the sector is climate change proofed as demanded under SDG13 on climate change action.

According to Barrera and Hertel (2020), "Sustainably in meeting the food demands of a growing population based on finite resources while protecting the environment is one of the great challenges of humanity in the coming decades." While the global food supply has been varied across the world, some countries are producing more than they require, while there are persistent shortages in some parts of the world. This could be attributed to matters surrounding food security including its distribution, access, and supply challenges. A study by Mc Carthy et al. (2018) notes that about 16% of the global population was suffering chronic hunger at a time when the world is producing enough food to feed everyone in the world. The causes for food insecurity across the world have been varied nonetheless with some academics and authorities complaining about excessive food waste which is particularly rampant in developed countries (Barrera and Hertel 2020).

While the SDGs were adopted in 2015 when hunger was highlighted as one of the key focus areas that warrant global attention, there has not been good progress regarding this goal with the situation of food insecurity worsening over the past couple of years. Although the world had witnessed some progress in reducing the number of hungry people, undernourished people seem to have picked up post-2015, mainly driven by setbacks in sub-Saharan Africa and Asia. According to FAO (2020), out of the 7.7 billion people in the world, about 5,7 billion were food secure, while about two billion people are food insecure. Of the food-insecure people, more than one billion are from Asia and 0,7 billion from Africa, while 0.2 billion people are from Latin America. Consequently, the number of undernourished people in the three continents is 381 million in Asia, 250 million in Africa, and about 48 million in Latin America.

Going with the trend, it shows that at a global level, the number of food-insecure people is increasing with the increasing incidents of extreme weather events with drought being one of the biggest drivers of global food insecurity (Fig. 1.7). Consequently, the world is moving further from its target of ending hunger by 2030. FAO projects that very few global regions are likely to meet their 2030 SDG 2 targets with most regions either having witnessed some progress (South East Asia, Southern Asia, and the Caribbean). In Africa and Latin America and other regions in Asia, the situation is worsening (FAO 2020).

In as much as the issue of ensuring sustainable food security cannot be linked to the hydrometeorological factors alone, this is one of the most significant threats that are facing the sector. Droughts, which have been noted as problematic in many regions across the world, call for tapping into the irrigation systems as an adaptation measure. However, the world is facing a real disaster as severe droughts tend to raise demand for water from various competing economic sectors, which also limits the capacity to allocate or provide more water for irrigation (Dube et al. 2020). In some regions, droughts have affected energy generation, which is crucial for ensuring irrigation schedules are completed even where water is available, resulting in diminished irrigation pro-

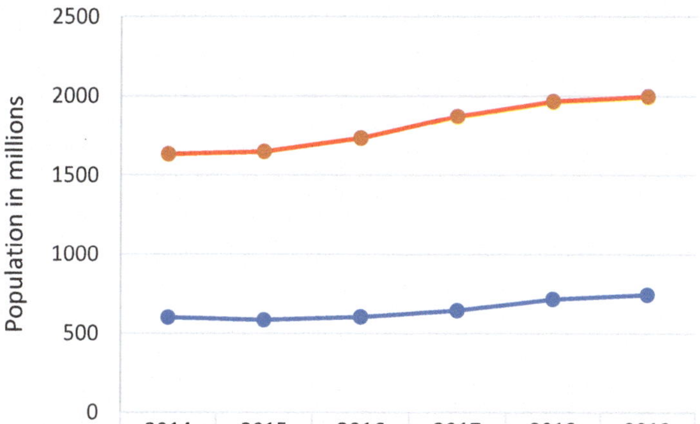

	2014	2015	2016	2017	2018	2019
severely food insecure	602	586	605	646.4	717.5	746
moderately or severely fod insecure people	1633.5	1649.5	1735.2	1874.5	1969.6	2001.1

Fig. 1.7 Global levels of food insecurity trends 2014–2019. (Source: Authors, data from FAO (2020))

duction (Dube and Nhamo 2020b). Accordingly, Nhamo et al. (2019) underscored that achieving the Agenda 2030 goals will be premised on ensuring water security. This is particularly challenging for the sub-Saharan Africa and impairs the capacity for the region to feed its population. Other factors that have been cited as problematic are political instability, which is one of the non-environmental drivers of food insecurity in many countries across the world (Brück and d'Errico 2019; Martin-Shields and Stojetz 2019). There is, therefore, a need to tackle SDG16 peace justice and strong national and international institutions to reduce the challenges and variables that the world has to deal with and face the global environmental challenges.

The next section presents the book and chapter outline. Methodologies are contained in every chapter. Overall, the mixed methods approach is used.

5 Book and Chapter Outline

This book comes in 5 major parts and 19 chapters. Part I is made up of a single chapter focusing on making SDGs work to end hunger, sustain energy, resolve climate change, and reverse biodiversity loss. Part II looks at food security and sustainable energy. From this part comes Chap. 2 that deals with the contribution of responsible leadership in raising funds to support the organizational mandate and the SDGs, with a focus on the Land Bank of South Africa. Chapter 3 deliberates on confronting poverty, hunger, and food insecurity in Malawi and Zimbabwe, while Chap. 4 considers preventing fall armyworm (*Spodoptera frugiperda* JE Smith) damage in maize by altering planting time and using varied genotypes. Chapter 5 discusses enhancing urban farming for sustainable development through SDGs, while Chap. 6 looks at water and sanitation access in the Shamva district for sustainability and development of the Zimbabwean smallholder farming sector. Chapter 7 presents responsible leadership and the implementation of SDG 7 with a focus on the UNDP Botswana biogas project. This is followed by Chap. 8 that harnesses elements of responsible leadership in driving climate action (SDG 13), while Chap. 9 narrows down to leadership capabilities for the successful implementation of SDG 7 targets at Energy Company X. Chapter 10 looks at designing effective social protection for food and nutrition security among farm workers from Masvingo, Zimbabwe.

Part III addresses climate action for SDGs. Chapter 11 looks at mitigating climate change through carbon sequestration for sustainable development in Cameroon's forest economy, while Chap. 12 considers the private sector SDGs' localization from Kruger Mpumalanga International Airport, South Africa. Chapter 13 deliberates on scaling up university engagement with the water SDG for general environmental stewardship and climate change resilience, and Chap. 14 addresses climate change in Zimbabwe's vulnerable communities using a case study of Supporting Enhanced Climate Action Project (SECA project) in Bulilima district. Chapter 15 presents climate resilience strategies and livelihood development in dry regions of Zimbabwe, while Chap. 16 looks at climate action at international airports focusing on the Airport Carbon Accreditation program.

Part IV discusses health, water, and biodiversity engagements. Chapter 17 considers protected areas interventions and SDGs using the case of the Bolsa Floresta program in Brazilian Amazon. Chapter 18 focuses on the implementation of SDGs through greening household responses to water, energy, and food shortages in Newlands West, Durban. Part (V) is made up of another single chapter that presents that book conclusion and policy recommendations.

References

World Meteorological Organization, 2019. *WMO confirms past 4 years were warmest on record.* [Online] Available at: https://public.wmo.int/en/media/press-release/wmo-confirms-past-4-years-were-warmest-record#:~:text=The%20year%202016%2C%20which%20was,C%20above%20pre%2Dindustrial%20levels. [Accessed 26 August 2020].

Addison, P. F. et al., 2020. Bringing sustainability to life: A framework to guide biodiversity indicator development for business performance management. *Business Strategy and the Environment,* pp. 1-11.

Aerts, R. & Honnay, O., 2011. Forest restoration, biodiversity and ecosystem functioning. *BMC ecology,* 11(1), p. 29.

Alova, G., Orihuela, J. C. & Karousakis, K., 2018. *Mainstreaming biodiversity and development in Peru,* s.l.: OECD Library.

Aydin, C., Esen, Ö. & Aydin, R., 2019. Is the ecological footprint related to the Kuznets curve a real process or rationalising the ecological consequences of the affluence? Evidence from PSTR approach. *Ecological indicators,* Volume 98, pp. 543-555.

Barlow, J. et al., 2016. Anthropogenic disturbance in tropical forests can double biodiversity loss from deforestation. *Nature,* 535(610), pp. 144-147.

Barrera, E. L. & Hertel, T., 2020. Global food waste across the income spectrum: Implications for food prices, production and resource use. *Food Policy,* p. 101874.

Betts, M. G. et al., 2017. Global forest loss disproportionately erodes biodiversity in intact landscapes. *Nature,* 7664(547), pp. 441-444.

Boadi, S. A. & Owusu, K., 2019. Impact of climate change and variability on hydropower in Ghana. *African Geographical Review,* 38(1), pp. 19-31.

Dube, K. & Nhamo, G., 2020a. Sustainable Development Goals localisation in the tourism sector: lessons from Grootbos Private Nature Reserve, South Africa. *GeoJournal,* pp. 1-18.

Dube, K. & Nhamo, G., 2020b. Vulnerability of nature-based tourism to climate variability and change: Case of Kariba resort town, Zimbabwe. *Journal of Outdoor Recreation and Tourism,* Volume 29, p. 100281.

Dube, K. & Nhamo, G., 2020. Greenhouse Gas Emissions and Sustainability in Victoria Falls: Focus on Hotels, Tour Operators and Related Attractions. *African Geographical Review,* pp. 1-16.

Dube, K., Nhamo, G. & Chikodzi, D., 2020. Climate change-induced droughts and tourism: Impacts and responses of Western Cape province, South Africa. *Journal of Outdoor Recreation and Tourism,* p. 100319.

European Union , 2020. *2030 climate & energy framework.* [Online] Available at: https://ec.europa.eu/clima/policies/strategies/2030_en [Accessed 22 August 2020].

FAO, 2020. *The State of Food Security and Nutrition in the World 2020: Transforming food systems for affordable healthy diets.* [Online] Available at: https://reliefweb.int/sites/reliefweb.int/files/resources/SOFI2020_EN_web.pdf [Accessed 17 August 2020].

Frankfurt School -UNEP, 2020. *Falling clean energy costs provide opportunity to boost climate action in COVID-19 recovery packages.* [Online] Available at: https://www.fs-unep-centre.org/global-trends-in-renewable-energy-investment-2020/ [Accessed 28 August 2020].

Iizuka, M. & Zanlungo, J. P., 2016. Environmental collapse and institutional restructuring: the sanitary crisis in the Chilean salmon industry. In: A. Hosono, M. Iizuka & J. Katz, eds. *Chile's Salmon industry.* Tokyo: Springer, pp. 109-135.

Infante-Amate, J. et al., 2018. Land embodied in Spain's biomass trade and consumption (1900–2008): Historical changes, drivers and impacts. *Land Use Policy,* Volume 78, pp. 493-502.

Intergovernmental Panel on Climate Change (IPCC), 2018. *The Special Report on Global Warming of 1.5 °C (SR15).* [Online] Available at: https://www.ipcc.ch/sr15/ [Accessed 22 August 2020].

Intergovernmental Science-Policy Platform on Biodiversity and Ecosystem Services-IPBES, 2019.

Global Assessment Report on Biodiversity and Ecosystem Services, s.l.: Intergovernmental Science-Policy Platform on Biodiversity and Ecosystem Services- IPBES.

International Renewable Energy Agency, 2020. *Investment Trends.* [Online] Available at: https://www.irena.org/Statistics/View-Data-by-Topic/Finance-and-Investment/Investment-Trends [Accessed 27 August 2020].

International Renewable Energy Agency, 2020a. *Solar Costs.* [Online] Available at: https://www.irena.org/Statistics/View-Data-by-Topic/Costs/Solar-Costs [Accessed 27 August 2020].

Janssens, C. et al., 2020. Global hunger and climate change adaptation through international trade. *Nature Climate Change,* pp. 1-7.

Jaramillo, P. & Nazemi, A., 2018. Assessing urban water security under changing climate: Challenges and ways forward. *Sustainable cities and society,* Volume 41, pp. 907-918.

Juutinen, A. et al., 2019. Trade-offs between economic returns, biodiversity, and ecosystem services in the selection of energy peat production sites. *Ecosystem Services,* Volume 40, p. 101027.

Kappelle, M., Van Vuuren, M. M. & Baas, P., 1999. Effects of climate change on biodiversity: a review and identification of key research issues. *Biodiversity & Conservation,* 8(10), pp. 1383-1397.

Kjaer, L. L. et al., 2019. Product/service-systems for a circular economy: the route to decoupling economic growth from resource consumption. *Journal of Industrial Ecology,* 23(1), pp. 22-35.

Klimenko, V. V., Fedotova, E. V. & Tereshin, A. G., 2018. Vulnerability of the Russian power industry to the climate change. *Energy,* Volume 142, pp. 1010-1022.

Liang, J. et al., 2016. Positive biodiversity-productivity relationship predominant in global forests. *Science,* Volume 354, p. 6309.

Lingbeek, B. J. et al., 2017. Arthropod diversity and assemblage structure response to deforestation and desertification in the Sahel of western Senegal. *Global ecology and conservation,* Volume 11, pp. 165-1.

Martin-Shields, C. P., & Stojetz, W. (2019). Food security and conflict: Empirical challenges and future opportunities for research and policy making on food security and conflict. *World development,* 119, pp. 150-164.

Marques, A. et al., 2019. Increasing impacts of land use on biodiversity and carbon sequestration driven by population and economic growth. *Nature ecology & evolution,* 3(4), pp. 628–637.

Mason-D'Croz, D. et al., 2019. Agricultural investments and hunger in Africa modeling potential contributions to SDG2–Zero Hunger. *World development,* Volume 116, pp. 38–53.

Mc Carthy, U. et al., 2018. Global food security–Issues, challenges and technological solutions. *Trends in Food Science & Technology,* Volume 77, pp. 11-20.

Met Office Hadley Centre, 2020. Global average temperatures have increased by more than 1°C since pre-industrial times. [Online] Available at https://www.metoffice.gov.uk/hadobs/hadcrut4/index.html [Accessed 11 July 2020].

Nhamo, G., Nhemachena, C., & Nhamo, S. (2019). Is 2030 too soon for Africa to achieve the water and sanitation sustainable development goal?. *Science of the Total Environment,* 669, 129-139.

Nhamo, G. et al., 2020. Concluding the Matter: Global Energy Gaps and Remedial Pathways to 2030. In: C. N. S. N. V. M. I. S. Godwell Nhamo, ed. SDG7–Ensure Access to Affordable, Reliable, Sustainable and Modern Energy. s.l.: Emerald Publishing Limited, pp. 233-262.

Nunez, S. et al., 2019. Assessing the impacts of climate change on biodiversity: is below 2° C enough?. *Climatic Change,* 154(3-4), pp. 351-365.

Nwilo, P. C. et al., 2020. Impacts of land cover changes on desertification in northern Nigeria and implications on the Lake Chad Basin. *Journal of Arid Environments,* Volume 181, p. 104190.

Opoku, A., 2019. Biodiversity and the built environment: Implications for the Sustainable Development Goals (SDGs). *Resources, conservation and recycling,* Volume 141, pp. 1-7.

Poore, J. & Nemecek, T., 2018. Reducing food's environmental impacts through producers and consumers. *Science,* 360(6392), pp. 987-992.

Sato, C. F. & Lindenmayer, D. B., 2018. Meeting the global ecosystem collapse challenge. *Conservation Letters,* 11(1), p. e12348.

Schaeffer, R. et al., 2012. Energy sector vulnerability to climate change: a review. *Energy,* 38(1), pp. 1-12.

SDG Tracker, 2018. *Sustainable Development Goals.* [Online] Available at: https://sdg-tracker.org/biodiversity [Accessed 22 August 2020].

Smale, D. A. et al., 2019. Marine heatwaves threaten global biodiversity and the provision of ecosystem services. *Nature Climate Change,* 9(4), pp. 306-312.

United Nations, 2015. *Agenda 2030 on Sustainable Development.* [Online] Available at: https://sustainabledevelopment.un.org/content/documents/21252030%20Agenda%20for%20Sustainable%20Development%20web.pdf [Accessed 11 July 2020].

Voisin, N. et al., 2016. Vulnerability of the US western electric grid to hydro-climatological conditions: How bad can it get?. *Energy,* Volume 115, pp. 1-12.

Weber, H. & Sciubba, J. D., 2019. The effect of population growth on the environment: Evidence from European regions. *European Journal of Population,* 35(2), pp. 379-402.

Wiedmann, T., Lenzen, M., Keyßer, L. T. & Steinberger, J. K., 2020. Scientists' warning on affluence. *Nature communications,* 11(1), pp. 1-10.

World Resources Institute, 2019. *CO2 Emissions Climb to an All-Time High (Again) in 2019: 6 Takeaways from the Latest Climate Data.* [Online] Available at: https://www.wri.org/blog/2019/12/co2-emissions-climb-all-time-high-again-2019-6-takeaways-latest-climate-data [Accessed 22 August 2020].

World Resources Institute (WRI). 2020. Global historical emissions. World Resources Institute. Wasington.

Yu, M. & Ruggieri, E., 2019. Change point analysis of global temperature records. *International Journal of Climatology,* 39(8), pp. 3679-3688.

Part II

Food Security and Sustainable Energy

The Contribution of Responsible Leadership in Raising Funding to Support Organisational Mandate and the SDGs: Case of the Land Bank of South Africa

2

Carolien Samson and Dawie (D.A.J.) Bornman

Abstract

Responsible leadership theory has become prominent in the past decade as businesses are expected to play a larger role through their public presence and business leaders are held accountable for the consequences of their enterprises' activities. This chapter focuses on the development of responsible leadership theory by testing the applicability of three models: (1) responsible leadership dimensions, (2) responsible leadership influence pathways and (3) multilevel outcomes of responsible leadership. Responsible leadership theory is applied to a state-owned entity to test the applicability of the theory beyond the private sector. The Land and Agricultural Development Bank of South Africa has a dual mandate to support both commercial and development agriculture and create a more inclusive and transformed sector in South Africa. Through its activities, the Land Bank has the ability to support multiple sustainable development goals (SDG) as its lending allows the agricultural sector to contribute to economic growth, employment creation and income generation. This case study analyses the leadership required during 2015–2018 to diversify the Land Bank's portfolio of investors and reduce its reliance on short-term lending. Consequently, Land Bank was able to raise funding from four multilateral institutions which could be applied specifically to support development farmers.

Keywords

Agriculture · Land Bank · Responsible leadership · SDGs · South Africa

1 Introduction and Background

Agriculture is a key element to ensure food security in South Africa by providing a sufficient quantity and quality of a range of food through field crops, horticulture and the raising of livestock. Although primary agriculture only contributed 2.4% to South Africa's Gross Domestic Product in 2018 (Statistics South Africa 2019a), the sector employs 5% of the South African workforce, mostly at a lower skilled level (Statistics South Africa 2019b), and is an important input into many other economic activities through agri-processing. The National

C. Samson
Grobank, Sandton, South Africa
e-mail: carolien.samson@me.com

D. (D.A.J) Bornman (✉)
University of Pretoria, Pretoria, South Africa
e-mail: dawie.bornman@up.ac.za

G. Nhamo et al. (eds.), *Sustainable Development Goals for Society Vol. 2*, Sustainable Development Goals Series, https://doi.org/10.1007/978-3-030-70952-5_2

Development Plan confirms the importance of the sector to create employment opportunities in rural communities characterised by inequality and poverty and identified the opportunity in agriculture to create one million jobs (National Planning Commission 2012). South Africa's current president Cyril Ramaphosa's economic recovery plan announced in his State of the Nation address in February 2018 prioritised inter alia employment creation, the revival of the agricultural sector and support for small business enterprises and youth (Land and Agricultural Development Bank of South Africa 2018a). Growth in the sector therefore contributes directly to increased food availability, job creation and income generation including foreign exchange earnings through exports.

Agriculture is a capital-intensive industry requiring significant investments in land, infrastructure and production inputs, and access to reliable, affordable sources of finance is a critical requirement for farmers. Thus, the government established the Land and Agricultural Development Bank of South Africa (Land Bank) in 1912 as a channel to provide finance to farmers. Up to 1994, Land Bank focused on supporting white commercial farmers through both a branch network and wholesale financing to co-operatives. In 1995, then President Nelson Mandela established the Strauss Commission of inquiry into the provision of rural financial services. The Commission recommended that Land Bank maintain its services to its existing commercial clients but should focus increasingly on providing financial services to new clients, especially beneficiaries of the land reform programme (Strauss et al. 1996). Over the next two decades, the Bank set out to become more developmental in its focus. However, the grants proposed by the Commission to enhance Land Bank's activities failed to materialise, and consequently it had to rely on commercial investors to support commercial and development farmers.

Land Bank's reliance on short-term commercial funding became a risk to its long-term financial sustainability and hampered its ability to support development farmers. In 2015, the Bank decided to change the structure and tenor of its funding and improve access to loans for development farmers. This required its senior management not only to understand the legitimate expectations of its investors and clients but also to respond appropriately.

This study firstly aims to contribute to the development of responsible leadership theory by testing three models proposed in the literature at an organisational level, namely, the dimensions of responsible leadership (Voegtlin 2016), responsible leadership influence pathways (Doh and Quigley 2014) and multilevel outcomes of responsible leadership (Voegtlin et al. 2012). Responsible leadership theory assumes that organisations operate mainly as private sector entities with private shareholders and a strong profit motive. This chapter intends to provide evidence that responsible leadership theory also applies within public sector entities with both a commercial and development mandate.

This chapter will continue with a literature review focusing on the objectives of Land Bank, responsible leadership theory and the sustainable development goals. This will be followed by the research design for the study focusing on the study's methodology and case study under investigation. Thereafter, the study's results and findings will be addressed before the chapter concludes.

2 Literature Review and Background

2.1 Objectives of Land Bank

In the aftermath of the South African War (1899–1902), farmers struggled to recover from the damage caused by the war as well as severe droughts in the early 1900s. As a result, the colonial government set up a number of so-called land banks and co-operatives to assist farmers with financial and other assistance (Jacobs 2013). It also passed various pieces of legislation to advance the interests of white farmers including the Land Settlement Act, 1912, which would result in approximately 14 million hectares being granted to successful applicants up to 1959 (Jacobs 2013) as well as the infamous Natives Land Act, 1913.

By the late 1980s, the Bank had become more profit-driven as it had to secure its balance sheet. In 1995, the Strauss Commission was appointed to assess rural financial services. The Commission recommended that Land Bank should focus on extending its activities beyond commercial agriculture to extend loans on wholesale and individual basis for development (Strauss et al. 1996:73). By the end of that decade, Land Bank had commenced with providing loans to Black farmers, many of whom were entering the agricultural sector at a commercial level for the first time.

In 2002, the Land Bank Act, 1944, was updated and replaced by the Land and Agricultural Development Bank Act 15 of 2002 (hereafter the Land Bank Act). Section 2 of the Land Bank Act sets out the objectives of the organisation, and although promulgated a decade prior to the publication of the National Development Plan and SDGs, the objectives are clearly aligned to both policies (Land and Agricultural Development Bank of South Africa 2017) in Table 2.1.

Despite its attempts to increase reserves in the 1990s, the Bank posted the first losses in its his-

Table 2.1 Objectives of the Land Bank Act

The Bank is obligated to promote, facilitate and support the following:	National development plan	Sustainable development goals
Equitable ownership of agricultural land, particularly increasing the ownership of agricultural land by historically disadvantaged persons Agrarian reform, land redistribution or development programmes aimed at historically disadvantaged persons for the development of farming enterprises and agricultural purposes Removal of the legacy of past racial, gender and generational discrimination in agriculture	Making land reform work to unlock the potential for a dynamic, growing and employment-creating agricultural sector	5. Gender equality 10. Reduced inequalities
Programmes that contribute to agricultural aspects of rural development and job creation Agricultural entrepreneurship Enhancement of productivity, profitability, investment and innovation in the agricultural and rural financial systems	Support job creation in the upstream and downstream industries Develop strategies that give new entrants access to product value chains and Support from better resourced players	1. No poverty 4. Quality education 9. Industry innovation and infrastructure
Commercial agriculture Land access for agricultural purposes Better use of land	Commercialise some under used land in communal areas and land reform projects Pick and support commercial agriculture sector and regions that have the highest potential for growth and employment	8. Decent work and economic growth 11. Sustainable cities and communities
Food security	Everyone should have access to sufficient, nutritious and safe food at all times	2. Zero hunger 3. Good health
Promote and develop the environmental sustainability of land and related natural resources	Expand irrigated agriculture; find creative partnerships between opportunities	6. Clean water and sanitation 7. Affordable and clean energy 12. Responsible consumption 13. Climate action 14. Life under water 15. Life on land 16. Peace and justice 17. Partnerships for the goals

Source: Authors, based on Land and Agricultural Development Bank of South Africa (2017:35)

tory between 2003 and 2006 based on poor credit decision-making and loan recovery processes. In addition, by 2007, it was subject to two forensic investigations relating to commercial property transactions that fell outside the ambit of the Land Bank Act as well as the misappropriation of money from the AgriBEE fund that it had been administering on behalf of the Department of Agriculture (Jacobs 2013:48). Following a period of managerial instability, then Deputy Director General in the National Treasury Mr P.S. Hadebe was appointed as CEO in 2008 to clean up and stabilise the organisation (Jacobs 2013:46).

In 2009, Land Bank was completely reliant on the Public Investment Corporation (PIC) and Corporation for Public Deposits (CPD) for its funding (Land and Agricultural Development Bank of South Africa 2016). These funds were all short term (1 year or less maturity dates) which resulted in a significant mismatch between the funding terms and client lending terms (typically mortgage loans with terms up to 20 years). It also limited Land Bank's ability to increase its loan book as its growth was constrained by the funding these two investors were willing to provide.

The steps taken by Mr Hadebe improved investor perceptions to such an extent that in 2012, the African Development Bank (AfdB) was willing to provide a government guaranteed 15-year loan of R1.0 billion to the Bank (Land and Agricultural Development Bank of South Africa 2016). This loan enabled Land Bank to find areas of growth.

Although Land Bank's financial situation had improved significantly by the time of Mr Hadebe's departure at the end of 2013, it still remained heavily reliant on a small group of investors willing to provide short-term funding to Land Bank. In addition, concerns on the financial sustainability of the organisation limited the scope of Land Bank to increase the transformational and developmental elements of its mandate. The proportion of its loan book deemed transformational increased from R790 million in 2005/2006 (Land and Agricultural Development Bank of South Africa 2008) to R2.3 billion in 20014/2015 (Land and Agricultural Development Bank of South Africa 2015). However, as a percentage of the overall loan book, it remained essentially flat at 5%–6%, and the lack of progress was challenged by a number of stakeholders including government and agricultural sector participants.

Mr T.P. Nchocho, who became CEO on 1 February 2015, set out to balance the financial sustainability with the development mandate of the organisation. He faced an immediate challenge as the demise of African Bank Limited resulted in a liquidity squeeze in the South African market which exposed the limitations of Land Bank's dependence on short-term funding. Another challenge followed in August 2016 when Futuregrowth, one of the large asset managers in South Africa, announced that it would cease to invest in six SOEs including Land Bank due to governance concerns (Futuregrowth Asset Management 2018).

In May 2015, Mr B.J. van Rooy joined Land Bank as Chief Financial Officer (CFO) and developed an investment strategy that would improve the stability of Land Bank's operations, crowd-in private sector funding to the sector and support Land Bank's development mandate. The actions related to this project and outcomes achieved are described in the next section as the case study.

2.2 Responsible Leadership Theory

Responsible leadership seeks to define the type of leadership required in a multi-stakeholder environment with complex demands from internal and external followers and is defined as a "values-based and thorough ethical principles-driven relationship between leaders and stakeholders who are connected through a shared sense of meaning and purpose through which they raise one another to higher levels of motivation and commitment for achieving sustainable values creation and social change" (Pless 2007:438). It recognises that leaders have responsibilities not only to internal followers, but broadens the scope to include external stakeholders (Pless and Maak 2011).

For the purpose of this study which focuses on the organisational and individual level, the definition of responsible leadership by Voegtlin, Patzer and Scherer (2012) was used, namely: "the awareness and consideration of the consequences of one's actions for all stakeholders, as well as the exertion of influence by enabling the involvement of the affected stakeholders and by engaging in an active stakeholder dialogue. Therein responsible leaders strive to weigh and balance the interests of the forwarded claims. Additionally, responsible leaders foster the public exchange of opinions and engage in public will formation".

The concept of responsible leadership within the leadership theory is interpreted in different ways. Waldman and Galvin (2008) identified two types of mindsets, namely, an economic perspective and a stakeholder perspective. The economic perspective revolves around the principle that responsible leaders act solely on behalf of the shareholders and that their focus should be on generating profits. The stakeholder perspective holds that leaders within an organisation are accountable to a broader range of interest groups including shareholders, clients, employees, society at large and the environment. This mindset acknowledges that leaders have a set of values that recognises the importance of the needs of multiple parties and understand the imperative to balance those needs. In addition, stakeholders expect leaders to "take active roles in fostering responsible behaviour, within and outside the organisation" (Pless and Maak 2011:4). This may take the form of an increased focus on corporate social responsibility, recognising performance across the triple bottom line and acting like a good corporate citizen.

Doh and Quigley (2014) consider the ways in which responsible leaders can influence the outcomes through stakeholder engagement. These authors identify two influence pathways, namely, psychological and knowledge based. The psychological pathway considers the evidence that a leadership style that is inclusive, stakeholder and community focused and based on trustful relationships and tends to have better outcomes in terms of follower perceptions related to effectiveness, fairness and honesty. The knowledge-based pathway recognises that leaders that are open and stakeholder-oriented may be well positioned to facilitate the flow of knowledge inside the organisation to improve effectiveness and support innovation.

2.3 Sustainable Development Goals (SDGs)

In light of linking up the SDGs and the reasonable leadership that can enable Land Bank to obtain the necessary funding through a number of partnerships in support of development and commercial agriculture in South Africa, it is paramount to provide a brief SDGs background. In September 2015, governments across the world adopted the 17 sustainable development goals (SDGs) and 169 targets as a plan of action to "free the human race from the tyranny of poverty and want and to heal and secure our planet" (United Nations 2015). These goals aim to direct action in five areas of critical importance over the next 15 years, namely, people, planet, prosperity, peace and partnership.

Governments, the private sector and civil society are aligning their policies and activities for meaningful contribution to the goals, of which a number are connected to agricultural activities, especially the improvement of smallholder agriculture. SDG 2 "Zero hunger" which aims to "end hunger, achieve food security and improved nutrition and promote sustainable agriculture" is the most obviously related to Land Bank's activities (United Nations 2015:15). Specific indicators with associated sub-indicators of importance are:

2.3 "By 2030, double the agricultural productivity and incomes of small-scale food producers, in particular women, indigenous peoples, family farmers, pastoralists and fishers, including through secure and equal access to land, other productive resources and inputs, knowledge, financial services, markets and opportunities for value addition and non-farm employment".

2.4 "By 2030, ensure sustainable food production systems and implement resilient agricultural practices that increase productivity and production, that help maintain ecosystems, that strengthen capacity for adaptation to climate change, extreme weather, drought, flooding and other disasters and that progressively improve land and soil quality".

2.a "Increase investment, including through enhanced international cooperation, in rural infrastructure, agricultural research and extension services, technology development and plant and livestock gene banks in order to enhance agricultural productive capacity in developing countries, in particular least developed countries".

Land Bank aims to create a more inclusive agricultural sector, and its financing activities of female farmers contribute to SDG 5 – "Gender equality" and specifically indicator 5.a "Undertake reforms to give women equal rights to economic resources, as well as access to ownership and control over land and other forms of property, financial services, inheritance and natural resources, in accordance with national laws".

In the light of the significant contribution of the sector to employment creation, Land Bank indirectly contributes to SDG 1 "No poverty", SDG 8 "Decent work and economic growth" and SDG 10 "Reduced inequalities". It is estimated that every R1 billion in new capital disbursed by Land Bank contributes to the creation of 3200 direct and indirect employment opportunities (Land and Agricultural Development Bank of South Africa 2017). It is also estimated that every R1 billion of Land Bank's capital employed results in an additional R2 billion of indirect capital being leveraged in the sector (Land and Agricultural Development Bank of South Africa 2017).

The agricultural sector is dependent on many other inputs for its success, including natural resources and financial and human capital. Therefore, partnerships are central to its sustainable development. In addition, Land Bank as an organisation is unable to meet its mandate without support from a range of external stakeholders, and partnerships are thus a key element of its strategy. SDG 17 "Partnerships for the goals" speaks directly to the need of both the sector and bank to develop partnerships that support the co-creation of a transformed and resilient sector. In this respect, Indicator 17.17 "Encourage and promote effective public, public-private and civil society partnerships, building on the experience and resourcing strategies of partnerships" is specifically relevant to this case study (United Nations 2015:27).

The case study in this chapter examines the impact of responsible leadership that enabled Land Bank to obtain the necessary funding through a number of partnerships in support of development and commercial agriculture in South Africa, which allowed it to contribute specifically to SDG 2 and SDG 17.

3 Research Design

3.1 Methodology Employed

This study relied on qualitative research to understand a single case in sufficient details. Qualitative research can be summarised as an inquiry process of understanding, based on distinct methodological traditions of inquiry that explore a social or human problem. Furthermore, the researchers build a complex, holistic picture, analyse words, report detailed views of informants and conduct the study in a natural setting. This allows the researchers to systematically analyse and organise large quantities of data contained in multiple data sources and formats in order to understand a situation from within (Creswell 1998; Evers and Van Staa 2012).

Over the period of April 2015 to March 2018, Land Bank as a specialist Development Finance Institution (DFI) significantly diversified and increased its sources of funding from external stakeholders to support the agricultural sector and pursue its development mandate. Land Bank was used as a case study to firstly examine the extent to which the leadership capacities described by responsible leadership theory were present in this transition and secondly assess the extent to which Land Bank's mandate affects the type of leadership required from its leadership.

Land Bank's activities are related to a number of SDG targets, and the organisation underwent significant change in the period under review, especially with regard to the structure of its funding. It will be critical for SOEs to crowd-in private sector investment to achieve their mandates within the fiscally constrained environment in which the South African government currently operates, and this case study highlights the contribution which responsible leadership may make in these circumstances.

The first step in analysis was to confirm the existence of responsible leadership capacities within the case study. This was tested against models proposed by Voegtlin (2016) regarding dimensions of responsibility. The next step was to test the influence pathways that resulted in the observed outcomes which was based on a model proposed by Doh and Quigley (2014). The final step confirmed the outcomes of responsible leadership based on another model proposed by Voegtlin, Patzer and Scherer (2012). Inductive analysis occurred during this phase as the combination of various data sources and models highlighted areas for additional research and consideration.

Focusing on the levels of analysis, within the field of responsible leadership, it is possible to analyse the conduct of leaders at various levels as discussed by Miska and Mendenhall (2018). Macro-level analysis includes linkages with broader cultural and societal structures which determine the environment within which the organisation operates and may impact the values held by the responsible leader. Meso-level analysis focuses on the linkages between responsible leadership and the organisational context, for instance, corporate governance and human resource management. This level of analysis may also focus on the outcomes of responsible leadership on, for instance, corporate social responsibility. Micro-level analysis focuses on individual leaders and specific capacities and competencies.

The research sample consisted of the Land Bank CFO and two senior managers (i.e. General Manager-Treasury and Manager-Investor Relations, both who interacted with investors in their roles who reported directly to the CFO and engaged with investors during the period under review). This case study examined how the Land Bank's funding profile changed to enable the organisation to meet its mandate and provide funding firstly to the agricultural sector and more importantly to emerging farmers. In a bank, the CFO is typically the person who is most involved with investors. For that reason, this case study concentrated on the leadership of the Land Bank CFO during the period of interest, although other managers were also considered to be linked to specific organisational outcomes.

In terms of the interview method and structure, the primary research took the form of semi-structured interviews. At the time of conceptualising the research, the CFO was still an employee of Land Bank. However, this CFO resigned prior to the interviews, which provided an opportunity to amend the interview structure. The interview with the CFO therefore focused on the specific case but also allowed the CFO to reflect on the outcomes of their decisions which related directly to the conceptual model used in the study.

The interviews with the followers were held at the Land Bank Head Office in Centurion and lasted about 20 minutes each. They were requested to reflect on the type of characteristics that they thought had been brought to the organisation by the former CFO but also to provide an idea of the leadership characteristics that a future CFO should bring to the organisation. These interviews specifically confirmed certain events that had not been published publicly by Land Bank. This allowed the researchers to identify certain differences between the period of the case study and prior period in the data analysis phase.

Once all the data had been collected, the interviews and documentation were analysed. The first phase entailed documenting the case study based on documentary evidence supplemented by the interview content. This involved analysing mostly written records from the period of interest and developing a timeline of events related to Land Bank's funding strategy and structure. The case study considered the funding structure at the start of the period, the partnerships and new funding obtained during the period and the steps taken to obtain the funding. This phase of the analysis relied on a combination of reports, press releases and content from the individual interviews to confirm certain elements that had not been clearly documented.

The next phase entailed an analysis of the facts of the case study in relation to leadership theory and the models as described. The first step was to confirm the existence of responsible leadership capabilities in the case study based on the understanding of responsibility as defined by Voegtlin (2016). This was based on the factual outcomes supported by exemplary quotes from various sources. Once the presence of responsible leadership had been confirmed, the influence pathways were considered as proposed by Doh and Quigley (2014). The final step was to assess the case study within the outcomes framework provided by (Voegtlin et al. 2012). The various outcomes at macro-, meso- and micro-level were identified, and the interviews, reports and other documents were used to identify changes that occurred during the time span of the case study and the relevance thereof to the model.

3.2 Case Study: Land Bank

Any bank requires funding to lend out to its clients. Funding may come in the form of equity from shareholders, deposits from clients, loans from other banks or funds from investors like asset managers or institutional investors. In this regard, Land Bank has limited options. Section 26 of the Land Bank Act specifies a number of sources of funds including share capital, income from operating activities, interest earned on investments, funds appropriated by Parliament, deposits, grants and donations as well as the proceeds of the loans that the Bank provide. Equity is provided by its shareholder on an occasional basis, and Land Bank ceased to take deposits in the 1990s.

Land Bank must raise funding in order to provide the loans required to meet its mandate, which allows it to support economic growth, employment creation, transformation and inclusivity in the sector and thus contribute to the achievement of various SDGs. Fundraising is therefore a key component of the Bank's strategy and business objectives. This section considers the impediments the Bank has to contend with and the actions required from leadership to improve the quantity, cost and tenor of funding available to Land Bank.

The section specifically considers the Bank's responses to liquidity concerns, comparability of financial information, corporate governance concerns, funding to support its development mandate and social and environmental standards. During 2015, Land Bank's shareholder, National Treasury, required it to restructure its balance sheet and develop a suitable funding model for an agricultural DFI (Land and Agricultural Development Bank of South Africa 2016). By May 2018, the Bank was able to issue an unguaranteed 10-year fixed rate bond in the commercial market and receive two unguaranteed 12-year loans from multilateral lenders (Land and Agricultural Development Bank of South Africa 2018a).

In order to provide investors with a consistent basis for comparison, it was decided to adopt a number of ratios used by the commercial banking sector based on the so-called Basel Accords set by the Basel Committee on Bank Supervision. Land Bank adopted the Capital Adequacy Ratio, Net Stable Funding Ratio and Liquidity Coverage Ratio (Land and Agricultural Development Bank of South Africa 2016). These had to be adapted to provide for the fact that Land Bank is not a deposit-taking institution and therefore required additional engagement with investors to reach agreement on the structure of the ratios. The standard only came into effect in 2018, but Land Bank had adopted the standard in its 2015/2016 financial year. This required the development of new financial models in line with global requirements which implied that the staff involved had to develop the necessary skills to comply with the new standards. Land Bank achieved its objective of reducing its reliance on a limited number of investors. It attracted both new commercial investors and additional multilateral investors (Land and Agricultural Development Bank of South Africa 2018a:53). In this manner, it created new partnerships to support both commercial and development agriculture and thus expand resources available to contribute to the SDGs.

Land Bank immediately set out to engage directly with Futuregrowth to address their concerns. The engagement allowed the CFO to identify specific investor concerns related to credit committee quorums and information sharing with investors. Within less than a month, Land Bank became the first entity where Futuregrowth resumed investing thus establishing "that you can never underestimate the power of negotiation. The past three weeks enabled both institutions to clarify any existing problems and misunderstandings they might have had" (Mchunu 2016). The public commitment required internal engagement starting with the Land Bank Board. The Board had to agree to the stringent requirements on corruption and increased oversight on risk and corporate governance. Consistent communication was required with staff to create awareness of the requirements for investors and the impact thereof. Transparency required that Land Bank publish some of its key policies publicly. These included its supply chain management policy, board charters, fraud and corruption policy, policy on politically exposed persons and conflict of interest policies (available on the Bank's website at https://landbank.co.za/About-Us/Pages/Corporate-Governance.aspx). At a time when other SOEs struggled to raise money in the capital and debt markets (Donnelly 2017), Land Bank's bond auctions were oversubscribed, and it was able to raise long-term unguaranteed debt (Land and Agricultural Development Bank of South Africa 2018a). By the end of its 2017/2018 financial year, Land Bank had significantly improved the funds available to the agricultural sector and specifically development farmers through its multilateral facilities. The Kreditanstalt für Wiederaufbau (KfW) and European Investment Bank (EIB) facilities combined provided 105 million EURO and the World Bank facility an additional US $93 million. These facilities are mainly aimed at supporting development farmers who make up a key element of SGD 2, specifically indicators 2.3 and 2.4. These funds are also to meet the requirements of indicator 2.2 which specifically refers to the availability of foreign aid and other flows to support agriculture.

By March 2018, Land Bank had screened ten transactions including its largest single debtor across six commodities (Land and Agricultural Development Bank of South Africa 2018a). The EIB was sufficiently satisfied with the Bank's progress in this regard that it ring-fenced a portion of its loan to Land Bank to promote climate change projects in the agricultural sector. More investors were participating and willing to provide funds over longer terms; by March 2018, the reliance on short-term funding had been reduced to 43.8%. The cost of funding had also reduced when compared to similar terms in 2015/2016. These developments improved the financial sustainability of Land Bank and increased the overall level of funding available to the sector.

4 Findings and Results

4.1 Responsibility Leadership Dimensions of Land Bank

The first step is to determine the extent, if any, to which responsible leadership was present in the case study. For the purpose of analysis, the four dimensions identified by Voegtlin (2016) are applied.

The first dimension relates to the fact that leaders do not operate in a vacuum and that leaders are frequently required to take decisions in a complex environment involving multiple stakeholders (i.e. not isolating the leader) (Voegtlin 2016). There are a number of examples in the case study where various stakeholders collaborated with Land Bank leadership to achieve a specific outcome. The shareholder expected Land Bank to extend the tenor of its funding to reduce its reliance on short-term funding in 2015. However, it recognised that the leadership would not be able to achieve this objective given its available resources at the time and therefore provided support in the form of a R2.7 billion government guarantee. The CFO was then able to identify and attract a number of private sector investors willing to invest for a period of 5 years in the organisation albeit at a relatively high interest rate. These investments were prepaid in

2018 in accordance to the loan agreements (Land and Agricultural Development Bank of South Africa 2019).

The second dimension recognises that there are often no clear moral norms to guide ethical decision-making especially when there are multiple stakeholders with valid but conflicting concerns (i.e. critically evaluating prevailing societal norms, rules and conventions) (Voegtlin 2016). Land Bank incorporated specific loan covenants related to fraud and anti-corruption activities in their loan agreements which allow investors to demand repayment of their loans if Board members or senior management are convicted of such activities. Land Bank was the first SOE to create these covenants and commit the organisation to such ethical standards (Van Rooy 27 June 2019).

The next dimension focuses on the forward-looking orientation of responsible leaders who should "anticipate the consequences of their actions, orient their thinking to the long-term, and consider the negative impact of doing business" (i.e. forward-looking rather than backward-looking) (Voegtlin 2016:590). During the period under review, Land Bank committed to report specific environmental events to investors that may have a material impact on their operations. In addition, it collaborated with the multilateral funders to improve its environmental and social assessment process of all loans through its direct network. As a result of the progress in implementing the environmental assessment in its loan process, the EIB extended funding to Land Bank that may be specifically used to support farmers to adapt to climate change.

The final dimension is closely related to the stakeholder-centric approach of responsible leadership and focuses on shared responsibility (i.e. shared responsibility and collective problem-solving) (Voegtlin 2016:590). Given the lack of government funding, Land Bank identified the need to raise specific funding to support development farmers and approached various multilateral funders. It subsequently raised funding through the World Bank, KfW and its MIGA facility that are earmarked for development farmers.

4.2 Influence Pathways Identified

The second model considered in the analysis relates to pathways that may influence outcomes as a result of responsible leadership. Doh and Quigley (2014) identified two possible pathways through which responsible leaders may influence stakeholders to achieve certain outputs. These influence pathways are not mutually exclusive and may co-exist within organisations.

The first pathway is based on the psychological benefits of responsible leadership where the authors cite a number of instances where a stakeholder inclusive approach is linked to higher levels of engagement with the organisation. The willingness of responsible leaders to include the needs, expectations and opinions of internal and external stakeholders appears to improve employee satisfaction, commitment and performance (Doh & Quigley 2014).

In terms of staff commitment, Land Bank's attrition rate of 7.89% in 2017/2018 was well below the market norm for the financial services sector of 12% and 92% of employees who responded to a climate and culture survey found meaning in their work (Land and Agricultural Development Bank of South Africa 2018a:79.81). This would appear to support the premise that responsible leadership is positively correlated with issues such as staff commitment and engagement.

The second influence pathway is knowledge-based where responsible leaders occupy positions that allow them to spread knowledge throughout the organisation and external stakeholders. Knowledge-sharing allows organisations to create new knowledge and is an important element of innovation. Responsible leaders extend the process of knowledge-sharing beyond the organisation through interactions with external stakeholders which allow them to incorporate external perspectives into the organisation (Doh and Quigley 2014).

The development of loan covenants that respond directly to investor concerns regarding governance and potential corruption at the organisation is an example of a knowledge-based influence pathway. Land Bank was invited to provide inputs in the JSE Bond Listing requirements review that is being undertaken in the face of criticism that these were not adequate to protect investors from fraudulent actions by debt issuers (Futuregrowth Asset Management 2018).

4.3 Responsible Leadership Outcomes

The third model to be confirmed in this case study relates to the expected eight outcomes of responsible leadership at macro-, meso- and micro-level in the organisation which were identified by Voegtlin, Patzer and Scherer (2012). The expected outcomes were converted to the research questions considered in this case study and the findings compared to the proposed outcomes.

4.3.1 Macro-level Outcomes

At a macro-level, the outcomes proposed by Voegtlin, Patzer and Scherer (2012) focus on stakeholder relations and the potential impact of positive stakeholder relationships for the organisation. Three aspects are highlighted in relation to the macro-level outcomes.

The first aspect, building and maintaining the legitimacy of an organisation through responsible leadership, was evident as Land Bank had to balance the need of development farmers who require access to patient capital with investor expectations regarding the performance of the organisation and returns on funding. Land Bank improved its corporate governance and strengthened investor relations in order to access longer-term funding by multilateral organisations for this purpose.

The effect on building trustful stakeholder relations through responsible leadership, as the second aspect, indicated that Land Bank engages regularly with investors through road shows and ad hoc engagements to share information on its financial performance and business strategy. During these engagements, the CFO recognised valid investor concerns and expectations, and actions were undertaken to respond to these concerns. Land Bank committed to public, transparent and timely communications through the Stock Exchange News Service (SENS) to keep investors informed of important developments at the Bank including changes in the leadership and financial performance.

The third aspect, enhancing the social environment through responsible leadership behaviour, was identified by the CFO's pro-active engagement and responsiveness with investor expectations which reduced Land Bank's reliance on a limited number of investors during the period. Land Bank attracted new asset managers as well as multilateral organisations as funders reducing its reliance on the PIC and CPD by 5% (Land and Agricultural Development Bank of South Africa 2018a).

4.3.2 Meso-level Outcomes

At a meso-level, the proposed model aims to consider how responsible leader's actions and conduct may influence organisational practices and culture. The first aspect was encouraging a culture of discursive conflict resolution and deliberative practices through responsible leadership in order to change the ethical culture of an organisation over time. Land Bank engaged the Ethics Institute to conduct an ethics assessment and review the perceptions of internal and external stakeholders on ethics in the organisation. In response to the findings of the assessment, it developed an ethics training programme for all employees and facilitated the certification of two senior employees as Ethics Officers of Land Bank (Land and Agricultural Development Bank of South Africa 2018a:104–105). This resulted in a number of policies being developed and published including the Code of Ethics and Business Conduct and Fraud and Corruption policies which are available on the Land Bank website at https://landbank.co.za/About-Us/Pages/Corporate-Governance.aspx.

Second, another aspect highlighted the effect of responsible leadership on the perceived importance of corporate social responsibility in an organisation. Land Bank is assessed against the Financial Services Sector Code and has a Level 4 B-BBEE rating (Land and Agricultural Development Bank of South Africa 2018a).

However, Land Bank's mandate, which requires support for development farmers and transformation in the sector, demands a CSR orientation from its leadership. A constant challenge for the CFO is to balance financial sustainability with the developmental mandate. During the period under review, Land Bank increased the percentage of its loan book defined as transformational to 12% of the total loan book (Land and Agricultural Development Bank of South Africa 2018a:38). Land Bank also forfeited approximately R174 million in profit to provide subsidised loans to development farmers.

The likelihood of responsible leaders to act as social entrepreneurs was identified as a third aspect. One example of innovation that Land Bank undertook in partnership with KfW during the period related to the disbursement of its loan facility which was done in South African Rand. This was the first loan in Africa that KfW disbursed in local currency which reduced the cost thereof and associated currency risk significantly for Land Bank (Land and Agricultural Development Bank of South Africa 2018b). This allows Land Bank to extend the funds to development farmers at an affordable rate to support its mandate.

Lastly, the aspect of the contribution that responsible leadership has on the performance of an organisation under the caveat of ethical or moral means was evident. During the period of the case study, Land Bank was profitable, achieving a total comprehensive income of R285.3 million in 2015/2016, R305.8 million in 2016/2017 and R182.3 million in 2017/2018. Furthermore, Land Bank achieved its goal of extending the terms of its funding liabilities and increased the number of investors in the organisation.

4.3.3 Micro-level Outcomes

In addition to the impact on external stakeholders and the broader organisation, responsible leaders will also have an impact on their direct reports where they will act as role models (Voegtlin et al. 2012). Therefore, it was evident that responsible leadership had an effect on followers' attitudes and cognitions in Land Bank.

The CFO summarised their leadership style in an external interview shortly after resigning, which follows:

> I believe that in retrospect that people won't necessarily remember what you may have told them, but they will always remember how you made them feel. Did you help them? Did you engage with them? Were they made to feel respected? I believe that we need to understand the psychology of people behaviour in order to get the best out of them. I think that I'm an intuitive leader and this is important in understanding people and what drives them (Anonymous 2019).

5 Conclusions

This chapter reflected on the type of leadership displayed in Land Bank during the period of March 2015 to March 2018. Land Bank is the only DFI in South Africa focused on primary agriculture and had a specific mandate to transform the agricultural sector. Primary agriculture as a sector is able to contribute specifically to SDGs 2, 5, 8 and 10 as it is a significant employer in rural areas. The Land Bank's mandate to support development farmers allows it to make a direct contribution to SDG 2 and, through the partnerships that it established to obtain funding for its development mandate, to SDG 17.

During the period under review, Land Bank's leadership had to respond to a few specific challenges raised by investors to lengthen the tenor of its funding and diversify its investor base. The case study illustrated the extent to which responsible leadership was present in the actions of the CFO and other senior managers. They identified legitimate concerns and expectations of investors and developed specific responses which exceeded regulatory requirements to address these issues. Their actions as responsible leaders resulted in a number of positive outcomes for the organisation, external and internal stakeholders.

Three models proposed by previous contributors to responsible leadership theory, namely, the dimensions of responsible leadership, responsible leadership influence pathways and multilevel outcomes of responsible leadership, were applied to the case study. These models were tested and found to be relevant at a macro-, meso- and micro-level to Land Bank as an SOE. The CFO and other executive managers at Land Bank displayed responsible leadership during the period under review, and these findings correlated to the multilevel outcomes model of responsible leadership. The presence of both a psychological-based and knowledge-based influence pathway in the case study was confirmed. Through its activities, the Land Bank has had (and can continue to have) the ability to support multiple sustainable development goals (SDG) as its lending allows the agricultural sector to contribute to economic growth, employment creation and income generation.

References

Anonymous (2019). Stepping stones in preparation for what follows (September 2019) [Online] Available from: https://www.bravura.net/stepping-stones-in-preparation-for-what-follows/ [Accessed: 4 November 2019].

Creswell JW (1998). Qualitative inquiry and research design: Choosing among five traditions, Sage Publications, Thousand Oaks, California.

Doh JP & Quigley NR (2014). Responsible leadership and stakeholder management: Influence pathways and organisational outcomes. The Academy of Management Perspectives, 28(3), 255-274.

Donnelly L (2017). Parastatals' poor record hits home, Mail & Guardian, 2 June 2017 [Online] Available from: https://mg.co.za/article/2017-06-02-00-parastatals-poor-record-hits-home [Accessed: 2 November 2019].

Evers JC & Van Staa A (2012). Qualitative Analysis in Case Study. In: Mills, A.J., Durepos, G. & Wiebe, E. (eds.). Encyclopedia of Case Study Research. Thousand Oaks, California: SAGE Publications, Inc.

Futuregrowth Asset Management (2018). SOE governance unmasked: a learning journey, Futuregrowth Asset Management, Cape Town, South Africa.

Jacobs J (2013). History of the Land Bank: financing agriculture for 100 years, Land Bank Public Information Centre, Centurion, South Africa.

Land and Agricultural Development Bank of South Africa (2008). Annual Report 2007/2008, Pretoria, South Africa.

Land and Agricultural Development Bank of South Africa (2015). Integrated Annual Report FY2015, Centurion, South Africa.

Land and Agricultural Development Bank of South Africa (2016). Annual Integrated Report 2016, Land and Agricultural Development Bank of South Africa, Centurion, South Africa. [Online] Available from: https://landbank.co.za/Shared%20Documents/Annual-Report-2015-2016.pdf.

Land and Agricultural Development Bank of South Africa (2017). Annual Integrated Report 2017, Centurion, South Africa. [Online] Available from: https://landbank.co.za/Shared%20Documents/Annual-Report-2016-2017.pdf.

Land and Agricultural Development Bank of South Africa (2018a). Integrated Annual Report 2018, Land and Agricultural Development Bank of South Africa, Centurion, South Africa. [Online] Available from: https://landbank.co.za/Shared%20Documents/Annual-Report-2017-2018.pdf.

Land and Agricultural Development Bank of South Africa (2018b). Land Bank partners with KfW Development Bank to grow local agricultural sector Press release issued on 1 February 2018 2018c. [Online] Available from: https://landbank.co.za/Media-Centre/Press%20Releases/2018/1.%20Press%20Release%20-%20Land%20Bank%20partners%20with%20kfW%20Development%20Bank%20to%20grow%20local%20agricultural%20sector.pdf.

Land and Agricultural Development Bank of South Africa (2019). Integrated Annual Report 2019, Centurion, South Africa. [Online] Available from: https://landbank.co.za/Shared%20Documents/Annual-Report-2018-2019.pdf.

Mchunu S (2016). Land Bank CFO hails Futuregrowth move. Business Report, 27 September 2016 [Online] Available from: https://www.iol.co.za/business-report/economy/land-bank-cfo-hails-futuregrowth-move-2073136 [Accessed: 28 September 2019].

Miska C & Mendenhall ME (2018). Responsible Leadership: a Mapping of Extant Research and Future Directions. Journal of Business Ethics, 148(1), 117-134.

National Planning Commission (2012). National Development Plan 2030: our future - make it work, Pretoria, South Africa.

Pless NM (2007). Understanding Responsible Leadership: Role Identity and Motivational Drivers. Journal of Business Ethics, 74(4), 437-456.

Pless NM & Maak T (2011). Responsible Leadership: Pathways to the Future. Journal of Business Ethics, 98, 3-13.

Statistics South Africa (2019a). Gross domestic product: fourth quarter 2018 P0441, Pretoria, South Africa. [Online] Available from: http://www.statssa.gov.za/publications/P0441/P04414thQuarter2018.pdf [Accessed: 20 October 2019].

Statistics South Africa (2019b). Quarterly labour force survey: second quarter 2019 P0211, Pretoria, South Africa. [Online] Available from: http://www.statssa.gov.za/publications/P0211/P02112ndQuarter2019.pdf [Accessed: 20 October 2019].

Strauss CB, Brand DT, Dolny HM, Goldblatt S, Jack BS, Joubert JSG, Memela BT, Motsepe DR, Nhlapo K, Raath JW, Vink N, Wrighton PGA & Coetzee GK (1996). Final Report of the Commission of Inquiry

into the Provision of Rural Financial Services (Strauss Commission), Pretoria, South Africa.
United Nations (2015). Resolution adopted by the General Assembly on 25 September 2015: Transforming our world: the 2030 Agenda for Sustainable Development. [Online] Available from: https://www.un.org/ga/search/view_doc.asp?symbol=A/RES/70/1&Lang=E [Accessed: 2 November 2019].
Voegtlin C (2016). 'What does it mean to be responsible? Addressing the missing responsibility dimension in ethical leadership. Leadership, 12(5), 581-608.
Voegtlin C, Patzer M & Scherer AG (2012). Responsible Leadership in Global Business: A New Approach to Leadership and its Multi-Level Outcomes. Journal of Business Ethics, 105(1), 1-16.
Waldman DA & Galvin BM (2008). Alternative Perspectives of Responsible Leadership. Organisational Dynamics, 37(4), 327-341.

3 Confronting Poverty, Hunger, and Food Insecurity: Lessons from Malawi and Zimbabwe

Mavis Thokozile Macheka
and Gift Wasambo Kayira

Abstract

Since 2016, Malawi and Zimbabwe have made strides to domesticate the 17 sustainable development goals (SDGs) that the United Nations formulated in 2015. This chapter analyses how Malawi and Zimbabwe address the goals of ending poverty and achieving hunger-free societies. What strategies exist at the district level for confronting challenges in agriculture and how are rural communities embracing them are the key questions this chapter seeks to answer. The chapter addresses these questions through a comparative analysis that draws experience from Rumphi and Chiredzi districts of Malawi and Zimbabwe, respectively. Using data obtained through interviews from the two districts, the chapter argues that although the SDGs are framed within the optimism of "African Rising Narrative," which holds that post-2010 Africa is on the right path to achieving development, experiences from Malawi and Zimbabwe do not inspire strong confidence. Although public-private partnerships and contract farming arrangements yield positive results for local farmers, the story of these communities is that of the same problems of the past, manifesting through poor access to agricultural inputs and markets and lack of active farmer participation in experts' interventions. Unless addressed through deliberate state intervention, these shortfalls are likely to compromise the fight against poverty and hunger.

Keywords

Africa Rising narrative · Hunger · Malawi · Poverty · SDGs · Zimbabwe

1 Introduction and Background

On 1 January 2016, Southern African countries and others adopted the 17 sustainable development goals (SDGs), which the United Nations identified in 2015 as part of the 2030 Agenda for Sustainable Development. Since then, both Malawi and Zimbabwe have made strides to domesticate these goals through country-specific frameworks such as the *2017–2022 Malawi Goals Development Strategy* (*MGDS*) *III* and the *2016–2018 Interim Poverty Reduction Strategy Paper* (*I-PRSP*) for Zimbabwe. This chapter

M. T. Macheka (✉)
Great Zimbabwe University, Masvingo, Zimbabwe
e-mail: mtmacheka@gzu.ac.zw

G. W. Kayira
University of Malawi, Chancellor College,
Zomba, Malawi
e-mail: gkayira@cc.ac.mw

G. Nhamo et al. (eds.), *Sustainable Development Goals for Society Vol. 2*, Sustainable Development Goals Series, https://doi.org/10.1007/978-3-030-70952-5_3

analyses how state intervention in Malawi and Zimbabwe translates into practice as the two countries strive to address the goals of ending poverty (goal 1) and achieving hunger-free societies (goal 2). It draws experience from Rumphi and Chiredzi districts of Malawi and Zimbabwe, respectively. Given that the societies in question are predominantly agrarian, what strategies exist at the district level for confronting challenges to agriculture, and how are rural communities embracing them? What role do the communities play in implementing such interventions? What challenges of implementation exist, and how have they been mitigated? This chapter addresses these questions. These are significant questions because they are raised against the optimism of the "Africa Rising" narrative, which assumes that the post-2010 African societies are on the right path to achieving the sustainable development.

First popularized by *The Economist* in the 3 December 2011 and 2 May 2013 issues, the Africa Rising narrative propagates the view that Africa is registering economic growth, which is a subject of sound democratic governance, inclusive politics, and better economic management (Taylor 2014:146). Africa also benefits from good economic relations with the rising powers in the Global South and others, notably Brazil, Russia, India, China, and South Africa (the BRICS). This narrative departs from an earlier view which *The Economist* advanced in its 13 May 2000 issue, which condemned Africa as a hopeless continent with lopsided economic policies and poor record of political governance that did little to spur progress (Pillay 2015:60).

Since then, the narrative has matured into a full-blown discourse, albeit one that is polarized around two opposing poles. For such scholars as Hanson et al. (2018), the improved economic environment which Africa Rising narrative highlights could provide an impetus to achieving the SDGs and other related frameworks such as the African Union Agenda 2063. Hanson and colleagues call upon African leaders to shepherd the Africa Rising narrative by crystallizing it with sustainable development, which the SDGs advance. Unlike in the previous decades, the post-2010 Africa has witnessed an improvement in its moribund infrastructure and improved and well-connected markets which have made goods accessible. Besides, "increased Africa's leverage in the global system has matched with a surge in other socio-economic developments such as the construction of new hospitals and schools, and investments in research and development programmes" (Naidu et al. 2009:1–4). With these achievements, Taylor (2012) like the *Economist* dismisses the previous pessimistic view which saw Africa locked up in unfair international dependence or internal patrimonial politics and other ills that retarded the progress of the continent. Therefore, African states can build on this progress and achieve the SDGs than otherwise was the case with the MDGs.

This reading of the post-2000 Africa's economic environment does not inspire confidence in other scholarly circles. Pillay (2015), first, argues that the Africa Rising narrative is misleading, not least because "it draws the people of Africa into a false sense of promise of 'development' and 'decent' jobs for all—that can never be delivered by the current economic growth paradigm" (Pillay 2015:71). More interesting are the premises of Pillay's work. He notes that scholars that celebrate this narrative base their arguments on statistics that do not give a comprehensive picture of Africa's economic performance. Second, the statistics on growth reflects activities under the extractive sectors with little regard to the net loss to Africa; third, and perhaps more significant to this chapter, Pillay argues that "it is a truism that narrow Gross Domestic Product (GDP) growth, loved by investors who only see returns for themselves, can mean little or nothing to the lives of ordinary people" (Pillay 2015:61). Pillay is not alone. Earlier, Taylor (2014) faulted the optimistic narrative for basing its arguments on GDP figures, which mask the inequality gap or the struggles the poor face within sub-Saharan African countries. Because this growth corresponds to the "rise of commodity prices, new discoveries of natural resources, or increase in sources of foreign capital," Taylor (2014: 145) adds it is not sustainable. Guchu (2018), Slattery and Stanton (2013), and Vickers (2013) share similar sentiments and criticize the Africa Rising narrative for projecting an optimistic picture of Africa's economic performance that does not speak of the specific concerns on poverty eradication.

This opposing view is important because it points to the plight of the African poor living in both rural and urban areas. It reminds us about the dangers of putting faith in GDP figures, hoping that national economic growth will trickle-down to the poor. Such shortfalls, however, should not warrant a dismissal of the changing economic climate in Africa. The Africa Rising discourse is significant because it highlights the age-old debate on the contradictions of the development process. As Cowen and Shenton (1996) have illustrated elsewhere, unless regulated, capitalist-induced growth has the tendency to create the dualism of accumulation and dispossession. While empowering some, whether at international, regional, or national levels, it also disempowers others. The SDGs recognize this anomaly and act as an instrument with which nations and the global world can regulate growth and ensure that it benefits the various sectors of the society. The SDGs, therefore, are a necessary guard against deepening inequalities and dispossession. If carefully implemented, we argue, SDGs 1 and 2 and other related goals can help to address the perceived shortfalls highlighted above.

Since the SDGs envisaged the shortfalls of the Africa Rising narrative, the challenge we have is to understand how the interventions African governments employ to domesticate these goals target the poor without sacrificing economic growth. We argue that experience from Malawi and Zimbabwe, relative to how the states in question translate into practice the goals on poverty and hunger, does not command strong confidence. Although public-private partnerships and contract farming arrangements are yielding some positive results, the overall impression suggests some gaps that would compromise the attainment of goals 1 and 2 of the SDGs. The story of Rumphi and Chiredzi communities is that of the same problems of the past, manifesting through poor access to agricultural inputs and markets and lack of active farmer participation in experts' interventions. As Puplampu et al. (2018) emphasize, the attainment of the SDGs partly depends on good governance that advances the principles of transparency, accountability, and local participation. Although the will to achieve local participation is strong in both Malawi and Zimbabwe, knowledge about the specific agricultural interventions is transmitted from above such that farmers are reduced into recipients or consumers of that knowledge. As a result, the communities do not fully appreciate state interventions targeted at improving their lives.

By bringing cases of Malawi and Zimbabwe on the same plane, this chapter speaks to SDG's attention to regional realities about poverty, hunger, and other related goals (Puplampu et al. 2018). Again, by comparing experience from the localized studies of the two countries, we can better appreciate the effort African governments make to domesticate the SDGs and identify common bottlenecks. Because SDGs call for international and regional cooperation as a strategy to achieving the set goals, moreover, equally significant is an approach that measures the successes gained or the challenges encountered through a similar approach. While the contribution African agriculture can make in raising the GDP of African countries remains debatable, the agricultural sector plays a significant role in cautioning African communities from poverty and malnutrition. Besides, agriculture remains a significant sector providing wage or self-employment to African communities, although barely 10% of sub-Saharan arable land is utilized. Understanding specific challenges confronting this sector is crucial to arresting poverty and hunger among rural communities.

We present our argument through four significant sections. The subsequent one discusses the methodology guiding the study and the sources of data used. The second discusses how Malawi and Zimbabwe have approached the problem of poverty since their independence in 1964 and 1980, respectively. This section further provides a context to the third, which specifically reflects on Rumphi and Chiredzi to assess how the two districts domesticated goals 1 and 2 of the SDGs. In the concluding section, we present some suggestions on how to re-orient the existing interventions in ways that would align them with the aims of the SDGs.

2 Materials and Methods

This study adopted a qualitative research methodology to understand the lived experiences of the farming communities in Rumphi (Malawi) and Chiredzi (Zimbabwe) and how they interact with state and non-state officials who enforce the SDGs' interventions. Our analysis draws on oral interviews conducted among selected communities in two districts. In both cases, we interviewed men and women through focus group discussion (FDGs) and one-on-one interviews. We purposively sampled state and non-state institutions, including NGOs working in communities under study. Here, we interviewed officials serving in the departments of Agriculture, Development, and Planning at the district level and those serving in selected non-state institutions. The choice of communities under examination is deliberate. Rumphi and Chiredzi communities predominantly engage in farming for their livelihood and mostly rely on rain-fed agriculture. The two areas are also vulnerable to droughts and floods.

In Rumphi, maize, tobacco, and legumes stand out as the primary economic crops. Maize is equally important among the people of Chiredzi, while cotton and irrigation-based sugar rank as significant economic crops. These communities also specialize in both subsistence (food) and cash crop production. To make sense of the interviews collected, we subjected the data to a thematic analysis, grouping them into themes such as those of the agronomic challenges farmers face, including how they access agricultural inputs. Other themes focused on farmers' interaction with agricultural and development experts and the specific interventions the latter groups bring to the communities. The detailed descriptions, narrative vignettes, and direct quotes from interviews and focus group discussions which we present follow these themes. In keeping with ethical considerations, we have anonymized the interview respondents and only partially revealed those who allowed us to do so through their official titles.

3 Presentation and Discussion of Findings

3.1 Fighting Poverty and Hunger in Malawi and Zimbabwe: A Historical Overview

The fight against poverty, hunger, and food insecurity in Malawi and Zimbabwe predates the more recent interventions of the MDGs and SDGs. After independence in 1964 (for Malawi) and 1980 (for Zimbabwe), African governments in these countries put in place strategies for reversing the economic disparities that colonial economies had created. In both cases, the new governments inherited a dual economy characterized by a prosperous white large-scale estate sector and impoverished peasant sector (Kydd and Christiansen 1982). Again, in both cases, colonialism had left countries with a neglected peasant sector and widespread poverty (Mhone 1993; Mlambo 1997). Such inequalities contrasted the anticipation people had for independence. Therefore, it was necessary to bridge the gap between the rich and the poor and raise the economic capacity of the countries.

With Zimbabwe, the immediate solution lay in introducing a "socialist policy" dubbed "Growth with Equity." Introduced in 1981, Growth with Equity set out to achieve economic growth and address the socioeconomic disparities (Chinake 1997). To improve the living standards of the poor, the policy conditions provided for free healthcare, education for all, and rural development. Increasing rural minimum wage was another aspect of this intervention (Balleis 1993). Between 1982 and 1990, the state raised two development plans aimed at regulating the development process. These were the 1982–1985 Transitional National Development Plan (TNDP) and 1986–1990 First Five-Year National Development Plan (FFYNDP). Both emphasized the need to achieve economic growth, create employment, and reduce poverty and inequalities. Some argue that between 1981 and 1990, the quality of life for the majority of Zimbabweans improved, although the country remained economically unstable (Mlambo 1997).

To stabilize the economy, Zimbabwe, like other African countries, implemented the WB and IMF-inspired Economic Structural Adjustment Programs (SAPs) between 1991 and 1995. The World Bank (WB) criticized Zimbabwe's policy on Growth with Equity. The Bank argued that the policy increased expenditure on social developmental programs without compensating it with similar growth in the economy. Moreover, such expenditure reduced national savings and investment capital (World Bank 1995).

Although the SAPs coincided with the drought of the early 1990s, which negatively impacted agricultural productivity, the reforms in question helped to reverse the previous optimism (Mlambo 1997; Sachikonye 2002). Under these SAPs, Zimbabwe was forced to abandon its socialist reforms and the economic protectionism of the 1980s. Market liberalization, devaluation of local currency, deregulation, and removal of subsidies from the public sector combined to impoverish Zimbabweans.

Even with the onset of the SAPs, the government did not completely abandon social programs. For example, in 1991, it introduced the Social Development Fund (SDF) to cushion the vulnerable poor communities against negative effects of the reform programs through targeted interventions (Human Development Report 1998). The SDF supported the "Employment and Training Program" for those that were retrenched and the "Social Welfare Program." The latter provided subsidized food or food money to the urban poor. The focus of SDF was too narrow to contain poverty meaningfully. Bureaucratic delays and limited resources further limited its success. As a result, the government developed a much broader approach to poverty alleviation under the Poverty Alleviation Action Plan or PAAP (Chinake 1997). The PAAP broadened the overall scope, coverage, and impact of selected social programs which the state retained despite implementing the SAPs (UNICEF 1994). For Chinake (1997), this program too failed because of inadequate funding. But the PAAP was the exact antithesis of what the WB and the IMF advocated.

Between 1998 and 2003, the government introduced several other strategies directed at mitigating poverty among the poor, most of which failed because of inadequate resources, over-politicization, and somewhat weak cooperation among stakeholders (Nyathi 2012). These included Zimbabwe Programme for Economic and Social Transformation of 1998, which aimed at replacing ESAP and was dubbed as a "home-grown" reform package; Fast Track Land Reform Program of 2000 whose aim was to improve the amount of land freed up for resettlement and to empower black communities; and 2001 Millennium Economic Recovery Programme which attempted to fight escalating economic inflation. The Rural Electrification program (2002–2003), National Economic Revival Programme (2003), and Short-term Economic Recovery program (2009) were among other programs the government launched to improve the living standards of the rural people. As noted, these performed meagerly because of the reasons cited above.

The will to improve the rural societies remained strong in the 2000s, and this coincided with the policy environment the MDGs created. It was clear, however, that the government would not halve poverty levels as prescribed under the MGDs. In 2013, the state introduced the Zimbabwe Agenda for Sustainable Socio-Economic Transformation (ZIMASSET). ZIMASSET had four clusters that later aligned with the SDGs on poverty and hunger. These were "food security and nutrition," "social services and poverty eradication," "infrastructure and utilities," and "value addition and beneficiation cluster." These clusters were later integrated into the 2016–2018 I-PRSP, which the government raised to domesticate the SDGs.

Similar dynamics played out in Malawi. The state's plan for improving the lives of citizens and raising the economic growth of the country had a strong agrarian bias. In general, the development plans (transitional development plan, 1962–1965; development plan of 1965–1969; and the Development Policy of 1971–1980) emphasized supporting both large-scale commercial and rural farmers simultaneously. To avoid replicating the colonial era economic imbalances, the planners ensured that both the estate and smallholder agriculture would thrive without prejudicing the other.

Although not much took place in the early years of independence, from the late 1960s, the state introduced four integrated rural agricultural projects and improved 16 colonial era irrigation schemes to allow farming communities grow food and cash crops. The four integrated projects, two in the Central Region and each of the remaining two in the Southern and Northern regions of the country, were large scale and capital intensive. They targeted the production of crops such as maize, cotton, and groundnuts. In one of these—the WB-supported Lilongwe Land Development Programme of Central Malawi—the state planned to create markets in land by providing farmers with secure registered land titles (Ng'ong'ola 1982). A Customary Land Registration Act was passed in 1967 to support these reforms. The government envisaged that once piloted in Lilongwe, the program would extend to the rest of the country. Such plans did not materialize as Lilongwe farmers rejected registering customary land under individual titles (Kayira 2020).

The state also established four Smallholder Authorities, which were quasi-commercial state-owned enterprises. These aimed to assist small-scale farmers to participate in cash crop production (Chirwa and Kydd 2005). Each of the Authorities targeted the production of one of the major cash crops of the country—tea, coffee, tobacco, and sugar. In all these interventions, farmers were provided with input credits with which to purchase subsidized fertilizer, seed, and other agricultural inputs. Besides, the government empowered its marketing board, the Agricultural Development and Marketing Corporation (ADMARC), to facilitate the marketing of farmer's produce. With over 90% of the five million population living in the rural areas by 1967, such interventions would potentially uplift the living standards of the rural poor while contributing to the economic growth of the country.

By 1979, these strategies had achieved mixed results. Malawi became self-sufficient in food, but the structural imbalances of the colonial era remained intact. The SAPs of the 1980s did little to reverse this trend (Mhone 1993; Pryor 1990). Like in Zimbabwe, the agricultural liberalization of the 1980s removed the subsidy scheme and reduced Malawi from a country that was self-sufficient in food (through maize production) to one experiencing constant food deficits and hunger (Chinsinga and Poulton 2014; Conroy et al. 2006; Devereux and Zoltan 2007). Besides, the bias against food crop production was and remains a big part of the problem.

In the early 1990s, poor weather conditions combined with the negative effects of a liberalized economy to produce the 1992 famine. Besides, the political developments of the early 1990s, whose climax was the removal from power of Dr. Hastings Kamuzu Banda (President 1964–1994) through the 1994 Multiparty General Election, helped to worsen the country's poverty levels (Conroy et al. 2006). Between 1998 and 2005, the state raised Poverty Reduction and Strategy Papers (PRSP), which included the Vision 2020—an instrument that was designed to guide the country toward sustained long-term economic growth. These plans emphasized domestic ownership of development initiatives (Chinsinga and Poulton 2014; Malawi Government Vision 2020 2000).

In keeping with aspirations on achieving domestic autonomy, the state reverted to farm input subsidies, first through the Starter Pack (SP) and second the Targeted Input Programme. The former was a country-wide program reaching out to all the smallholder farmers. The latter targeted the poorest groups in the society. In both cases, the state provided farmers with free fertilizer and seed with which they cultivated a maximum of 0.1 hectares of land (Chinsinga and Poulton 2014:127). The 2000/2001 famine reminded the state about the weaknesses of these strategies. Coming in the wake of the MDGs, the state needed to do more than what the existing strategies could achieve. As Conroy et al. (2006:22) note, between 1976 and 2006, the country's per capita income had grown only by a meagre 0.2%.

Given these realities, the new regime of Dr. Bingu wa Mutharika implemented a robust Fertiliser Input Subsidy Programme (FISP) and rolled it across the country from 2005 (Chinsinga and Poulton 2014). Against donor advice, the government subsidized over 90% of fertilizer cost price and allowed rural farmers to purchase one 50 kg bag of maize fertilizer for basal and top dressing. Farmers also purchased subsidized seeds. Under the first phase of the FISP (2005–2009) Malawi was praised within the region for having achieved food self-sufficiency. Despite having a high expenditure budget on subsidies within sub-Saharan Africa, poverty levels in the country remain high (Masanjala 2020).

With the dawn of the 2015 SDGs, the state raised the 2017–2022 MGDS III, which outlines the specific ways for domesticating the SDGs. As noted, Malawi's economic trajectory does not depart from that of Zimbabwe. Over the years, the two countries have not run short of development interventions, all of which have achieved similar results. When the continent is celebrating the Africa Rising narrative, Malawians and Zimbabweans are poorer today than they were 20 years ago. In fact, in 2016, the World Food Programme listed Malawi and Zimbabwe, countries which have had a relatively successful history at food production, as ranking among those to experience hunger for a foreseeable future (OECD/FAO 2016). To what extent the SDGs will reverse this trend is an aspect to which we turn.

3.2 Domesticating SDGs in Malawi and Zimbabwe: A Critical Analysis

As noted earlier, the 2017–2022 MGDS III and 2016–2018 I-PRSP are critical instruments Malawi and Zimbabwe have used to domesticate the SDGs (Malawi Government, 2017; Government of Zimbabwe, 2016). Here, we focus on what the two countries planned to achieve to address goals 1 and 2 of the SDGs, particularly as this relates to the local agricultural communities of Rumphi and Chiredzi Districts. For the most part, the 2017–2022 MGDS III and 2016–2018 I-PRSP put particular emphasis on agricultural production as an avenue for eliminating hunger while leading societies out of the poverty trap. Among the seven "priority sectors" for achieving development in Zimbabwe, the I-PRSP ranks on first position the "agriculture productivity, growth and rural food security" sector. Special emphasis on farming is not surprising because agriculture contributes over 10% to the GDP of Zimbabwe, and about 60% of raw materials feeding agro-industries come from the agricultural sector. Over 70% of the country's population, moreover, relies on this sector (I-PRSP 2016:72). In responding to the specific problem of food security, the I-PRSP proposed to implement a "Special Maize Production Programme" between the years 2017 and 2020. The program aimed at supporting willing farmers with agricultural inputs, irrigation facilities, and mechanized farm technology to produce a yearly output of 2 million tons of maize on 400,000 hectares of land (I-PRSP 2016:75).

Like Zimbabwe, Malawi's 2017–2022 MGDS III is the crucial document operationalizing the SDGs by integrating them into priority areas of the country. Here, too, "agriculture, water development, and climate change" is ranked first among the five priority areas. This ranking reflects the contribution the agricultural sector makes to the economy. The sector contributes 28% to the GDP of the country and supports about 64% of Malawi's total workforce. Besides, 80% of the national export earnings of the country come from agriculture. To increase agricultural production and productivity, the MGDS III proposes several strategies. These include promoting agricultural extension services, increasing agricultural mechanization and diversification, and coordinating and facilitating the development of new structured markets (2017–2022 MGDS III 2017:32–35). The Malawi government, like that of Zimbabwe, also encourages public-private partnerships as a strategy of addressing the SDGs. These are excellent plans. Interviews with Rumphi and Chiredzi communities, however, point to some gaps that make one wonder as to whether the two countries are on the right path toward arresting poverty and achieving hunger-free societies. The shortfalls are detailed below.

3.3 Access to Agricultural Inputs and Markets

To begin with, the farming communities in the two countries are experiencing the same age-old problems of access to agricultural inputs and markets. Save for sugar and tobacco growers in Chiredzi and Rumphi, whose case will be analyzed later, maize growers lack adequate fertilizer, seed, and other related inputs for their crop (FGD with Chozoli farmers, Rumphi, 22/032020; Masivamele and Village section 6 farmers, Chiredzi 25/03/2020). State officials at District level corroborated farmers' concern regarding limited access to maize fertilizer and seed (District Agriculture Development Officer (DADO), Rumphi, 23/03/2020; District Development Fund Coordinator (DDFC), Chiredzi, 23/03/2020). The problem of access to resources does not correspond to the scarcity of agricultural inputs on the market. Farmers do not have credit to purchase the products. Using conservative figures, a farmer needs a minimum of six 50 kg bags of fertilizer to produce one hectare of maize yielding a minimum of 3 tons (Mutegi et al. 2015). A farmer would need about US$184 to buy fertilizer on the market. For farmers living on less than US$2 per day, it is difficult to accumulate enough savings for agricultural inputs.

Farmers' concerns are surprising given that both Malawi and Zimbabwe run agricultural input subsidy programs. As noted, the FISP continues in Malawi, just like the Presidential Input Support Scheme with Zimbabwe. These programs claim to provide farm inputs to the poorer segments of society. However, farmers expressed a dislike of the programs. "We do not want subsidy because it is there only in name. How can two people share one bag of fertiliser? You are lucky if you receive a coupon to buy fertiliser," complained one farmer in Rumphi (FGD with Chozoli farmers, Rumphi, 22/03/2020). Farmers in Chiredzi expressed similar sentiments, with one noting that "the presidential input scheme is not benefiting every farmer here. Only those who are politically connected get the 10 kg bag of seeds. If you are lucky to get the inputs, it will be very late into the farming season" (FGD with Masivamele farmers, Chiredzi, 25/03/2020). Farmers prefer input loans to subsidy.

One reason explaining farmers' concerns centers on what Chinsinga and Poulton (2014) earlier described in Malawi as the "politicization" of the subsidy scheme, a view that echoes that of Dawson and Kelsall (2012) and several others in Zimbabwe. With Malawi, FISP proved successful between 2005 and 2009 because President Mutharika was committed to using it to galvanize popular political support from among the masses. Once he had found alternative means of sustaining his power in his second tenure in office, which ended in 2012 following his death, the FISP program was less rigorously pursued (Chinsinga and Poulton 2014). Over time, politicians use the program to galvanize votes from among the electorate. In Zimbabwe, Shonhe (2018) criticizes the government for assigning the President's office powers of administration over the Input Subsidy Programme (ISP) because such an arrangement undermines the effective execution of the program.

These subsidy schemes, it appears, serve what Ferguson (1994) refers to as the "anti-politics machine"—the idea that state programs serve political motives of extending state power in the rural areas. As such, even though studies doubting the efficacy of these subsidies are on the increase (Jayne et al. 2013; Moyo et al. 2014), they cannot be eliminated. "Politics" forestalls any possibilities of rethinking the input subsidy programs in the two countries because leaders use them to project themselves as benign actors with concerns for the poor at heart.

Political problems are one aspect affecting the effectiveness of the subsidy programs. Farmers also appear to sell the coupons/vouchers they use to access the subsidized inputs. Some farmers sell fertilizer they obtain through the subsidized scheme. Officials blame farmers for ignorance and often complain that the "inputs land in the wrong hands. We need to identify productive farmers who are willing to change their status and improve their lives" (Interview with the Director of Planning and Development (DPD), Rumphi, 20/03/2020). But this behavior only points to the dilemma farmers face. Caught up in poverty cycles, and lack of adequate returns from agriculture, they prefer to address their most immediate needs rather than use the inadequate inputs for a less promising agricultural season.

Because of the limitations associated with the subsidy program, the private sector stands out as an alternative to state resources. In both Malawi and Zimbabwe, non-state institutions have taken an active role to address the problems farmers face in accessing agricultural credit. In Rumphi, for example, the National Smallholder Farmers Association of Malawi (NASFAM), Action Aid, and Development Action for Marginalized Rural Area (DAMRA) closely work with farmers and facilitate farmers' access to agricultural inputs. NASFAM is a farmer's organization with branches in all the agricultural districts of the country. It initially aimed at assisting tobacco farmers to access input credit. With time, its activities extended to maize growers. DAMRA is a relatively new NGO that, among others, assists farmers to embrace smart agricultural technologies. It also provides farmers with small amounts of fertilizer and seed, just as does Action Aid, another NGO. In Zimbabwe, NGOs such as Plan International and CARE International provide farm inputs to vulnerable groups who are into "small grains farming" (such as sorghum and millet) in Chiredzi North, East, and South.

The presence of these non-state institutions in rural areas is a step toward achieving public-private partnerships crucial to the attainment of the SDGs (SDG 17). NASFAM activities in Malawi, however, might limit farmers' access to credit. For instance, NASFAM extends its credit to farmers who pay membership fees, valued at MK25,000 (US$33) per year and a deposit fee of about the same amount. The latter serves to prove farmer's creditworthiness before accessing the loans (Interview with NASFAM Manager, Rumphi, 23/03/2020). Most farmers are incapable of raising these sums of money and, therefore, unable to access NASFAM credit. That said, these non-state institutions complement well the government by augmenting the state's staff responsible for agricultural extension services. Both Malawi and Zimbabwe have low extension service providers-farmer ratio. To mitigate these shortfalls, government officials celebrate "demand-driven" approach to agricultural extension—the idea that farmers need to request for such services. "We do not have adequate extension workers," acknowledged the Rumphi Director of Planning and Development: "the few that are available have mobility problems such that we have resorted to recruiting lead farmers who work as volunteers within the farming areas hoping these would help their fellow farmers" (Interview with DPD, 20/03/2020).

3.4 Contract Farming in Rumphi and Chiredzi

If the story of maize farmers presents a bleak future relative to access to agricultural inputs, that of tobacco and sugar growers, most of which also grow maize, is different. It is here that contract farming touted in both the 2017–2022 MGDS III and 2016–2018 I-PRSP addresses many of the challenges cited above. According to the two documents, contract farming, manifesting through a partnership between small-scale and large-scale farmers, presents several advantages. First, less skilled small-scale commercial farmers can receive agricultural credit from large scale-farmers whose creditworthiness allows them to access huge loans from the banks. Second, local farmers can benefit from the extension skills of large-scale farmers rather than relying on the rather thin government agricultural extension staff. In Chiredzi, local sugar producers, grouped in a farmers' association, have used to their advantage the Tongaat Huletts, a sugar-growing/sugar-producing company. The company extends input credit such as fertilizer and herbicides to local farmers who are guaranteed markets for their sugar. The company recoups this credit with interest once farmers sell their sugar. Again, farmers access the best extension services from the well-qualified team of agricultural officers the company recruits. Although the contract gives farmers little freedom to bargain for better interest rates on agricultural input loans, sugar farmers rate this partnership favorably. One farmer stated that:

> The contract arrangement is very beneficial. We receive the input credit in good time and also receive our payment even before they [the company] sell our sugar. We also have an option of getting part forex [in US dollars] and part RTGs [Real-time Gross Settlements] or 100% RTGs on our revenue. During harvesting time, they sometimes offer haulage services such as the use of tractors to ferry our sugar to the mills. We get advice from the research station in terms of soil sampling. They also offer free training in sugar production to willing sugar farmers who receive diplomas after finishing training. (Interview with a member of Sugar Growers Association, 26/04/2020)

Its weaknesses notwithstanding, the case of Chiredzi sugar producers provides a workable model and one worth replicating. But it is also similar to that which plays out among burley tobacco farmers in Malawi, although the latter yields contrasting results. Like in Zimbabwe, burley tobacco farmers in Rumphi and other districts of Malawi engage in contract farming with international tobacco buyers who work through their local representatives—Alliance One Malawi and Limbe Leaf. To ensure that they buy better tobacco leaf, these companies provide agricultural inputs to farmers, who are grouped in clubs. Besides, the companies provide extension services to local farmers. Like in Zimbabwe, the companies recoup expenses made on farmers at the point of sale. This arrangement guarantees markets for tobacco farmers. Ironically, it is these same companies that buy tobacco at the auction floors. Over time, the tobacco industry has not yielded its intended benefits for farmers. The price of tobacco at the auction floor is unnecessarily low. Interests charged on input credit and transport expenses further drive down farmers' proceeds.

More worrisome is how the contract system potentially ties farmers to the credit system without providing an exit option. Farmers on contract have a preference in selling their tobacco at the auction floors. Those growing tobacco using their capital are regarded as "second-class farmers," and their tobacco rarely gets better prices at the auction floors. As the Manager of NASFAM noted in Malawi:

> It is just difficult for farmers with their own capital to grow burley tobacco. Tobacco buyers are interested in the tobacco of their sponsored farmers. What they want is to claim back their money. So, even farmers that have their resources prefer to get a loan from these buyers in order to have better tobacco market prices. (Interview with NASFAM Manager, 23/03/2020)

The government is aware of this problem. During the 2019 General Election, presidential candidates campaigned on bringing sanity in the tobacco marketing system by ensuring that international buyers do not engage in tobacco production to avoid conflict of interest. It is such anomalies in the international trade system that critics of the Africa Rising narrative often cite as representing nothing other than the same old exploitative tendencies. If not carefully crafted and regulated, these contract arrangements have the potential to sentence nations to conditions akin to the tenancy system and hence frustrate the achievement of sustainable development goals on poverty.

3.5 Community Participation in Experts' Interventions

Finally, it is important to discuss on how communities participate in the various interventions state and other non-state actors make to better the lives of the societies under study. In order to mitigate the problems of hunger, food insecurity, and poverty, officials from government departments and other institutions highlighted above engage the communities in several areas, chief among them being challenges brought about by "climate change." With some exceptions, the Rumphi and Chiredzi communities mostly rely on rain-fed agriculture. Although the two districts have had a relatively good history of rainfall patterns, droughts and floods typically characterize their agricultural seasons in more recent times. For example, farmers highlighted that dry spells and increasing variations in rainfall patterns, coupled with frequent seasonal droughts, threatened the livelihoods of Rumphi and Chiredzi communities in the years 2015, 2017, 2018, and 2019. Interviews with both farmers and officials corroborate these difficulties. "Floods affected operational works of the community, and they led to the destruction of land, houses, and death of livestock," complained one farmer in Chiredzi while reminiscing the devastating effects of the 2019 Cyclone Idai (Interview with Chilonga farmer, 09/04/2020). Similar sentiments were evident in Rumphi, with one farmer affirming that "in 2020 we have experienced more than enough rains in some parts of the district and fertiliser has been washed away, and more crops have been washed by the flooding waters" (Interview with farmer at Kayiwale Village, 24/03/2020).

As destructive as the problem of flooding has been, dry spells have challenged the traditional methods of growing maize in ridges and the practice of solely relying on rains. In response, both state departments and the NGOs have launched a concerted effort to bring irrigation farming to the communities in question. In Rumphi, DAMRA has taken the lead. It has supplied communities with mechanical and treadle pumps to help pump water from the main Lunyina river to farmers' maize fields. Together with the Department of Agriculture, DAMRA is also teaching farmers to adopt smart technologies in production, which manifests through zero tillage (Interview with DAMRA Executive Director, 22/03/2020). In Zimbabwe, the Department of Mechanisation and Irrigation has teamed up with Southern Alliance for Indigenous Resources, Enhancing Community Resilience Against Shocks, and Plan International, to provide irrigation water to the communities through small dams. Through the District Development Fund, Chiredzi is also supplied with several solar-powered water points which allow the communities to irrigate their crops. The completion of Tokwe-Mukosi Dam in Chivi complements well the existing effort.

Undoubtedly, these are success stories that show the effort the two states make to mitigate poverty and create hunger-free societies through public-private partnerships. It appears, however, that farmers do not positively complement this effort. Officials frequently complained about farmers' low adoption of these technologies. In some areas of Chiredzi, solar-driven water pumps are stolen. "A lot of farmers take farming as a part-time activity, not as a serious business," complained DARMA Executive Director, further stating that "the mentality of rural farmers often militate against the achievement of the food security and poverty reduction in these marginalised areas" (Interview with DARMA Executive Director, 22/03/2020). In Chiredzi, farmers resent similar methods of farming that are aimed at mitigating effects of climate change despite the many workshops that the DDFC and Plan International personnel hold with the communities (Interview with DDFC, 12/04/2020).

Such complaints point to how the expert-farmer relationship plays out on the ground and the contribution farmers make toward these interventions. Both state and non-state officers consult farmers before introducing new interventions. A closer analysis shows that this level of consultation aims at collecting some baseline data among farmers, the aim of which is to clear out obstacles that would affect the adoption of new agricultural technologies and ideas. There is little attempt to tap on farmer's own local knowledge and how it could further inform experts' interventions. In Malawi, for example, farmers doubted the necessity of zero tillage when, in the immediate past, experts insisted on the need to make ridges. Some questioned DARMA's reasoning in installing petrol-powered water pumps, which required farmers to consolidate their earnings and buy fuel to run them. Farmers appear to reject the somewhat patronizing and condescending consultative meetings that do little than convince them about the need to adopt experts' proposals and interventions. Many of these interventions are conceptualized from above and shallowly speak to the sociological aspects of the communities. There is little consideration regarding how new technology would change household economics. Such interventions usually reshape the allocation of existing family labor and resources such that farmers are likely to resist them to maintain a status quo. While grassroots participation in the development process or development interventions is crucial to achieving goals on poverty and hunger, equally important to consider is the nature of that participation. The cases of Rumphi and Chiredzi suggest that farmers' participation is crafted within a hierarchical and prescriptive framework, which development scholars have criticized for decades.

From a slightly different angle, external intervention in these communities should also aim at reorganizing farmers in ways that would help them reduce overreliance on agriculture. This direction is a necessary step for societies prone to adverse climatic conditions. When asked about their major sources of income, both areas cited seasonal agriculture, with some exceptions. "Once we have harvested our maize and tobacco," lamented one member of the FGD in Rumphi, "we wait for the next growing season. Life becomes unbearable if the harvest is poor." A good poverty mitigation strategy, we argue, should be one that slowly rechannels farmers' proceeds toward off-farm investments either through value addition or direct investment in other areas. Off-farm investments could raise farmers' creditworthiness and allow them to access more agricultural credit. Besides, such a strategy would cushion farmers from the shocks of climate change like frequent dry spells and floods.

4 Conclusion

This chapter set to understand how Malawi and Zimbabwe are domesticating the SDGs in the areas of alleviating poverty and achieving hunger-free societies by 2030. Our interest was to examine how government plans surrounding SDGs 1 and 2 translate into practice among rural agricultural communities of Rumphi and Chiredzi districts in Malawi and Zimbabwe, respectively. Although the two countries have well-defined instruments to mitigate poverty and hunger among rural communities, there are some gaps that need addressing to achieve these objectives. In both cases, communities grapple with the same old challenges of limited access to agricultural inputs, extension services, and imperfect markets for their agricultural produce. Again, although the various organizations and government departments work with farmers to mitigate the problems of unpredictable climatic conditions, there is little attempt to tap on farmers' agency and develop interventions that align with aspirations of peasant households. Strategies of helping farmers to adopt smart agricultural technologies, for example, remain hierarchical and patronizing and, therefore, minimally command farmer's confidence.

These challenges notwithstanding, the two countries have made strides to forge public-private partnerships and make them work for the good of society. Such partnerships are clearly articulated in the specific national documents that operationalize the activities of local departments at district levels. The case of Chiredzi contract sugar farmers in Zimbabwe presents a workable model of how these partnerships can benefit societies. They can act as channels through which the private sector can extend agricultural inputs and extension services to farmers in more beneficial arrangements. In contrast, contract farming among tobacco farmers in Malawi has yielded negative results because of poor tobacco prices on the market. We argue that if not regulated, such contract arrangements can perpetuate conditions similar to the tenancy system and tie farmers to economic exploitation, which is the very antithesis of the optimism inherent in the Africa Rising narrative. Addressing the highlighted bottlenecks would smoothen the path toward achieving SDGs 1 and 2 in the two countries.

References

Balleis, P. (1993). *A Critical Guide to ESAP*. Mambo Press, Gweru, Zimbabwe.

Chinake, C. (1997). Strategies for poverty alleviation in Zimbabwe. *Journal of Development Studies in Africa* 12, 1, 39-51.

Chinsinga, B, Poulton, C (2014). Beyond Technocratic Debates: The Significance and Transience of Political Incentives in the Malawi Farm Input Subsidy Programme (FISP), *Development Policy Review*, 32, 2,123-150.

Chirwa, E.W. & Kydd, J. (2005). Farmer organization for market access: Study on farmer organization in smallholder tea in Malawi. Retrieved from: https://pdfs.

semanticscholar.org/0709/13593a59a7bfd43074de85d59e512f2b1466.pdf. Accessed 28 Dec. 2019.

Conroy, A.C., Blackie, M.J., Whiteside, A. Malewezi, J.C. & Sachs, J. (2006). *Poverty, AIDS and Hunger: Breaking the Poverty Trap in Malawi.* New York, Palgrave Macmillan.

Cowen, M. & Shenton, W.R. (1996). *Doctrines of Development.* London, Routledge.

Dawson, M. & Kelsall, T. (2012). Anti-Developmental Patrimonialism in Zimbabwe, *Journal of Contemporary African Studies* 30, 1, 49–66.

Devereux, S. & Zoltan, T. (2007). The Malawi first famine, 2001–2002, In S. Devereux (ed) *The New Famines: Why famines persist in an era of globalization* (pp 143-177), London, Routledge.

Ferguson, J. (1994). *The Anti-Politics Machine: 'Development,' depoliticisation, and bureaucratic power in Lesotho,* Minneapolis, University of Minnesota Press.

Government of Zimbabwe. (2016). *Zimbabwe Interim Poverty Reduction Strategy Paper (I-PRSP) 2016–2018,* Harare, Fiscal Policy & Advisory Services Department, Ministry of Finance and Economic Development.

Guchu, W. (2018). *The Gods Sleep through It All: A collection of Essays*, Harare, Mwanaka Media and Publishing.

Hanson, K.T., Puplampu, K.P. & Shaw, T.M. (2018). Crystallising the Africa Rising narrative with sustainable development, In K.T. Hanson, K.P. Puplampu, T.M. Shaw (eds), *From Millennium Development Goals to Sustainable Development Goals: Rethinking African Development*, (pp. 167-177), London and New York, Routledge.

Human Development Report. (1998). Harare, Zimbabwe.

Jayne, T., Mather, D., Mason, N. & Ricker-Gilbert, J. (2013). How Do Fertiliser Subsidy Programs Affect Total Fertiliser Use in Sub-Saharan Africa? Crowding Out, Diversion, and Benefit/Cost Assessments, *Agricultural Economics* 44, 6, 687–703.

Kayira, G.W. (2020). Wealth Creation, Poverty, and Social Welfare in Malawi, 1940s-1980. In M. Khan & G. Mkodzongi (eds), *Africa History and Culture* (pp 11-28), Dubuque, IA, Kendall Hunt.

Kydd, J. & Christiansen, R. (1982). Structural Change in Malawi since Independence: Consequences of a Development Strategy Based on Large-scale Agriculture, *World Development* 10, 5, 355–375.

Malawi Government. (2017). *The Malawi Growth and Development Strategy (MGDS) III, 2017-2022: Building a Productive, Competitive and Resilient Nation.* Lilongwe, Ministry of Finance, Economic Planning and Development.

Malawi Government, 2000. *Vision 2020: The Long-term Development Perspective for Malawi*, Lilongwe, National Economic Council.

Masanjala, W. (2020). Findings and Lessons from Vision 2020, Presentation made at the launch of Envisioning Process, BICC, Lilongwe, Malawi.

Mhone, G.C.Z. (1993). The Political economy of Malawi: An Overview. In Mhone G.C.Z. (ed), *Malawi at the Crossroads: The Post-colonial political Economy*, (pp 1-33), Harare, Sapes Books.

Mlambo, A.S. (1997). *The Economic Structural Adjustment Programme: The Case of Zimbabwe, 1990–1995.* University of Zimbabwe Publications, Harare.

Moyo, S., Chambati, W. & Siziba, S. (2014). Agricultural Subsidies Policies in Zimbabwe: A Review, Study commissioned by the World Bank, Mimeo, Harare.

Mutegi, J., Kabambe, V., Zingore, S., Harawa, R. & Wairegi, L. (2015). *The Fertiliser Recommendation Issues in Malawi: Gaps, Challenges, Opportunities and Guidelines.* Lilongwe, Soil Health Consortium of Malawi.

Naidu, S., Corkin, L. & Herman, H. (2009). Introduction, *Politikon* 36, 1,1-4.

Ng'ong'ola, C. (1982). The design and implementation of customary land reforms in Central Malawi. *Journal of African Law* 26, 2,115–32.

Nyathi, D. (2012). 'An Evaluation of Poverty Alleviation Strategies Implemented by NGOs in Zimbabwe: A Case of Binga Rural District' Masters Dissertation submitted to Department of Development Studies, University of Fort Hare, South Africa.

Pillay, D. (2015). The Global Economic Crisis and the Africa Rising Narrative, *Africa Development* 40, 3, 59-75.

Pryor, F.L. (1990). *The Political Economy of Poverty, Equity and Growth: Malawi and Madagascar.* Oxford: Oxford University Press.

Puplampu, K.P., Hanson, K.T. & Shaw, T.M. (2018). From MDGs to SDG's: African development challenges and prospects. In K.T. Hanson, K.P. Puplampu, T.M. Shaw (eds), *From Millennium Development Goals to Sustainable Development Goals: Rethinking African Development*, (pp. 1-10), London and New York, Routledge.

Sachikonye, L. (2002). Whither Zimbabwe? Crisis and Democratisation *Review of African Political Economy*, 29, 91, 13-20.

Shonhe, T. (2018). The Political Economy of Agricultural Commercialisation in Zimbabwe, APRA Working Paper 12, Future Africulture Consortium, https://opendocs.ids.ac.uk/opendocs/handle/20.500.12413/13817.

Slattery, M. & Stanton, L. (2013). Africa Rising? *Foreign Policy*, 201, 88-89.

Taylor, I. (2014). Is Africa Rising? *The Brown Journal of World Affairs* 21, 1,143-161.

Taylor, S. (2012). *Globalisation and the Cultures of Business in Africa: From Patrimonialism to Profit,* Bloomington, Indiana University Press.

UNICEF. (1994). A Situational Analysis of Women and Children in Zimbabwe, UNICEF, Harare.

Vickers, B. (2013). Negotiating the rise of new powers, *International Affairs*, 89, 3, 673-693.

World Bank. (1995). *World Summit for Social Development,* Copenhagen, World Bank.

4 Preventing Fall Armyworm (*Spodoptera frugiperda* JE Smith) Damage in Maize by Altering Planting Time and Using Varied Genotypes

Leonard Nyabanga, Ronald Mandumbu, Joyful T. Rugare, Never Mafuse, Emmanuel Zivenge, Handsen Tibugari, George Nyamadzawo, and Christopher T. Gadzirayi

Abstract

Fall armyworm (*Spodoptera frugiperda* JE Smith) is a polyphagous pest indigenous throughout the Americas which has invaded the African continent since 2016 from South America. It has had devastating effects on cereals with yield losses sometimes reaching 100%, and yet, there are no confirmed means of managing the pest. This militates against the achievement of sustainable development goal (SDG) number 2 which seeks to end hunger, achieve food security and improved nutrition and promote sustainable agriculture. It also impedes the achievement of SDGs 1 and 3 which, respectively, seek to end poverty and ensure good health and the wellbeing of all people, particularly people in resource-constrained societies. Thus, the experiments conducted in the study were aimed at determining the effect that maize genotypes and planting time had on fall armyworm (FAW) damage in a bid to minimise yield losses and concomitantly ensure food security. The results indicated that late planting resulted in a significant 40% increase in the incidence of FAW attack compared to early and medium planted maize crops. Consequently, early planting gave a significantly higher yield by 122% compared to late planting. In contrast, there were no significant effects of genotypes on the number of plants affected by FAW. The study concluded that early planting is effective in reducing FAW

The original version of this chapter was revised. The correction to this chapter is available at https://doi.org/10.1007/978-3-030-70952-5_20

L. Nyabanga · R. Mandumbu (✉)
Department of Crop Science, Bindura University of Science Education, Bindura, Zimbabwe

J. T. Rugare
Department of Crop Science, University of Zimbabwe, Harare, Zimbabwe

N. Mafuse · E. Zivenge · C. T. Gadzirayi
Department of Agricultural Economics, Extension and Education, Bindura University of Science Education, Bindura, Zimbabwe

H. Tibugari
Department of Crop Science, Faculty of Life Sciences, Gwanda State University, Gwanda, Zimbabwe

G. Nyamadzawo
Department of Environmental Science, Bindura University of Science Education, Bindura, Zimbabwe

G. Nhamo et al. (eds.), *Sustainable Development Goals for Society Vol. 2*, Sustainable Development Goals Series, https://doi.org/10.1007/978-3-030-70952-5_4

attack on maize and results in increased maize productivity in FAW endemic areas.

Keywords

Fall armyworm · Genotypes · *Spodoptera frugiperda* · *Zea mays* · Early planting

1 Introduction

Maize (*Zea mays* L.) is the single most important staple crop in sub-Saharan Africa (SSA) (FAO 2018). It constitutes about 75% or more of cereal area in countries like Kenya, Malawi, Zambia and Zimbabwe (Jayne 2003). Yet, maize as a vital source of food has been threatened by the invasive fall armyworm (FAW) that invaded the African continent since the year 2016.There are speculations that this pest might have come to Africa through imported or food aid consignments. Hruska (2019) reported that FAW has been a consistently important insect pest for a number of crop species especially maize in the United States. In the invaded range, FAW is projected to constitute a lasting threat to several important crops as the region provides diverse host sources and favourable climatic conditions for consistent reproduction in many areas (Midega et al. 2018).

FAW threatens to undermine the efforts to achieve sustainable development goals (SDGs) for millions of poor people in sub-Saharan Africa (SSA). The invasion of Africa by FAW is causing a serious threat to food security and adds to the insecurity caused by many other factors including extreme weather conditions such as the El Nino phenomenon and tropical cyclones. This in turn negatively impacts on the realisation of the second SDG which seeks to end hunger, achieve food security and improved nutrition and promote sustainable agriculture (United Nations 2015). Biotic constraints such as FAW have been reported to cause significant yield losses which affect food security among communities that rely on subsistence farming. In addition, the FAW scourge is a threat to SGD number 3 which seeks to double the agricultural productivity and increase the incomes of small-scale food producers, in particular women and resource-constrained smallholder farmers whose livelihoods solely depend on subsistence farming. The invasive nature of FAW also threatens the achievement of SDG 15 target number 8 which seeks to prevent the introduction and significantly reduce the impacts of invasive alien species on land and water ecosystems (United Nations 2015).

The potential economic impacts of FAW in Africa are huge. FAW has the potential to cause yield losses of 8.3 million to 20.6 million metric tonnes per year in just 12 of Africa's maize-producing countries in the absence of proper control methods (CABI 2017). The countries include the southern and eastern African countries such as Zimbabwe, Mozambique, Botswana, Zambia, Namibia, Swaziland, Malawi, Tanzania, Kenya and Uganda. The level of losses represents 21–53% of annual production of maize averaged over a 3-year period in these countries, and their value is estimated to be between US$2.48 billion and US$6.19 billion (CABI 2017). FAW could have serious regional and international trade through strict phytosanitary regulations which may reduce the movement of crop produce among different countries, particularly grain. Anecdotal evidence suggests that FAW has been intercepted at quarantine points in Africa and Europe, suggesting the potential for phytosanitary trade issues inside and outside Africa despite the pest being capable of migrating long distances on prevailing winds making natural migration a possibility (Huesing 2018). The polyphagous nature of FAW presents challenges in management due to the presence of numerous alternative hosts outside the production season of many crops (Hruska 2019).

2 Literature Review

2.1 FAW Distribution in Africa

FAW has been confirmed present in all the sub-Saharan African countries except Lesotho (Fig. 4.1). The environmental conditions in sub-Saharan African countries are suitable for the multiplication of the pest because the countries are in the tropical and sub-tropical regions (FAO 2019).

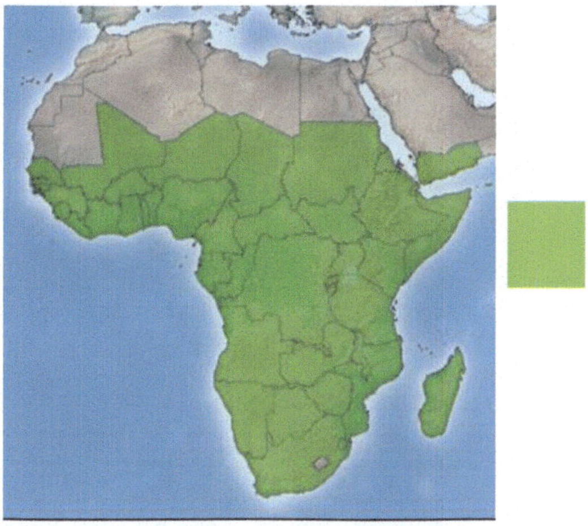

Fig. 4.1 FAW distribution in sub-Saharan Africa. (Source: FAO (2019))

2.2 Crop Losses

Damage by FAW is manifested through photosynthetic area loss, lodging, impaired reproduction and direct damage to grain (Chimweta et al. 2019). Young larvae mainly feed on epidermal leaf tissue and also make holes on the leaves, which is the typical damage symptom of FAW (Sisay et al. 2019). Given the importance of maize in Africa as a primary staple food crop, the recent invasion of FAW threatens the food security of millions of people in the sub-Saharan Africa region that will likely have an aggravated drought due to climate change or El Nino (Sisay et al. 2019).

2.3 Life Cycle

Following emergence, the adult moths feed on suitable flowers in the dark for up to 2 h, before females start their mating call by emitting pheromones that attract males to mate. Adults fly at night and are attracted to light, especially those with a strong ultraviolet component. The use of pheromone traps has been used to monitor populations in integrated pest management programmes (Cruz et al. 1999). Oviposition starts later on in the same night that mating takes place. Eggs are laid as ‘egg masses’ in batches of 100–200 eggs and hatch in 2–4 days in optimum temperatures (FAO 2019). Oviposition is usually on the underside of leaves, but as the density of moths increases, oviposition becomes increasingly indiscriminate on other parts of the host plant, other non-host plants and inanimate objects. Adult moths mostly live for 2–3 weeks. Females will mate multiple times during this period and lay multiple egg masses with a potential fecundity of up to 1000 eggs per female. There are six larval instars: it is the final instar which consumes the most plant material (77%) and causes the most damage. The developing larvae eat different parts of the host plant, depending on the crop, the stage of crop development and the age of the larvae. On maize, young larvae usually feed on leaves creating a characteristic windowing effect. This and moist sawdust-like frass near the funnel and upper leaves can be an easily spotted sign of larva feeding. The larvae also feed on the developing cob (Fig. 4.2).

2.4 Household Level

FAW is likely to directly affect capital costs through increased labour needed and the type of knowledge required to deal with the pest, through yield losses and the ability of agricultural lands to respond to shocks and, financially, through increasing the cost of production due to costs of control (defined as the cost of technology and its application) and its effect on income. It will also indirectly affect households’ social and physical capital (the household’s assets) (Harrison et al. 2019).

Fig. 4.2 *Spodoptera frugiperda* feeding on a developing maize cob. (Source: Authors)

2.5 Host Range

Field crops are frequently injured, including alfalfa, barley, Bermuda grass, buckwheat, cotton, clover, maize, oat, millet, peanut, rice, ryegrass, sorghum, sugar beet, Sudan grass, soybean, sugarcane, timothy, tobacco and wheat. Among vegetable crops, only sweet maize is regularly damaged, but others are attacked occasionally. Other crops which are sometimes injured are apple, grape, orange, papaya, peach, strawberry and a number of flowers. Weeds known to serve as hosts include bent grass, *Agrostis* ssp.; crabgrass, *Digitaria* spp.; Johnsongrass, *Sorghum halepense*; morning glory, *Ipomoea* spp.; nutsedge, *Cyperus* spp.; pigweed, *Amaranthus* spp.; and sandspur, *Cenchrus tribuloides* (Prasanna 2018).

2.6 Management of FAW in Zimbabwe

Crop losses due to FAW may be prevented or reduced by deploying effective crop protection measures which to a large extent depends on farmers' knowledge and behaviour towards pest management and the availability and effectiveness of crop protection methods (Kansiime et al. 2019). Elements of smallholder maize integrated pest management need to be carefully studied across ecosystems of Africa to better understand the conditions in which they work best and their mechanisms and to develop mechanisms that can be scale up (Abate et al. 2000; Hruska 2019). The sustainable approach is the integrated pest management approach which employs a multi-tactic approach to lower pest populations so that they go below economic injury levels which take into cognisance the economic status of the smallholder farmers (Mandumbu et al. 2011). Technically, these farmers have two ways of managing the pest: they can use either improved technology or traditional sustainable approaches which can be accommodated by their resources.

The ability of maize to compensate for foliar damage depends on genetics, nutrition and water availability of the plant. This therefore raises the need to test the currently available maize genotypes against FAW. This could provide the first line of defence to the maize crop and probably the cheapest way of managing the pest under smallholder agriculture. Many smallholder farmers grow maize under unsuitable conditions and as a result suffer yield losses due to inadequate nutrition and moisture stress. There is a need to develop sustainable FAW management techniques that fit into the economics of the smallholder maize producer for sub-Saharan Africa. Some farmers use sand, soap or unregistered chemicals. Most of these methods used by farmers and reasons underlying their use are anecdotal and lack scientific backing (Kansiime et al. 2019). The use of plant resistance and early planting are possible management methods that can be sustainable for resource-poor small-scale farmers. Early planting after effective rainfall usually provides better growing conditions for maize by making use of more heat units at the beginning of the cropping season. Early planted crops escape pest pressure as the crop life cycle will escape the time of pest abundance. According to Prasanna (2018), early planting creates asynchrony between the pest and critical crop stages. These options are especially key to SSA countries such as Zambia, Zimbabwe, Malawi and Mozambique where the majority of the farmers have limited access to safe and affordable FAW control options.

2.7 Cultural Pest Management

Cultural pest management strategy comprises of those methods that involve the manipulation of agrosystems in order to decrease the success of a pest species within it. Many cultural techniques form the basis of preventive pest management. From the arthropod pest management strategy, the primary aim of cultural management techniques are (i) to reduce the colonisation of a crop by a pest and to decrease the pest dispersal from that crop and (ii) to reduce the reproduction and survival of a pest in a crop once colonisation has occurred (Thacker 2002). Since the invasion of sub-Saharan Africa by the FAW, there has been a resurgence in the interest of cultural pest management strategies. However, private companies which are 'product' based do not have much interest in these strategies because of the absence of saleable products.

2.8 Planting Date Manipulation and Arthropod Pest Management

Many have periods during the year when they are most dispersive and can colonise plants more easily. For Zimbabwe, *Spodoptera frugiperda* does not appear in large numbers very early in the season probably because of the shortage of hosts from the dry summers. Planting date manipulation is therefore necessary to avoid the peak period for crop colonisation. Plants will therefore be able to establish themselves before pests arrive.

For Zimbabwe, early planting means the crop will get more heat units and therefore grow very fast compared to later planted crops. The early planted crop is likely to escape the periods of heavy infestation by *Spodoptera frugiperda*. This is a principle that was tested for other indigenous bollworms in cereals.

2.9 Effect of Maize Genotypes on the FAW Pandemic

The use of resistance is an attractive option particularly if the resistance is complete in the sense that the attacking organism is no longer able to cause economic damage. According to Thacker (2002), host plant resistance is a collective heritable characteristic by which plants may reduce the possibility of its utilisation as a host for the pest.

Sometimes, secondary plant substances are responsible for resistance. The main function of secondary plant metabolites is to act as a defence against herbivore attack by acting as repellents, inhibitors and toxins. Other compounds may deter feeding of pest, disrupt development, provide barriers from attack and assist with wound healing and to provide many other neurotoxins to herbivorous insects. Plants can also use a range of morphological features to defend against insect attack like the use of trichomes.

3 Materials and Methods

3.1 Description of Trial Location

The trial was conducted in Kandava village under Seke District in ward 4 in Mashonaland East Province which falls under Zimbabwe's Agro-ecological Region 2. The site coordinates are 18.1279° S and 31.2701° E. Site is 8 km from Marondera/Hwedza junction along Chitungwiza-Marondera road. Altitude is 1600 metres, and total annual rainfall ranges from 600 to 800 mm. Most of the rainfall is received from November to April.

3.2 Planting Preparations

Land preparation was done using an ox drawn plough. The maize varieties were planted in plots measuring 4 m × 5 m. Plant spacing was 0.75 m × 0.25 m, and distance between plots was 1 metre, while that between blocks was 3 metres. Two seeds were planted per station and thinned to one at 2 weeks after crop emergence (WACE) to leave a plant population of 53,333 plants per hectare. Compound D Fertilizer (7% N/14% P/7% K) was side dropped in planting stations at the rate of 350 kg ha^{-1}, and top dressing was split applied at 150 kg ha^{-1} at 4 and 8 WACE, giving a total of 300 kg ha^{-1} ammonium nitrate (34.5% N). The trial was under drip irrigation to safeguard against total crop failure due to drought. Irrigation was applied as a supplementary measure. Weeding was done three times at 3, 8 and 12 WACE using a hand hoe, and the site was kept weed-free at most times.

3.3 Monitoring and Observations Done

Although the trial was under rain-fed conditions, two survival irrigations were applied. A total of 20 mm was applied on each occasion using drip irrigation system.

Four mid-season maize hybrids were selected for evaluation. The selected hybrids are among the most commonly grown by villagers, and these were SC649, SC637, SC633 and DKC8053. The four maize hybrids were planted over three planting dates of 15 November, 15 December and 28 December signifying early, mid-season and late planting, respectively. FAW infestation was natural, and the site was previously grown to maize resembling the monoculture practises by the community which also ensured that sufficient infestation to the treatments occurred.

The trial was held under no chemical spray conditions for all treatments. The experimental design was a 4*3 factorial which was laid down in randomised complete block design (RCBD) giving 12 treatments replicated three times to give a total of 36 treatments. Blocking was done to reduce the effect of environmental factors such as soil texture and slope.

A plant was randomly selected to be the 1st of five consecutive plants in each treatment as from 2 WACE for leaf score measurements. The first row was omitted for edge effect, and sampling was done from the second row of each plot. The first metre on the sampling row was also omitted for the same reason. Each of the five consecutive plants was examined to determine the number of damaged leaves and leaf lesion size. Records of foliar damage ratings were done using a nine 9-point visual rating scale (1, no damage, to 9, severe foliar damage) (Davis and Williams 1992) (Table 4.1). A score was given for each plant, and an average was recorded for the treatment.

The number of affected plants was also recorded at each sampling for particular growth stages and on the same plants sampled for leaf damage score. Whole plot counts were done to determine the number of plants affected by FAW. The number affected for each treatment was expressed as a percentage of the total. One hundred plants were sampled per each treatment to determine exit holes. Fifty plants were sampled per treatment to determine kernel score at maturity stage, and whole plots were harvested and cobs sun dried to 12% moisture to determine yield. Yield per plot was adjusted per hectare basis. Discard as a result of FAW damage and subsequent secondary infection of grain was also recorded following harvesting.

3.4 Data Collection and Analysis

Data collected were leaf damage score as explained above, exit holes, kernel score (modified Davies scale), Yield (t ha^{-1}), and percent discard yield (t ha^{-1}) as a result of FAW damage. The data were subjected to analysis of variance

Table 4.1 Scale for assessment of foliar damage due to FAW in maize germplasm entries. Score damage symptoms/description response

Scale	Description	Resistant/susceptible
1	No visible leaf-feeding damage	Highly resistant
2	Few pinholes on 1–2 older leaves – resistant	
3	Several shot-hole injuries on a few leaves (<5 leaves) and small circular hole damage to leaves	Resistant
4	Several shot-hole injuries on several leaves (6–8 leaves) or small lesions/pinholes, small circular lesions and a few small elongated (rectangular-shaped) lesions of up to 1.3 cm in length present on whorl and furl leaves	Partially resistant
5	Elongated lesions (>2.5 cm long) on 8–10 leaves, plus a few small- to mid-sized uniform to irregular-shaped holes (basement membrane consumed) eaten from the whorl and/or furl leaves	Partially resistant
6	Several large elongated lesions present on several whorl and furl leaves and/or several large uniform to irregular-shaped holes eaten from furl and whorl leaves	Susceptible
7	Many elongated lesions of all sizes present on several whorl and furl leaves plus several large uniform to irregular-shaped holes eaten from the whorl and furl leaves	Susceptible
8	Many elongated lesions of all sizes present on most whorl and furl leaves plus many mid- to large-sized uniform to irregular-shaped holes eaten from the whorl and furl leaves	Highly susceptible
9	Whorl and furl leaves almost totally destroyed and plant dying as a result of extensive foliar damage	Highly susceptible

Source: Modified from Davis and Williams (1992)

using Genstat version14 after testing for normality and other assumptions of analysis of variance. Where there were significant differences the means were separated using the least significant differences at 0.05 probability level.

4 Presentation and Discussion of Findings

Leaf average score, number of maize plants infested, average exit holes, kernel average score and yield were not significantly ($p > 0.05$) affected by maize genotype (Table 4.2).

Maize genotypes did not show any significant differences on FAW damage across all the growth stages. This could be because the maize hybrids used could have emanated from similar or related parental lines. It is important to note that three of the maize varieties used emanated from the same seed producing company and the possibility of them emanating from the same or related parents are high. The results concur with findings by Harrison et al. (2019) who articulate that there are a few corn varieties which can withstand earworm or FAW attack.

The results indicated that FAW has no oviposition and feeding preferences among the genotypes used which means that the pest is likely a generalist pest. Goergen (2016) also reported that breeding for FAW resistance has only been initiated in Africa, and therefore, there are no current varieties which can withstand FAW attack. Prasanna (2018) also confirmed that Africa has no adapted varieties with scientifically validated resistance to FAW. This gap is however being currently addressed by the International Maize and Wheat Improvement centre (CIMMYT) who is evaluating several germplasm for resistance to FAW (Prasanna 2018). Again, in other seed companies which are privately owned, breeding has been initiated as it might provide a sustainable management of the pest in the face of impoverished African farmers who do not afford chemicals and associated application devices.

Therefore, to date, there are no confirmed maize varieties that can withstand the effects of FAW which signifies the absence of antixenosis, tolerance, antibiosis and apparent resistance in the tested genotypes. Given that volatiles from plants differ quantitatively and qualitatively, FAW moth and lar-

Table 4.2 Effect of variety on LAS, number affected, AEH, KAS and maize yield

Variety	Leaf average score	Maize plants affected	Average exit holes	Kernel average score	Yield (t ha^{-1})
SC627	1.93	49.5	0.28	3.33	2.57
SC639	1.99	56.3	0.358	2.93	2.76
SC633	2.02	48.9	0.56	3.44	2.65
DKC8053	2.16	59.9	0.527	3.15	2.8
P-value	>0.05	>0.05	>0.05	>0.05	>0.05
Sed	NS	NS	NS	NS	NS

vae are presented with differences in olfactory cues when making choices among hosts (Carroll et al. 2006). Varieties would differ in the quality and quantity of volatiles and that may influence the oviposition behaviour of the moth and the feeding behaviour of the larvae in different sorghum varieties. However, it was not the case in this study as the varieties were uniform in influencing FAW moth oviposition. Efforts are currently under way in the international research organisations such as CIMMYT and the International Institute of Tropical Agriculture where they are evaluating a lot of germplasm for determination of resistance. It is hoped that the African breeding community may take a coordinated approach to develop elite varieties with relevant traits for smallholder farmers in Africa. It also has to be noted that breeding is a continuous process with no finishing line to the perpetual race between the host and the evolving pest.

4.1 Effect of Time of Planting on Maize Damage by FAW

Time of planting had a significant ($p < 0.05$) effect on maize damage by FAW. Late planting had the highest score of 3.3, while early planting and mid-season planting were the same with means of 2.57 and 2.53, respectively (Fig. 4.3).

4.2 Effect of Time of Planting on Leaf Damage

Time of planting significantly affected the number of maize damages by FAW as measured by the Davies scale (Fig. 4.4). The results show that the late the planting, the more the damage to a maize crop across all varieties. Time of planting has for long been recognised as a cultural management technique for smallholder farmers in traditional agriculture.

4.3 Effect of Time of Planting on Number of Plants Damaged at Reproductive Stage

Time of planting had a significant ($p < 0.05$) effect on number of maize plants damaged at reproductive stage. Early planting had significantly lower numbers of plants affected with a mean of 21%. On the other hand, late planted maize had significantly higher numbers of damaged plants with a mean of 83.4% compared to mid-season planted maize which had a mean of 56.5% (Fig. 4.5).

Visual leaf score damages showed that the late planted crop was severely damaged by FAW compared to late planted crop (Plates 4.1 and 4.2).

4.4 Effect of Time of Planting on Maize Yield

Variety had no significant ($p < 0.05$) effect on maize yield. However, time of planting significantly ($p < 0.05$) affected yield of maize. Early planting had the highest yield of 3.93 t ha^{-1}, while late planting had the lowest yield of 1.77 t ha^{-1}. Yield for mid-season planting was 2.39 t ha^{-1}. Late and mid-season planting performed the same in terms of yield (Fig. 4.6).

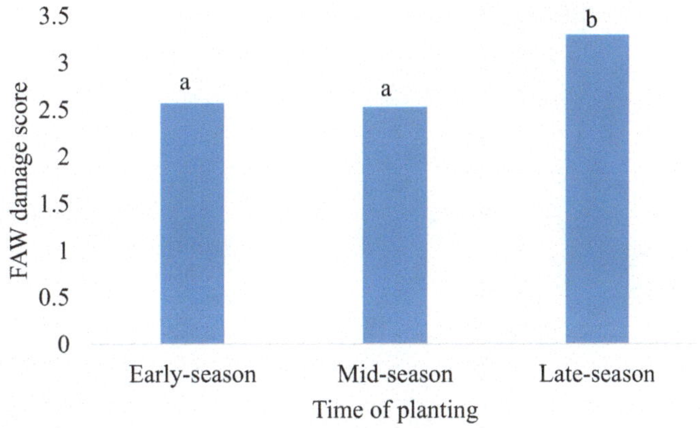

Fig. 4.3 Effect of time of planting on FAW attack. (Source: Authors)

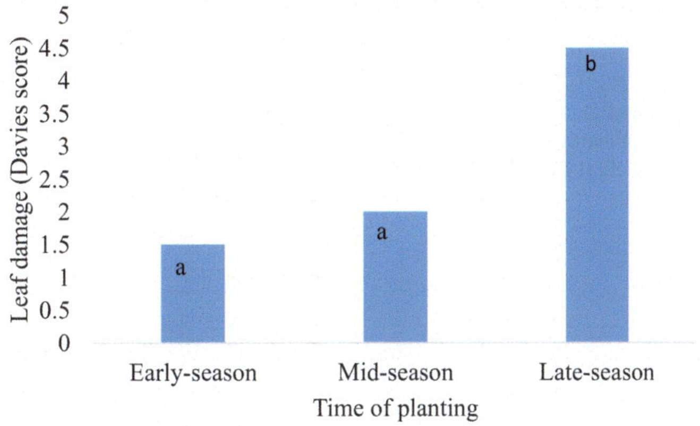

Fig. 4.4 Effect of time of planting on leaf damage (Davies scale). (Source: Authors)

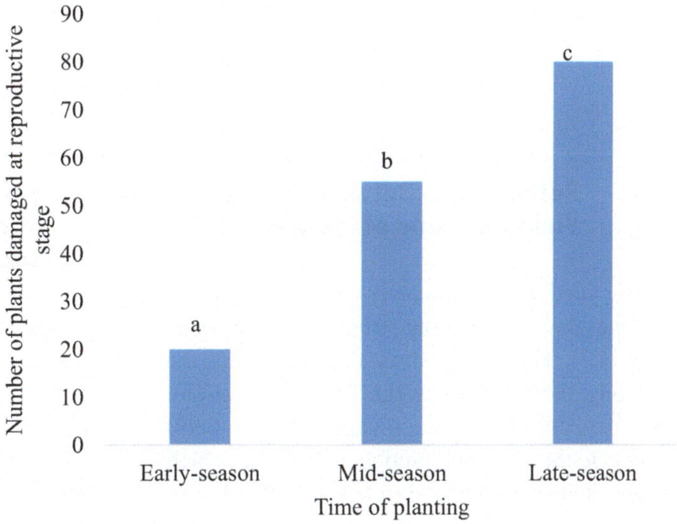

Fig. 4.5 Percent number of plants affected by FAW at reproductive stage. (Source: Authors)

Plate 4.1 FAW damage on late planted crop

Plate 4.2 FAW damage on early planted crop

Fig. 4.6 Effect of the time of planting on maize yield (t ha^{-1})

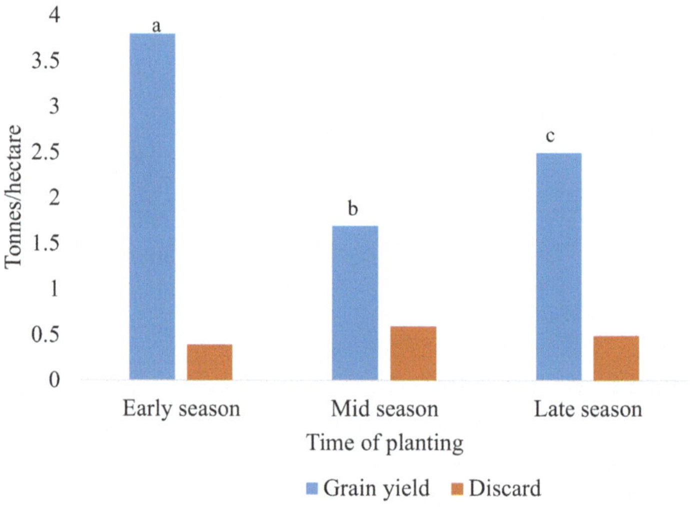

4.5 Relationship Between Maize Yield and Time of Planting

Maize yield was significantly ($p < 0.05$) affected by the day of planting from the first rains (Fig. 4.7). The results show that the earlier the planting, the higher the yield. There was a strong positive correlation coefficient ($R^2 = 0.99$) between the day of planting and yield which shows a strong relationship between the day of planting and yield.

4.6 Relationship Between Day of Time of Planting and Percent Maize Discard

There was a significant ($p < 0.05$) effect of time of planting on the discarded grain at the shelling stage (Fig. 4.8). The percentage of discarded grain grew with increase from the day of the first rains. A correlation coefficient of 0.961 showed a very strong relationship between the two.

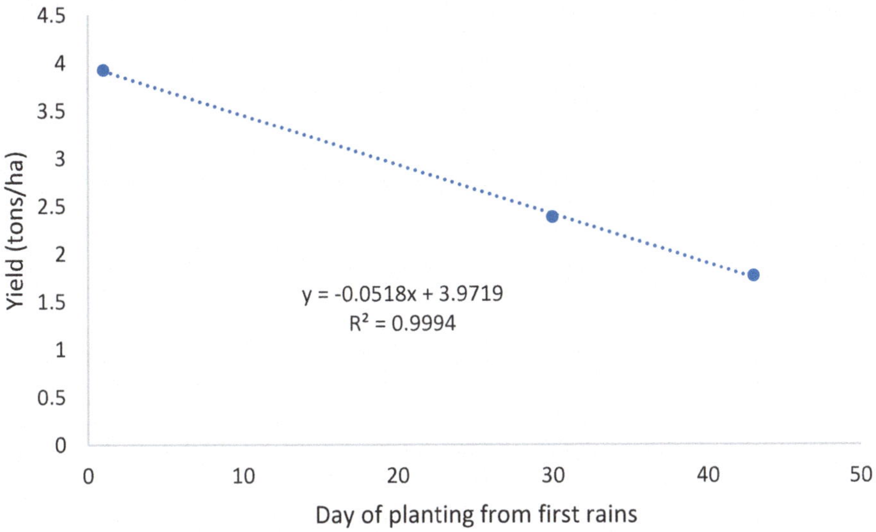

Fig. 4.7 The relationship between maize yield and time of planting. (Source: Authors)

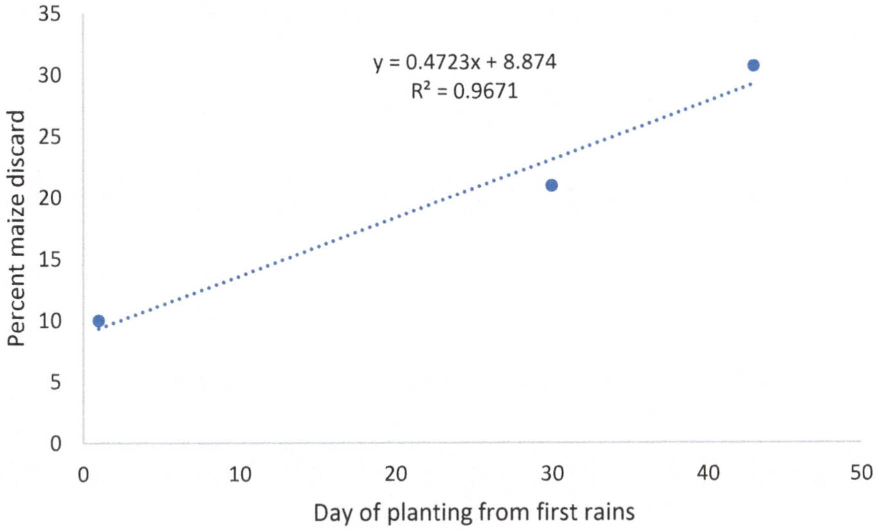

Fig. 4.8 The relationship between time of planting and percent maize discard. (Source: Authors)

5 Relationship Between Time of Planting and Leaf Average Score

Leaf average score and day of planting from the day of the first rains had a correlation coefficient of $R^2 = 0.662$ which showed a close relationship between the two. The results showed that leaf damage score increased with days of planting from the first rains (Fig. 4.9).

There was, however, no significant difference between mid-season planting and late season planting on kernel score. There was a significant difference between late and early planting (Fig. 4.10). Early planting had the least score of 1.22. Mid-season planting had the highest score of 4.38. Early planting was different from the rest, while mid- and late planting were ranked the same.

The following plate shows results in picture form for observed parameters (Plates 4.3, 4.4, 4.5, and 4.6). The results show that late planting

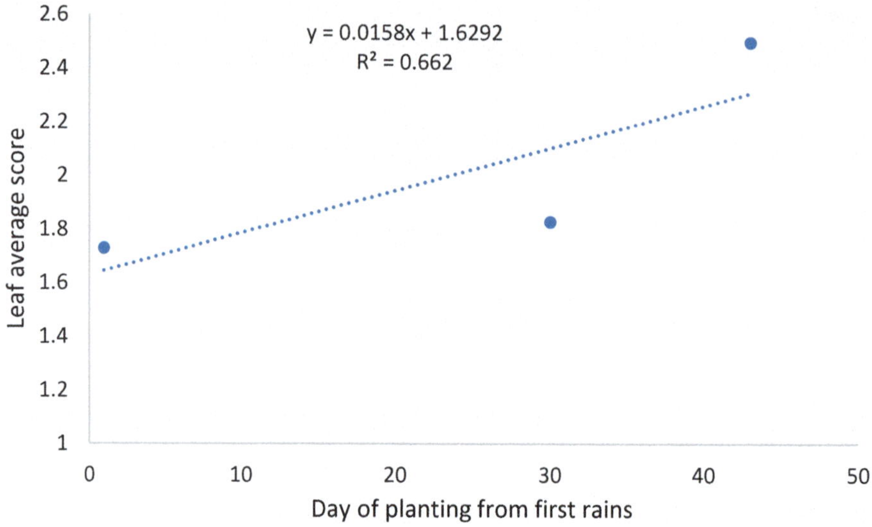

Fig. 4.9 The relationship between leaf average score damage and day of planting from first rains. (Source: Authors)

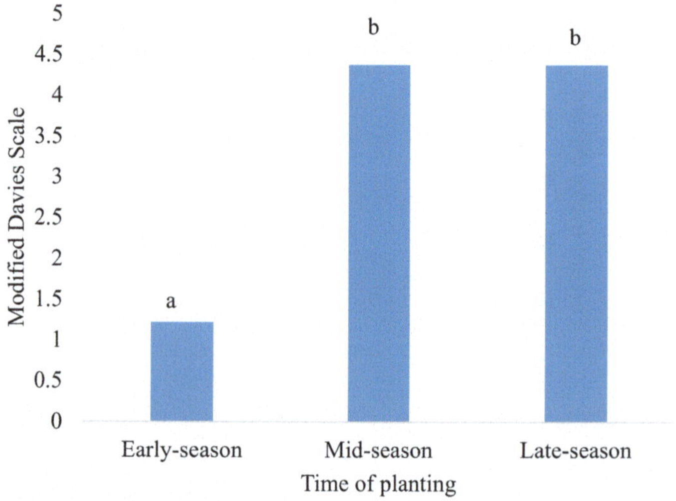

Fig. 4.10 Average kernel score (modified Davies scale)

results in serious damage to grain cobs by FAW, a phenomenon that is avoided by early planting.

Early planting led to significantly less percentage of plants being affected by FAW. According to Harrison et al. (2019), early planting is done to avoid peak migration of the FAW adult moth. According to Gebre-Amlak et al. (1989), early planting has worked for other stem borers because when planted early, crops have higher chances of escaping pest infestation compared to late planting. Mitchell et al. (1991) also noted that early planting is the most important cultural practice employed widely in the southern states of the United States to avoid the pest and early maturing varieties to mitigate against FAW. Recently CABI (2017) recommended early planting to escape FAW attack.

Midega et al. (2018) also noted that early planting after the first effective rainfall usually

Plate 4.3 Effect of FAW attack on kernel damage on late planting (damaged grain) compared to early planting (undamaged grain) for DKC8051

Plate 4.4 Late crop (left) and early crop (right)

Plate 4.5 Early crop (left) and late crop (right) for SC 633)

provides better growing conditions for maize making better use of more heat units at the beginning of the growing season. The resultant healthy plants may be capable under good moisture and nutrition conditions to compensate for foliar damage that may appear alarming to farmers (Abate et al. 2000). However, many smallholder farmers grow maize under unsuitable conditions and hence suffer greater losses due to poor nutrition and moisture stress (Hruska 2019).

Maize yields decreased with delay in time of planting from the first rains. Generally, early planted crops receive better heat units and generally grow faster than later planted crops. However, the main reason for differences between the pictures shown was because of FAW infestations. For similar varieties, there were striking differences between early planted and late planted maize. Hruska (2019) reported that late planting is often infested with high levels of FAW as

Plate 4.6 Late crop (left) and early crop (right) for SC 637. (Source: Authors)

moths increase as the season progresses and seek vegetative maize. Cobbing of late planted maize coincides with peak FAW period, and hence, grain damage increased. Soft dough stage cobs are more preferred as they are softer, and that results in substantial grain damage resulting in a higher percentage in discard.

As can be noted from plates 3, 4, 5, 6 and 7, the damage done to the crop due to late planting was extensive. The late crop coincided with the peak pest stages, and that could have resulted in extensive infestations with subsequent serious yield losses as was indicated on the yield. Therefore, manipulating host plant development through altering planting dates relative to the pest creates asynchrony between the pest and the critical crop stages (Thacker 2002).

6 Conclusion and Future Implications

The study concludes that the maize genotypes used did not contain any form of partial or complete resistance to FAW. Although new breeding was initiated across all the breeding houses in Zimbabwe and the regions so far no genotypes was found to be resistant to the pest. However, the results from this study also showed the effectiveness of early planting as a means of managing the pest. Early planting was shown to reduce the number of plants affected by FAW per plot. The cob quality tended to be lower the later the maize plantings. Evidence from other agronomic management practices for other stem borers showed that these agronomic techniques which are relatively affordable and friendlier to the environment are good for adoption. There is a need for promoting these low-cost methods for FAW management. This may form the basis for developing an integrated pest management programme which seeks to use all the suitable techniques in a compatible manner to maintain population levels below the economic injury level. Early planting affects the larvae whose effect tends to lessen as the time of infestation relative to the age of the plant delays. The sensitivity of the crop to FAW tends to get reduced with delay on the time of infestation. Further studies need to be conducted to test the effects of other agronomic techniques such as irrigation planting density, manuring, fertiliser application, crop rotation and clean cultivation on the FAW pandemic.

References

Abate, T., Van Huis, A. and Ampofo, J.K.O. (2000). Pest management strategies in traditional agriculture: an African perspective. *Annual Review of Entomology 45*, 631 – 659.

CABI. (2017). Fall armyworm: impacts and implications for Africa. Available at http://www.cabi.org/isc/datasheet/2981. (Accessed on 16 July 2019).

Carroll, M.J., Eric A., Schmelz, Robert. L., Meagher, P. and Teal, E. A. (2006). Attraction of *Spodoptera frugiperda* Larvae to Volatiles from Herbivore-Damaged Maize Seedlings. *J Chem Ecol* 32, 1911–1924.

Chimweta, M., Nyakudya, I.W., Jimu, L. and Mashingaidze, A.B. (2019). Fall armyworm (*Spodoptera frugiperda* (JE Smith)) damage in maize, management options for flood recession cropping small holder farmers. *International Journal of Pest*

Management. https://doi.org/10.1080/09670874.2019 1577514

Cruz, I., Figueiredo, M.L.C., Oliveira, A.C. and Vasconcelos, C.A. (1999). Damage of *Spodoptera frugiperda* (Smith) in different maize genotypes cultivated in soil under three levels of aluminium saturation. *International Journal of Pest Management 45*: 293-296.

Davis F.M. and Williams F. (1992). Visual rating scales for screening whorl-stage corn for resistance to fall armyworm. Mississippi Agric and Forestry Experiment station. Technical Bulletin 186. Mississippi State University, USA.

FAO. (2018). Integrated management of FAW on maize a guide to farmer field schools in Africa. Available at http://www.fao.org/faostat/en/. Accessed 12/09/2019

FAO. (2019). Brief note on FAO Actions of fall armyworm. FAO. Rome. Available at http://www.fao.org/ note. (Accessed on 20 September 2019).

Hruska, A.J. (2019). Fall armyworm (*Spodoptera frugiperda*) management by small holders. *CABI Reviews 14*, 1 – 11.

Huesing, J.E. (2018). Integrated Pest Management of the Fall Armyworm in Africa: An Introduction. *Stakeholders Consultation Workshop on Fall Armyworm* (p. 13). Nairobi, Kenya: USAID, CIMMYT, CGIAR.

Gebre-Amlak, A., Sigvald, R. and Petterson, T. (1989) the relationship between sowing date, infestation and damage by stalkborer, *Busseola fusca* (Noctuidae) on maize in Awassa, Ethiopia. *Tropical Pest Management 35*, 143 – 145.

Goergen, G. I. P. (2016). First report of outbreaks of the fall armyworm (*Spodoptera frugiperda* (JE Smith) Lepidoptera, Noctuidae), a new alien invasive pest in west and central Africa. PLoS One. https://doi.org/10.1371/journal.pone.0165632.

Harrison, R.D., Thiefelder, C., Baudron, F., Chinwada, P., Midega, C., Schiffner, U.and van Den Berg, J. (2019). Agro-ecological options for FAW management: providing low cost small holder friendly solutions to an invasive pest. *Journal of Environmental Management 243*, 318- 330

Jayne, M. S. (2003). Maize in Eastern and Southern Africa "Seeds" of Sucess in Retrospect. Washington DC: Technical Division International food Policy Research Institute, 2033 K Street, N.W, Washington DC.

Kansiime, K.M., Mgambi, I., Rwomushana, I., Nunda, W., Larontagne-Godwin, J., Rware, H., Phiri, N.A., Chipabika, G. and Ndhlovu, M., Day, R. (2019). Farmer perception of fall armyworm (*Spodoptera frugiperda*) and farm level management practices in Zambia. *Pest Management Science*. https://doi.org/10.1002/ps.5504.

Mandumbu, R., Jowah, P., Karavina, C. and Tibugari, H. (2011). Integrated weed management in Zimbabwe's smallholder sector: where are we? A review. *Morden Applied Sciences. 5 111–115*

Midega, C.A.O., Pittchar, J.O., Picket, J.A., Halu, G.W. and Khan, Z.R. (2018) A climate –adapted push-pull system effectively controls fall armyworm, *Spodoptera frugiperda* in maize in East Africa. *Crop Protection. 105*, 10 – 15.

Mitchell, E.R., McNeil, J.N., Westbrook, J.K., Silvian, J.F., Lalanne-Cassou, R.B. and Chalfant, S.D. (1991). Seasonal periodicity of fall armyworm (Lepidoptera: Noctuidae) in the Carribean Basin and northward to Canada. *Journal of entomological Science 26*, 39 – 50.

Prasanna, B.M. (2018). Fall Armyworm in Africa: A guide for Integrated Pest Management First Edition. *Stakeholders Consultation workshop on fall armyworm*. Nairobi: USAID, CIMMYT, CGIAR, FEED THE FUTURE.

Sisay, B., Tefera, T., Wakgari, M., Ayalew, G. and Mendesil, E. (2019). The efficacy of selected synthetic insecticides and botanicals against Fall armyworm (*Spodoptera frugiperda*) in maize. *Insects10*, 45. https://doi.org/10.3390/insects/10020045

Thacker, J.R.M. (2002). *An introduction to arthropod pest control*. Cambridge University Press. United Kingdom.

United Nations. (2015). *Transforming our world: the 2030 Agenda for Sustainable Development*. Resolution adopted by the General Assembly 25 September 2015. Available at: http://www.un.org/ga/search/view_doc.asp?symbol=A/RES/70/1&lANG=e (Accessed 02 July 2019).

5 Enhancing Urban Farming for Sustainable Development Through Sustainable Development Goals

Nyasha Chaminuka, Ernest Dube, Itai Kabonga, and Smart Mhembwe

Abstract

This chapter discusses how enhanced urban farming can be a strategy for sustainable development by localising the sustainable development goals (SDGs). Many urban households in Gweru urban in the Midlands province of Zimbabwe live in high levels of poverty, extreme hunger and malnourishment, hence the need for sustainable development. This study sought to unpack the role of urban farming as a strategy to achieve sustainable development through the SDGs; analyse the prospects of urban farming to urban households' welfare; and discuss the challenges encountered by urban households in achieving sustainable development and domesticating the SDGs through urban farming. This qualitative research was based on a purposive sample of 30 participants, with data gathered through interviews and on-site observations from 2018 to 2019. Findings show that urban farming was an option towards achieving sustainable development and domesticating the SDGs. Also, urban farming has the potential to reduce poverty and hunger and can contribute to good health and wellbeing of households. Furthermore, urban households encountered challenges such as shortage of farming land, lack of farming inputs (crop seed, fertiliser and pesticide), lack of extension services and lack of supportive policies in their urban farming endeavours. The study concluded that urban farming can either reduce or escalate poverty and hunger. Hence, the practice should be supported by stakeholders, including local authorities to achieve sustainable development and domesticate the SDGs. The study has the potential to assist the policymakers with information for better policy formulation and implementation. This study recommends the assistance to urban farmers in the form of improved access to farming land, inputs and skills capacitating to enhance the practice.

Keywords

Poverty · Hunger · Urban farming · Sustainable development · SDGs

N. Chaminuka
Department of Development Studies, Zimbabwe Open University, Gweru, Zimbabwe

E. Dube (✉)
Department of Development Sciences, Marondera University of Agricultural Sciences and Technology, Marondera, Zimbabwe
e-mail: edube@muast.ac.zw

I. Kabonga · S. Mhembwe
Department of Humanities and Social Sciences, Ezekiel Guti University, Bindura, Zimbabwe

G. Nhamo et al. (eds.), *Sustainable Development Goals for Society Vol. 2*, Sustainable Development Goals Series, https://doi.org/10.1007/978-3-030-70952-5_5

1 Introduction

The concept of urban farming has not been fully embraced, despite its potential to achieve sustainable development for urban populations. Urban populations have been steadily swelling in recent decades (Chevo 2018). This resulted in the need to increase activities towards the realisation of sustainable development goals (SDGs) in addressing issues of poverty, hunger and other needs of urban households. Many urban households have adopted urban farming activities to address poverty, food shortages and lack of income, amongst other basic needs deprivations. Evidence suggests that about one-quarter of the developing world's poor populations are found in urban areas and that food insecurity has become worse in urban areas (Ravallion et al. 2007). As such, urban farming has been one of the strategies used by urban populations in achieving SDG1 (no poverty) and SDG2 (zero hunger).

Statistics have highlighted the high dependency of populations on farming activities in dealing with poverty and hunger issues. About 800 million people globally have practised urban farming as a strategy to eliminate poverty and hunger, whilst about 200 million people have benefited from employment created through the practice (Zezza and Tasciotti 2010). Therefore, the large populations benefiting from urban farming suggest that the farming activity might be a worthwhile endeavour for reducing poverty, eliminating hunger and achieving other SDGs. Studies have also shown that about 40 per cent of urban households worldwide are impoverished (FAO 2012); hence, their welfare is at stake. Also, it is believed that many poor urban households spend between 60 and 80 per cent of their incomes on food, and this further exposes them to more food price volatility (Nhapi 2019).

The practice of urban farming by urban households usually occurs in backyards of urban residences, on public land as well as other public places like roadsides. This study argues that urban farming should be practised at properly designated places where there is no interference with the built environment and other developmental projects. Therefore, urban farming is a concept that also needs to be highly regulated so that it does not impact negatively on the environment. The negative impact of urban farming on the environment may negate sustainable development and the realisation of SDGs such as SDG11 (achieving sustainable cities), SDG12 (responsible consumption and production) and SDG13 (climate action) (United Nations Development Programme (UNDP) 2015). Thus, when urban farming activity is well practised, it can be replicated and extended in other public places, such as school grounds, unused open spaces and riversides, provided the practice does not contribute to environmental poverty. Environmental poverty occurs when people are deprived of a healthy environment that is suitable for human habitation.

In some instances, those in positions of authority in urban areas have also not supported the domestication of the SDGs through urban farming. This may defeat the aim of achieving sustainable development through the SDGs through a multi-stakeholder approach since no one should be left behind. Some authorities have regarded urban farming as an activity that disturbs urban development processes and that it does not support the goal of *sustainable cities and communities* (SDG11). Besides, research and policy debates surrounding urban farming show some doubts about its contribution as a poverty reduction and hunger elimination strategy (Crush et al. 2011; Lee-Smith 2013). However, the discussion argues that urban farming is not always a threat to urban development, but instead, it is a practice that has the potential to improve the lives and wellbeing (SDG3) of urban households.

In the African context, it has also been proved that urban farming has played an important role in reducing poverty to urban households (Chaminuka and Dube 2017; Goddard et al. 2009), hence improving people's wellbeing. For instance, 40 per cent of urban households in Africa are said to be engaged in farming activities for poverty reduction (Zezza and Tasciotti 2010). Also, many urban households in Zimbabwe are living with high levels of poverty and serious hunger as they cannot access basic needs such as food, medical care and education (Moyo 2013).

According to Nhapi (2019), the poverty situation of urban households in Zimbabwe is getting worse, as 38.2 per cent of urban households are living in poverty, with 4 per cent in extreme poverty. Therefore, the practice of urban farming in Gweru urban cannot be ignored as it can lead to the achievement of sustainable development by urban households. Thus, this study sought to (1) unpack the role of urban farming as a strategy to achieve sustainable development through the SDGs; (2) analyse the prospects of urban farming to urban households' welfare; and (3) discuss the challenges encountered by urban households in achieving sustainable development and domesticating the SDGs through urban farming. This study argues that urban setups have vast portions of land which urban farmers can utilise to produce crops to supplement food for households. Moreover, urban farmers do possess the necessary human capital in the form of farming skills and knowledge and social capital in the form of networks and teams, which they can use to produce crops in improving their livelihoods.

Urban farming is a practice that can help achieve food security for many poor households, resulting in sustainable development in urban areas. Nonetheless, despite the potential benefit of this farming activity, it seems that the concept has been overlooked in Gweru urban, in the Midlands province of Zimbabwe. This is despite the assets that urban populations have at their disposal including human capital, which is the ability to provide labour, farming skills and knowledge (Chaminuka and Dube). In Gweru urban, scenarios do show that urban farming has been under-practised, leading to increased food insecurity, poverty and hunger to many urban households. As such, the progress towards the achievement of sustainable development and the domestication of SDGs in Gweru urban has been very slow, with 2030 fast approaching. Different dimensions of poverty (food poverty, income poverty and education poverty) and hunger have severely impacted households in Gweru urban, hence the need to re-examine the practice of urban farming as a gateway to realising development through the SDGs. If the practice of urban farming in Gweru urban is not addressed, many households would continue to face food insecurity, poverty and hunger – failing to achieve sustainable development. There is a need to enhance urban farming for sustainable development through the SDGs. The next section provides literature about urban farming, the SDGs and theoretical framework informing the study – the sustainable livelihoods framework.

2 Literature Review

2.1 Urban Farming and the Sustainable Development Goals

Discussions around the subject of urban farming have been conducted by many scholars. Whilst linking urban farming with sustainable development and SDGs, a positive correlation has been identified. Urban farming has been understood as the production of crops and livestock within cities and towns (Zezza and Tasciotti 2010) for sustainable development. A broader conceptualisation of urban farming is given as the growing of plants and the raising of poultry and animals within or around urban areas for consumption and or commercial uses (Chaminuka and Dube 2017). This discussion considered these definitions and acknowledges that urban farming can play a crucial role in improving urban food security and alleviating urban poverty, supporting sustainable development and improving livelihoods (Prain and Dubbeling 2011). As such, urban farming is identified to have a huge contribution to the accomplishment of SDGs. For instance, SDG1 (*no poverty*) and SDG2 (*zero hunger*) are the two main goals that are urban farming outputs. In the context of this study, urban farming is the growing of crops and the keeping of livestock in urban areas for sustainable development. The study chose this definition to emphasise the purpose of urban farming to achieve sustainable development by reducing poverty and hunger at the household level, hence improving the wellbeing of urban residents. Therefore, this study assumes that urban farming is a potential strategy for achieving sustainable

Box 5.1 The United Nations SDGs

- *Goal 1. No poverty:* End extreme poverty in all forms by 2030.
- *Goal 2. Zero hunger:* End hunger, achieve food security and improved nutrition and promote sustainable agriculture.
- *Goal 3. Good health and wellbeing*: Ensure healthy lives and promote wellbeing for all at all ages.
- *Goal 4. Quality education*: Ensure inclusive and equitable quality education and promote lifelong learning opportunities for all.
- *Goal 5. Gender equality:* Achieve gender equality and empower all women and girls.
- *Goal 6. Clean water and sanitation*: Ensure availability and sustainable management of water and sanitation for all.
- *Goal 7. Affordable and clean energy*: Ensure access to affordable, reliable, sustainable and modern energy for all.
- *Goal 8. Decent work and economic growth:* Promote sustained, inclusive and sustainable economic growth, full and productive employment and decent work for all.
- *Goal 9. Industry, Innovation and Infrastructure*: Build resilient infrastructure, promote inclusive and sustainable industrialization and foster innovation.
- *Goal 10. Reduced inequalities:* Reduce inequality within and among countries.
- *Goal 11. Sustainable cities and communities*: Make cities and human settlements inclusive, safe, resilient and sustainable.
- *Goal 12. Responsible consumption and production:* Ensure sustainable consumption and production patterns.
- *Goal 13. Climate action*: Take urgent action to combat climate change and its impacts.
- *Goal 14. Life below water*: Conserve and sustainably use the oceans, seas and marine resources for sustainable development.
- *Goal 15. Life on land*: Protect, restore and promote sustainable use of terrestrial ecosystems, sustainably manage forests, combat desertification, and halt and reverse land degradation and halt biodiversity loss.
- *Goal 16. Peace, justice and strong institutions*: Promote peaceful and inclusive societies for sustainable development, provide access to justice for all and build effective, accountable and inclusive institutions at all levels.
- *Goal 17. Partnerships for the goals:* Strengthen the means of implementation and revitalize the global partnership for sustainable development.

Source: Authors, based on Unite d Nations Development Programme (UNDP) (2015: 14)

development through SDGs. The concept can reduce dimensions of poverty, especially food and income poverty commonly faced by urban households.

In understanding the SDGs, it should be highlighted that there is not much rich history behind them as there is one for urban farming. SDGs are part of a development plan made up of 17 sets of goals, with the anticipation to a future that is poverty-, hunger- and climate change-free as projected for 2030 (OECD 2016; United Nations Development Programme (UNDP) 2015). The 17 sets of SDGs as adopted from the United Nations (United Nations Development Programme (UNDP) 2015) are shown in Box 5.1.

The goals are meant to foster sustainable development within and amongst all nations. Therefore, there is a need to domesticate the goals as a way of fostering sustainable development at the local level. This study understands that by domesticating SDGs through urban farming, the possibility of achieving sustainable development by households exists. Hence, the study identified the sustainable livelihoods framework (SLF) as a way of achieving urban livelihoods outcomes. The next subsection discusses the SLF, showing its connection to urban farming, SDGs and sustainable development.

2.2 Sustainable Livelihoods Framework Adopted for the Study

In achieving sustainable development through domesticating SDGs for urban farming, the SLF is the fitting approach for the study context. Mancini et al. (2007) define the SLF as an evaluation tool that can be used to study and understand the livelihoods of poor communities, for instance, Gweru urban communities facing food and income poverty. The framework uses opportunities and constraints that can impinge in livelihoods. Thus, the SLF seeks to explain how urban farming as a livelihood in Gweru urban tends to be encountering challenges and opportunities in achieving sustainable development through SDGs. According to the Food and Agriculture

Organization (FAO) (2013), a livelihood is sustainable when it can cope with and recover from stresses and shocks and can maintain or enhance its capabilities and assets both now and in the future whilst not compromising the livelihoods of others. Urban farming in Gweru urban is a livelihood of poor urban households. The livelihood can be sustained by livelihood assets such as proper farming skills, allocation of land and assistance with farming inputs and equipment and income support to urban farmers.

In this study, urban farming is the opportunity identified to improve livelihoods of urban households in Gweru urban, as well as sustainable development by embracing SDGs 1 and 2 on eliminating poverty and hunger. In this case, sustainable development through urban farming and SDGs in Gweru urban is the positive livelihood outcome. Mancini et al. (2007) emphasise that the SLF (Fig. 5.1) has been understood as a theoretical tool that can be used to analyse and understand the livelihoods of the poor.

Figure 5.1 depicts the Gweru urban community which is living under vulnerable conditions characterised by shocks, trends and seasonality hunger, high prices of farming inputs and inflation. However, the community possesses livelihood assets such as natural assets (land for farming), human assets (labour, farming skills and knowledge) and social assets (social networks). The study argues that the Gweru urban community can be assisted with other important assets such as financial (income) and physical assets (farming equipment). The assets (Fig. 5.1) can be manipulated through combination by the households to support urban farming so that sustainable development is achieved and SDGs domesticated (Forsyth 2007; Momentum 2012). Manipulating the livelihood assets may lead to the attainment of goals of no *poverty* (SDG1), *zero hunger* (SDG2) and the enjoyment of *good health and wellbeing* (SDG3), amongst others in Gweru urban.

Households in Gweru urban area need assistance to eliminate poverty and hunger and to achieve good health and wellbeing. By eliminating poverty and hunger, and improving the wellbeing of urban households, urban farming has a potential of domesticating not only SDG1, SDG2 and SDG3 but also SDG11 – achieving sustainable cities and communities. Hence, the situation of households in Gweru urban is better explained through the SLF, which depicts them in the vulnerability context under the threats of climate change and seasonal droughts, amongst other issues. The situation created by these threats can disturb urban farming, thereby affecting food production and income levels of urban households and affecting the attainment of the SDGs. Therefore, the study emphasises the need for

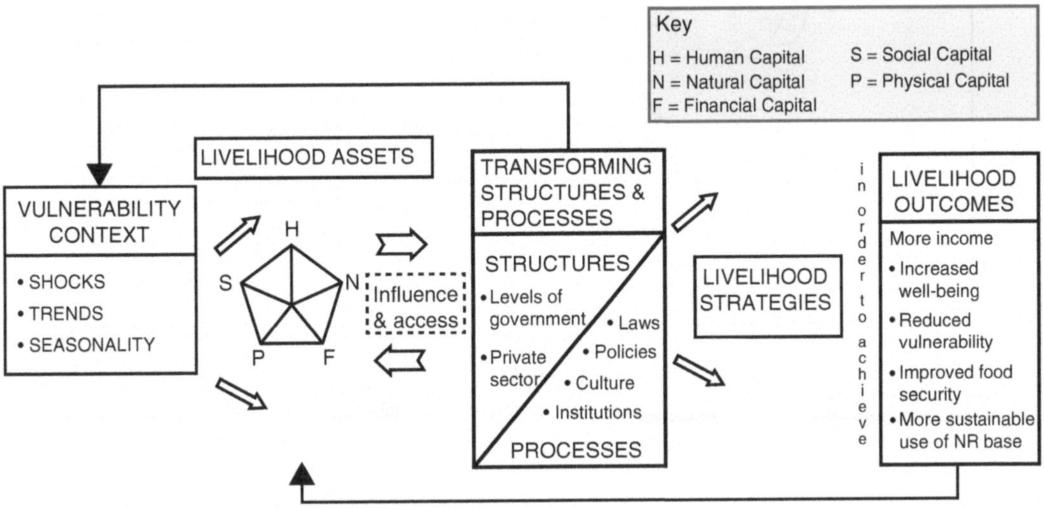

Fig. 5.1 The sustainable livelihoods framework. (Source: DFID (2010))

Gweru urban households to use their assets to break away from poverty and extreme hunger and to attain good health, amongst other SDGs aims.

However, reliance on the livelihood assets does not guarantee to achieve sustainable development and success towards the attainment of SDG targets, hence a proper combination of assets to effectively fight poverty (Forsyth 2007; Momentum 2012). This study assumes that achieving SDG1 and SDG2 will lead to the attainment of the other 15 SDGs because they are intertwined. Furthermore, if the Gweru urban populations are free from poverty and hunger, they would be able to work for themselves and come up with innovations. Urban farming in Gweru also needs to be mediated by the transforming structures – policies, institutions and processes (Fig. 5.1). Urban authorities through bylaws and enforcement on urban farming can either limit or promote the attainment of sustainable development through the SDGs. The following section discusses the materials and methods that were adopted to execute the study.

3 Materials and Methods

The research is a case study of Gweru urban in the Midlands province of Zimbabwe (Fig. 5.2). Gweru urban has a population of approximately 157,865 people (Zimstat 2012). The population is likely to increase when the next national census is conducted in 2022. Statistics show that the rates of urban population growth in sub-Saharan Africa are amongst the highest in the world (Chaminuka and Dube 2017; Chevo 2018). Any increase in the Gweru urban population might affect sustainable development and progress

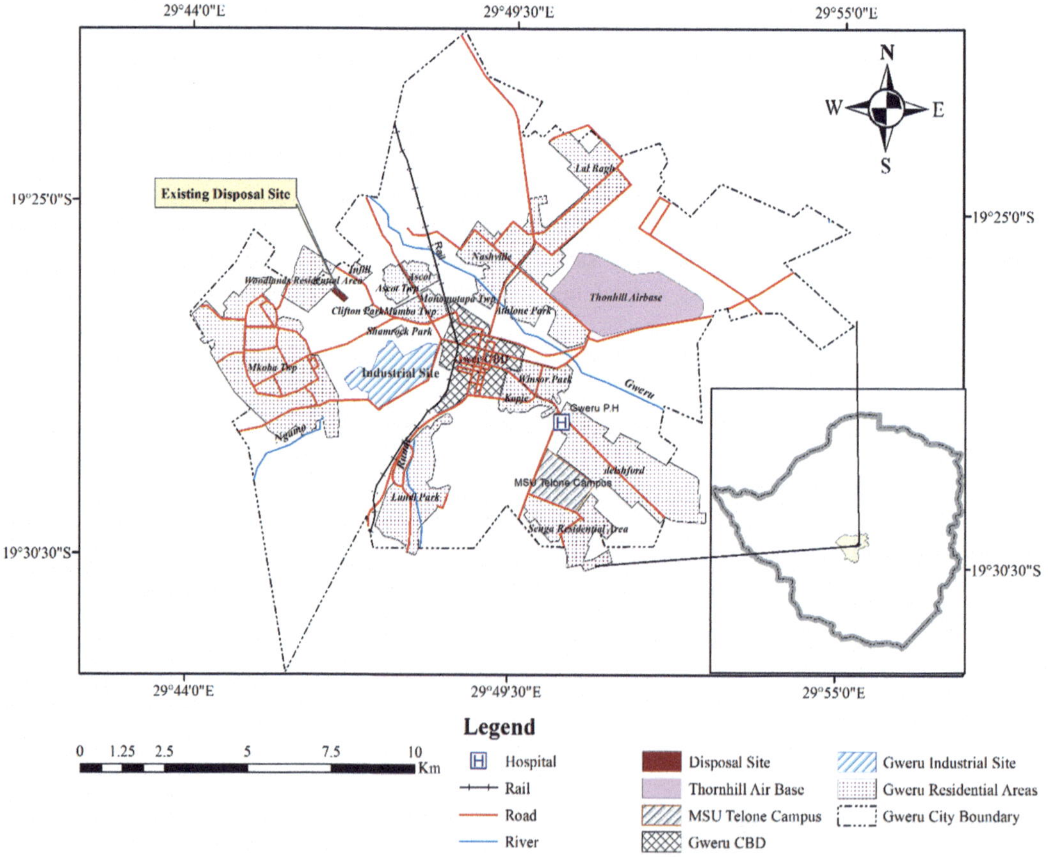

Fig. 5.2 Map of Gweru urban in the Midlands province of Zimbabwe. (Source: Authors)

towards SDGs, especially SDG1 (no poverty), SDG2 (zero hunger) and SDG3 (good health and wellbeing). Most households in Gweru urban earn a living through farming and vending and a few through formal employment.

The discussion adopted a qualitative research approach to understand and learn from the experiences of urban households. A qualitative approach is naturalistic and allows for the study of events in their natural settings (Lodico et al. 2006). A sample of 30 purposively selected participants was reached through saturation. Data were gathered from specifically targeted urban households through interviews conducted from October 2018 to April 2019. Gweru urban is dominated by ChiShona-language speaking residents, with a few residents speaking isiNdebele language. The participants were chosen on the criteria that they were information-rich since urban farming is part of their livelihood. The interview guide had open-ended questions to allow the researchers to probe further and gathered large volumes of data about urban farming in achieving sustainable development, as well as urban farming's contribution to households' welfare. Further, the study sought to gather data that interrogated the challenges faced by households in their urban farming endeavours. The interviews were complemented with researcher observations, in which the researchers sought to observe the nature of crops grown by urban households. The researchers acted as instruments to enhance dependability and credibility of the study through observation techniques, skills, knowledge and effectiveness.

Data analysis involved research themes, which were derived from the qualitative thematic categorisation so that the themes resonated with the objectives of the discussion. The analysis entailed collecting and categorising volumes of raw data to obtain meaningful information. Since the discussion is qualitative, data analysis aimed to make sense of the participants' 'views and opinions of situations, corresponding patterns, themes, categories and regular similarities' (Cohen et al. 2007:461). All the data were translated from ChiShona and isiNdebele to the English language. The following analysis procedure was adopted in analysing data collected from the 30 respondents. For data analysis, urban households are represented by the abbreviation UH. The respondents were coded as follows: 30 urban households (UH) – coded as UH01 – UH30. After coding, the categorisation process followed as indicated below (Table 5.1).

Table 5.1 Profiles of the households interviewed in Gweru urban

Interview code	Description
Category 1	*Categorisation according to the farming activity*
UH01 – UH12	Maize crop production
UH13 – UH21	Poultry production – broilers
UH22 – UH26	Poultry production – layers
UH27 – UH30	Sweet potato production
Category 2	*Categorisation according to the purpose of farming*
UH01 – UH18	Household consumption and selling of surplus
UH19 – UH30	For selling at the market place

Source: Authors

Table 5.1 represents the different stages the researchers followed in the analysis of the data collected from the 30 urban households which formed the unit of analysis. From Table1, there are two sets of categories, with Category 1 representing the farming activities practised in Gweru urban, whilst Category 2 shows the purpose of engaging in the farming activities. The household grouping was ranged from UH01 to UH30. From the 30 interviewed respondents in the category showing the nature of the farming activity they do to enhance their livelihoods, 12 (UH01–UH12) respondents indicated that they grew maize, nine (UH13–UH21) households practised poultry for broilers, five (UH22–UH26) households were into poultry production for eggs and three (UH27–UH30) households were farming sweet potatoes. Additionally, data was manipulated to fit into Category 2 which is about the purpose of the farming activity conducted. Under Category 2, 18 (UH01–UH18) households practised urban farming for household consumption, and they also sold surpluses to neighbours and other residents in need of fresh produce, to generate money to pay children's school fees. Twelve (UH19–UH30) households indicated that they practised farming so that they sell their products at the market. From the data analysis, the following themes were developed: (1) urban farming as a tool that can enhance household food security and livelihood enabler, (2) urban farming as an undertaking that can either reduce or escalate poverty and (3) urban farming as a practice that can promote agribusiness and entrepreneurship.

4 Presentation and Discussion of Results

The study findings presented in this section are based on the research objectives which were to (1) unpack the role of urban farming as a strategy to achieve sustainable development through the SDGs, (2) analyse the prospects of urban farming to urban households' welfare and (3) discuss the challenges encountered by urban households in achieving sustainable development and domesticating the SDGs through urban farming. Furthermore, the literature was consulted in the discussion of the results to generate new knowledge from the study. The results are presented in three thematic areas as mentioned in the data analysis section above.

4.1 Urban Farming in Achieving Sustainable Development

Urban farming in Gweru urban has been endorsed by many households as an option to achieve sustainable development and localising SDGs. According to the respondents, the practice had gained acceptance over the years, because of its contribution to urban livelihoods. According to the participants, urban farming contributed to poverty reduction and the elimination of hunger through improved food security and increased the general wellbeing of many households. By reducing poverty and hunger, urban farming also contributed to improving households' standard of living. This is an indication that the practice supports and domesticates SDG1, SDG2 and SDG3, by fulfilling the goals of *no poverty*, *zero hunger* and *good health and wellbeing*. This study argues that those who have practised urban farming as a livelihood strategy in Gweru urban have been food secure, hence the theme of urban farming as a tool to enhance household food security and livelihood enabler. The findings support a study by Peñalba (2019) who found that farming has the potential to deal with climate change effects by enhancing food security and livelihood. This theme of urban farming as a tool to enhance household food security and livelihood enabler is discussed in detail below.

4.2 Urban Farming as a Tool to Enhance Household Food Security and Livelihood Enabler

Most of the participants narrated that they increasingly relied on urban farming for sustenance. According to the respondents, they have resorted to food production through farming due to the rising cost of basic food commodities. It this way, urban farming seemed a viable strategy for dealing with food insecurity in urban areas, as it provided the needed food and nutritional needs to the households through improved access to fresh produce. As a tool to enhance household food security and livelihoods, urban households have mastered the skill and knowledge to grow their crops to achieve the goals of *no poverty*, *zero hunger* and *good health and wellbeing*. The nature of crops that urban households grow, according to Gweru urban participants, included green vegetables, sweet potatoes, groundnuts and maize. The study argues that urban farming is a livelihood enabler as it helped households to produce different farming produces for survival. Through the ability and willingness to produce different types of crops, urban farming is also seen as a tool that can enhance household food security. Hence, the practice fosters the domestication of the SDGs through eliminating poverty and hunger and achieving good health and wellbeing of Gweru urban households. Katsaruware et al. (2014) concur with this study's findings, as their study done in the city of Kadoma in Zimbabwe found that maize and vegetables were amongst the main crops produced through urban farming.

The research participants further mentioned that they produced such crops because the crops relied on rain-fed water. They indicated that the farming of rain-fed crops had some advantages since rainwater is available for free and this saves them from paying for water in urban areas. They further mentioned that at times they harvested and stored rainwater, which they would later use for their crops. The SLF recognises rainwater as a natural asset. The respondents opined that households in Gweru urban can combine water with human assets, namely, farming knowledge and labour for sustainable crop production – their desired livelihood outcomes. Besides, the respondents narrated that using rain-fed water was cost-effective as it does not require physical infrastructures, like irrigation equipment and hosepipes which infringe the city by-laws. Researcher observations also revealed that urban farming also took the dimension of allotment gardens. Some verification done with the local authorities revealed that the allotment gardens were legal and approved by the Gweru municipality. One such garden was observed in Nehosho suburb, where vegetables such as cabbage, tomatoes and beans amongst others were grown. Production of the variety of vegetables further reinforces the theme of urban farming as a tool to enhance household food security and livelihood enabler. It was also observed that the vegetable project was sustained by a borehole sunk by donors inside the garden to alleviate water problems. From the onsite visits, the authors further observed that maize was the most common crop grown by urban households in Gweru urban.

Apart from the attempt to overcome food poverty and hunger and achieve other SDGs through farming, the participants further indicated that they sold surplus farm produce, for instance, sweet potatoes as a way of supplementing their income. The following excerpt encapsulates one of the participant's lived experiences:

> Since I am not employed, I also rely on urban farming as part of my livelihoods. I usually harvest sweet potatoes in large quantities, such that I would sell some to other households as a way of supplementing my income. I would then use the money to buy other household necessities, to pay rent and school fees for my children. (Female adult, Senga suburb, Gweru)

The above narration by one of the participants shows that apart from providing food (SDG1 and SDG2), urban farming can also contribute to sustainable income generation (SDG8) and education poverty (SDG4) within societies. As such, through generating income for school fees, the practice is also seen as supporting the goals of ensuring quality education and promoting lifelong learning (SDG4). The next subsection discusses the theme of urban farming as a practice that can either reduce or escalate poverty and hunger.

4.3 Poverty and Hunger Reduction or Escalation via Urban Farming

The above verbatim from a resident in Senga suburb in Gweru shows the respondent as a hard-working person, who is capable of escaping poverty and hunger through combining one's farming knowledge with natural capital – land. Hence, urban farming in Gweru urban has the potential to transform people's lives in urban settings by reducing different dimensions of poverty and hunger faced by urban households. At the same time, the practice can worsen the poverty and hunger status of urban households if the intervention is not properly carried out (Orsini et al. 2013). Therefore, urban farming is an undertaking that, when well supported with proper assets, can reduce poverty and hunger, whilst it can also escalate the phenomenon if it is not properly supported. In reducing poverty and hunger, urban farming is indeed seen as a viable strategy for achieving sustainable development and domesticating SDGs of *no poverty* and *zero hunger*. The study, therefore, finds that a unique relationship is created when urban farming is supported with appropriate livelihood assets (Fig. 5.3). With proper assets, urban farming can reduce poverty (SDG1), eliminate hunger (SDG2) and contribute to the general wellbeing

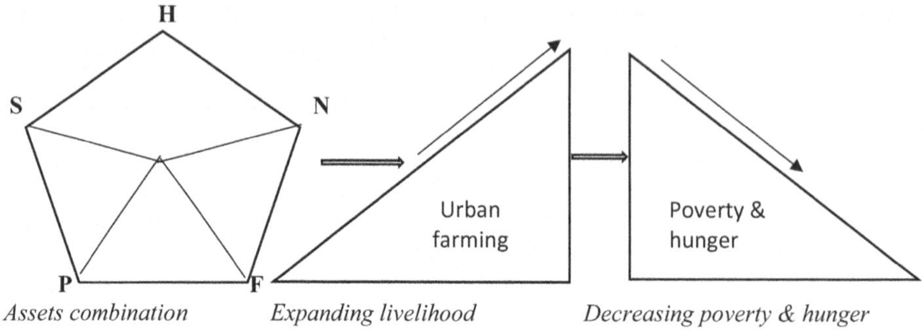

Fig. 5.3 Livelihood assets, urban farming and poverty and hunger relationship. (Source: Authors)

(SDG3) of urban households. The relationship showing the preferred condition of urban farming reducing poverty and hunger is illustrated below (Fig. 5.3).

Figure 5.3 explains how livelihood assets can influence the direction of urban farming in Gweru urban in a bid to achieve sustainable development and to domestic the SDG of *no poverty* and zero *hunger*. The explanation shows the preferred condition of urban farming, that of influencing the desired direction of poverty and hunger – reducing it. The relationship between livelihoods assets, urban farming and poverty and hunger (influence on SDG1 and SDG2) is illustrated by the horizontal directional arrows, which show the progression from one stage to another (Fig. 5.3). Hence, urban farming in Gweru urban can be supported with one or a combination of livelihood assets (H, human assets; S, social assets; P, physical assets; F, financial assets; N, natural assets) to expand or grow bigger, resulting in decreasing poverty and hunger.

Through the expansion of urban farming practice as shown by an upward arrow, urban farming can be a viable strategy in Gweru urban capable of reducing poverty and hunger. However, the study assumes that if urban farming has failed to reduce poverty or hunger, the arrow would face upwards, indicating the growth of the phenomenon and a challenge to sustainable development and domestication of the SDGs. Whilst livelihood assets can expand and strengthen urban farming (Fig. 5.3), urban farming may become stronger or weaker enough to reduce poverty and eliminate hunger for many urban households. According to some respondents, poverty and hunger can be reduced when urban households have access to food and education and have increased income amongst other household deprivations. Their views corroborated the narration by the unemployed Senga urban farmer (in the above quotation), who is using farming to generate income, pay bills and send her children to school. As such, most households indicated that they are likely to be more food secure (SDG1 and SDG2), obtain a quality education (SDG4) and increase their buying power (SDG3) if urban farming was well supported with assets such as training, farming clubs, farming equipment, funds and enough land.

These findings agree with the previous studies, which found that urban farming may lead to the expansion of local food production experiences (Dieleman 2016; Zainal and Hamzah 2017). Hence, the findings portray urban farming as a good strategy that can contribute to achieving sustainable development through SDGs. The findings also concur with a study done in the city of Bulawayo in Zimbabwe, which found that urban farming can improve food security and livelihoods through providing revenue collected from the selling of produce and through providing income resulting from employment created (Mudzengerere 2012). Therefore, urban farming can also fulfil SDG8 – which seeks to foster decent work and economic growth. Ultimately, urban households may depend on urban farming as one way to realise sustainable development.

4.4 Prospects of Urban Farming to Households' Welfare

From the interviews conducted, it seems that urban farming is a potential strategy to improve urban households' welfare. One of the themes derived from data analysis is that urban farming is a practice that can promote agribusiness and entrepreneurship. This theme is discussed below.

4.4.1 Urban Farming as a Practice to Promote Agribusiness and Entrepreneurship

According to 18 of the interviewed 30 participants (60%), they expected urban farming to grow into a large-scale endeavour capable of becoming an agribusiness venture. They stated that urban farming has efficient market potential and can finance their needs and contribute to infrastructure development – hence they regarded urban farming as a practice that can promote agribusiness and entrepreneurship. This view also portrays urban farming as an endeavour towards domesticating SDG3, SDG8, SDG9 and SDG11. According to the respondents, the practice is currently done at a low scale because of the lack of chemicals, seed inputs and financial challenges. However, the respondents indicated that with enough support from the government and stakeholders, the practice can improve their welfare. Previous studies also found that urban farming is frequently practised at small scale and that it is seasonal due to its reliance on rain-fed water (Webb 2011). Therefore, should the practice grow into a large agribusiness venture, the participants opined that it can generate employment for many unemployed urban residents. As such, this discussion argues that urban farming in Gweru urban can be sustained so that it can grow into a sustainable agribusiness sector. One urban farmer mentioned that:

> We are all aware that these days, there is a lack of employment in the country. Therefore, we envisage a situation where in the future, the practice of urban farming will expand into a business venture to enhance employment chances for the many jobless urban residents. In that sense, the practice will also be our major source of income generated through the business. (Male farmer, Sundowns, Gweru)

By employing the jobless urban population, urban residents have the potential of getting the much needed income from urban farming. As such, the authors reiterate that urban farming in Gweru urban, if expanded into an agribusiness and entrepreneurship, might provide employment and sustainably reduce poverty and hunger for many households. Views of other participants' group: 12 of the interviewed 30 (40%) mentioned that urban farming has the prospects of becoming a major supplier of green vegetables for Gweru urban markets. They stated that if adequately supported with enough land and farming inputs, the practice can be competitive with the potential to supply many urban markets with vegetables. Supplying vegetables in abundance and beyond the household level, to wider urban communities, would be a welcome development and a good strategy for achieving sustainable development and domesticating the SDGs. The provision of decent work and economic growth (SDG8) and creating sustainable cities and communities (SDG11) are some of the SDGs that can be domesticated. The Gweru urban scenario also shows urban farming's potential to achieve development through the SDGs by generating income, eradicating poverty and hunger and improving the general wellbeing of urban households. The findings agree with results from the previous study conducted by Drechsel et al. (2006) which found that above 60 per cent of fresh and perishable food consumed in urban areas in Africa were produced within the cities. Thus, by becoming a major supplier of vegetables to urban markets, urban farming is a sustainable household livelihood strategy that can promote agribusiness and entrepreneurship. Additionally, these results are linked to the SLF and SDGs. The SLF ends with livelihood outcomes in the form of increased wellbeing (SDG3), improved food security (SDGs 1 and 2) and sustainable use of natural resources base (SDGs11 and 12) (DFID 2010). The next subsection discusses challenges that are associated with urban farming practice in achieving sustainable development and domesticating the SDGs.

4.5 SDGs Domestication Challenges

The practice of urban farming in Gweru urban has not always been a smooth endeavour as many households have been facing challenges. Some of the key identified challenges are associated with availability of land, lack of access to farming inputs, careless use of chemicals and the need for capacitation for improved farming skills.

According to 15 of the interviewed 30 participants (50%), one major challenge is the small size of farming land. Through observations, the researchers verified that people were practising farming on small pieces of land, a scenario which seemed to not support food production endeavours for urban households. These findings concur with Lovell (2010) who found that the competition for land spaces made urban farming seem not an ideal practice in urban areas. Competing for bigger pieces for farming is against the efforts of achieving SDGs, as it results in overcrowding and poor sanitary conditions in urban areas. The participants further stated that the municipal authorities at times destroyed their crops, as the practice is illegal. The participants opined that if urban authorities supported their endeavours through land allocation, poverty and hunger would be reduced, thereby domesticating SDG1 and SDG2. The destruction of crops does not support the goals of *no poverty*, *zero hunger* as well as SDG3 (*good health and wellbeing*), SDG8 (*decent work and economic growth*), SDG11 (*sustainable cities and communities*) and SDG12 (*responsible consumption and production*) in Gweru urban. As such, the participants believed that urban farming should also be supported with appropriate policies for practice and that the authorities should always engage them. This view supports the SLF, which regards enabling processes, policies and structures as a good ground for realising positive livelihoods outcomes.

Another major challenge for urban households in the practice of urban farming towards sustainable development through SDGs was the lack of access to farming inputs. Ten out of the interviewed 30 participants (33%) indicated that they lacked access to farming inputs such as crop seed, fertiliser and pesticide. For urban farming to be sustainable, households should be supported with farming inputs, skills capacitation through extension services and farming equipment. Whilst many households indicated that they needed to be supported with the farming inputs, 33% of the respondents stated that they had no means for acquiring the inputs. As a result, they confined their activities to small pieces of land which did not produce enough food for an average family of six members. If not supported by farming implements and inputs, achieving sustainable development and SDGs targets would remain a challenge for urban households. These findings are significant, as Nasr et al. (2010) found that the costs of land and inputs in an urban setup can be a big challenge to farmers. The findings are also linked to the SLF, which understands natural assets (land) and physical assets (farming inputs) as important components for sustaining livelihoods and achieving positive outcomes. If urban households have access to these assets, the practice has the potential to achieving sustainable development and domesticating the SDGs.

Although urban farming plays a significant role as a means of survival, some participants (3 of the interviewed 30–10%) acknowledged that the practice can also pose environmental and health threats. They highlighted the carelessness of some urban farmers in the handling of harmful chemicals, such as pesticides and fertilisers. This compromises the sustainable practice of urban farming as a poverty and hunger reduction strategy (SDG1, SDG2). Further, the careless handling of the chemicals affects the environment and impacts against realising the goals of *clean water and sanitation* (SDG6), *sustainable cities and communities* (SDG11), *responsible consumption and production* (SDG12) and *life on land* (SDG15). Therefore, urban farming is not undertaken properly, and it can militate against achieving sustainable development and SDGs targets. This study emphasises that environmental preservation should be a priority so that urban farming can achieve SDGs' aims. Any act that disrupts or contaminates the environment should be avoided as that affects the sustainability of the environment. In the case of Gweru, the chemicals have the potential to contaminate water sources, resulting in health and environmental risks. One male respondent aged 54 years, from Mkoba suburb, felt that those embarking on urban farming should be provided with extension services, including services on the use of various pesticides to reduce associated chemical risks. He had this to say:

> Extension services for skills capacitation and better productivity should be put in place for the training of urban farmers on the use of pesticides. By so doing, the farmers would be equipped with proper knowledge which will help boost production, at the same time avoiding health risks from pesticides. (Male, aged 54 years, Mkoba suburb, Gweru)

The above excerpt advances the theme of urban farming as a tool to enhance food security and livelihood enabler for urban households. By equipping urban farmers with livelihood assets such as farming knowledge, skills, land and policies as enshrined in the SLF (Fig. 5.1), the households in Gweru urban are empowered to sustain their lives, hence domesticating the SDGs. As an empowerment practice, urban farming can be made to cascade to wider urban setups so that many urban households benefit from the practice and domesticate the SDGs. Some participants (2 of the interviewed 30; 7%) indicated that careless farming activities such as the use of large farming equipment and tractors also resulted in the damage to water and sewerage pipes. The use of heavy equipment such as tractors is not a good practice, as the damage caused by the tractors impinge against sustainable development in urban areas. Heavy equipment should only be used where permission has been granted by relevant authorities. These findings concur with another study by D'Alessandro et al. (2016), whose results indicated that there should be a balance between preserving the environment of the urban landscape and the promotion of livelihoods through urban farming. These finding also supports the goal of *responsible consumption and production* (SDG12).

However, despite all these challenges, the authors argue that urban farming is a worthwhile endeavour that should be allowed to continue as it has the potential to reduce poverty and improve urban livelihoods – hence this study views it as a tool to improve food security and livelihoods. Besides, previous research has shown that urban farming can improve people's livelihoods through maximum land utilisation, resulting in the practice being a commercial venture with a potential to generate household income (D'Alessandro et al. 2016). As such, urban farming can play a pivotal role in achieving sustainable development and foster domestication of the SDGs. The section that follows outlines the policy implications of the study.

5 Policy Implications of the Study

By outlining the way urban farming can achieve sustainable development through SDGs, this study has policy implications. The study has the potential to contribute to policy formulation by the government of Zimbabwe and local authorities on urban farming. Sound policies are needed to support urban farming for improved food security and livelihoods. Hence, the study may act as a guide that influences policymakers to come up with appropriate and implementable policies for urban farming.

The study recommends that urban farmers should be assisted with farming land, inputs and skills capacitation so that the practice is enhanced. More so, there is a need for policies that support the practice to achieve sustainable development.

6 Conclusions

There are major conclusions on the concept of urban farming as a strategy to enhance sustainable development through the SDGs. Urban farming can improve food security, can reduce poverty and can support agribusiness and entrepreneurship. As such, the practice succeeds through the proper combination of livelihood assets to attain sustainable development. However, urban farming can either reduce or escalate poverty and hunger, depending on the situation. Although the practice is perceived as being incompatible with urban development, urban farming is an important undertaking that can contribute to food security and livelihoods for many households.

The challenges of the small size of farming land, lack of extension services and lack of inputs can impinge urban farming endeavours, failing to realise sustainable development and to domestic the SDGs. Further, there is no guarantee that urban farming will always be a successful development strategy. The practice has at times failed to support the concept of sustainable development because of its negative effects on the environment.

References

Cohen, L., Manion, L. & Morrison, K. (2007). Research methods in education. *British Journal of Education Studies, 55*(4), 1- 638.

Chaminuka, N. & Dube, E. (2017). Urban agriculture as a food security strategy for urban dwellers: A case study of Mkoba residents in the city of Gweru, Zimbabwe. *PEOPLE: International Journal of Social Sciences, 3*(2), 26–45.

Chevo, T. (2018). *The construction of household livelihood strategies in urban areas: the case of Budiriro, Harare, Zimbabwe*. Unpublished PhD Thesis. Grahamstown: Rhodes University.

Crush, J., Hovorka, A., & Tevera, D. (2011). Food security in Southern African cities: The place of urban agriculture. *Progress in Development Studies, 11*(4), 285–305.

D'Alessandro, C., Hanson, K.T. & Kararach, G. (2016). Peri-urban agriculture in Southern Africa: Miracle or mirage? *African Geographical Review37(1):49-68.* https://doi.org/10.1080/19376812.2016.1229629.

DFID. (2010). *Sustainable livelihoods guidance sheets.* London: DFID.

Dieleman, H. (2016). Urban agriculture in Mexico City: Balancing between ecological, economic, social and symbolic value. *Journal of Cleaner Production163,*S156-S163. https://doi.org/10.1016/j.jcleopro.2016.01.082.

Drechsel, P., Graefe, S., Sonou, M. & Cofie, O.O. (2006). Informal irrigation in urban West Africa: An overview. *IWMI Research Report 102,* 40.

FAO. (2012). *Growing greener cities in Africa. First status report on urban and peri-urban horticulture in Africa.* Rome: FAO. Retrieved from: http://www.fao.org/ag/agp/greenercities/pdf/GGC-Africa.pdf. (Accessed 20 February 2020).

FAO. (2013). *Resilient livelihoods – disaster risk reduction for food and nutrition security framework programme.* Rome: FAO.

Forsyth, T. (2007). Sustainable livelihood approaches and soil erosion risks: Who is to judge? *International Journal of Social Economics, 34*(1/2), 88-102.

Goddard, M.A., Dougill, A.J. & Benton, T.G. (2009). Scaling up from gardens: biodiversity conservation in urban environments. *Trends in Ecology & Evolution, 25*(2), 90–98.

Katsaruware, R.D., Katanha, A. & Nyakujara T.J. (2014). The nexus between urban agriculture and poverty alleviation in Zimbabwe's cities: A case of Kadoma city. *Research on Humanities and Social Sciences, 4*(28), 60 – 67.

Lee-Smith, D. (2013). Which way for UPA in Africa? *City, 17*(1), 69–84.

Lodico, M.G., Spaulding, D.T. & Voegtle, K.H. (2006). *Methods in educational research: from theory to practice.* San Francisco: Jossey-Bass.

Lovell, S.T. (2010). Multifunctional urban agriculture for sustainable land use planning in the The United States. *Sustainability, 2,* 2499-2522. https://doi.org/10.3390/su2082499.

Mancini, F., Van Bruggen, A.H.C. & Jiggins, J.L.S. (2007). Evaluating cotton integrated pest management farmer field school outcomes using the sustainable livelihoods approach in India. *Experimental Agriculture, 43*(1), 97–112.

Momentum. (2012). *An asset approach to building sustainable livelihoods.* Eko Nomos: Canadian Women's Foundation.

Moyo, P. (2013). Urban agriculture and poverty mitigation in Zimbabwe: Prospects and obstacles in Bulawayo townships. *Journal of Human Ecology, 42*(2), 125-133.

Mudzengerere, F.H. (2012). The contribution of women to food security and livelihoods through urban agriculture in the city of Bulawayo, Zimbabwe. *Zimbabwe Journal of Science and Technology, 7*(6), 1–15.

Nasr, J., MacRae, R.M. & Kuhns, J. (2010). *Scaling up urban agriculture in Toronto: Building the infrastructure.* Toronto: Metcalf Foundation.

Nhapi, G.T. (2019). Socio-economic barriers to universal health coverage in Zimbabwe: Present issues and pathways toward progress. *Journal of Developing Societies, 35*(1), 153-174.

OECD. (2016). *Sizing up the SDGs: What is the importance of the United Nations Sustainable Development Goals for agriculture?* Paris: OECD.

Orsini, F., Kahane, R., Nono-Womdim, R., & Gianquinto, G. (2013). Urban agriculture in the developing world: A review. *Agron. Sustain. Dev.*33(4):695-720. https://doi.org/10.1007/s13593-013-0143-z.

Peñalba, E.H. (2019). Adaptation to climate change among farmers in Bulacan, Philippines. *The Journal of Rural and Community Development*, *14*(2), 1–23.

Prain, G., & Dubbeling, M.Ir. (2011). *Urban agriculture: A sustainable solution to alleviating urban poverty, addressing the food crisis, and adapting to climate change*. Leusden: RAUF Foundation.

Ravallion, M., Chen, S. & Sangraula, P. (2007). *New evidence on the urbanization of global poverty.* Washington DC: World Bank.

United Nations Development Programme (UNDP). (2015). *Sustainable development goals*. Geneva: UNDP.

Webb, N. (2011). When is enough, enough? Advocacy, evidence and criticism in the field of urban agriculture in South Africa. *Development Southern Africa*, *28*(2), 195-208.

Zainal, M., & Hamzah, S.R. (2017). Urban agriculture: The role of knowledge among farmer in Malaysia. *International Journal of Academic Research in Business and Social Sciences, 7*, 77 – 85.

Zezza, A., & Tasciotti, L. (2010). Urban agriculture, poverty, and food security: Empirical evidence from a sample of developing countries. *Food Policy*, *35*(2010), 265–273.

Zimstat. (2012). *Census 2012 National Report.* Retrieved from: http://www.zimstat.co.zw/documents/census. (Accessed 6 November 2019).

Water and Sanitation Access in the Shamva District for Sustainability and Development of the Zimbabwean Smallholder Farming Sector

6

Theresa Tendai Rubhara
and Olawaseun Samuel Oduniyi

Abstract

Over the past three decades, there have been tremendous developments relating to clean water and sanitation supply in sub-Saharan Africa. About 60% of households have access to clean water, while 40% have access to sanitation facilities. However, it is a different situation in rural communities where high concentrations of people are partaking in farming activities. Thus, this research seeks to examine access to water and sanitation by smallholder farmers in Zimbabwe and the factors determining that. Data was collected in two wards in the Shamva district through multistage random sampling procedure whereby 187 smallholder farmers were identified. In describing water and sanitation access, mean and frequency tables were used. The bivariate probit model was used to identify the determinants of access to water and sanitation facilities. Results revealed that 71% of the households had access to water and about 60% had access to sanitation facilities. The empirical results revealed that the marital status of the household head, household size and farm equipment were determinants affecting access to water, while land tenure, education level of the household head, livestock ownership and farm equipment affected access to sanitation. The study concludes by noting that improved access to water sources can be attributed to government and non-governmental efforts to drill community boreholes. Thus, a need exists for sustained efforts for water and sanitation infrastructural development by the government and non-governmental organisations to sustain the farming activities of smallholder farmers.

Keywords

Water source · Sanitation · Bivariate model · Smallholder farmers · SDGs

1 Introduction

The sixth sustainable development goal (SDG 6) of the United Nations (UN) General Assembly aims to "to ensure availability and sustainable management of water and sanitation for all by

T. T. Rubhara (✉) · O. S. Oduniyi
Department of Agriculture and Animal Health, University of South Africa, Florida, South Africa
e-mail: erubhatt@unisa.ac.za

G. Nhamo et al. (eds.), *Sustainable Development Goals for Society Vol. 2*, Sustainable Development Goals Series, https://doi.org/10.1007/978-3-030-70952-5_6

2030" (United Nations General Assembly 2015 p. 18). The goal stresses the importance of clean drinking water, sanitation facilities, water resources management and the reduction of pollution in waterbodies in ensuring social, economic and environmental sustainability. An adequate supply of clean drinking water is vital for the physical wellbeing of the body, thus, the inclusion of access to clean, safe drinking water and sanitation services as a basic human right in the 2015 United Nations Agenda. It is important to note that SDG 6 is interlinked with two other SDGs, namely, health (SDG 2) and food security (SDG 3). The micronutrients found in drinking water are essential for human survival (Dinka 2018), and for food security to be achieved, there should be enough quantities of quality water for health, livelihoods and sustainable food production (Food and Agriculture Organization [FAO] 2014). Therefore, well-calculated decisions for the development and management of water sources are required to attain global food security and health. According to the World Health Organization (WHO) and United Nations Children's Emergency Fund (UNICEF) (2017), "a household is considered as having access to water if they are using a safely managed source which is near their premises, available when needed and free from contamination" (p. 11). As such, when estimating access to water sources in a community, the proportion of people who have access to improved water sources is used as a proxy. An improved water source is a source of drinking water constructed in such a way that it is adequately protected from outside contamination (WHO and UNICEF 2017). Boreholes, taps, protected wells, protected springs and rainwater are the common examples of protected water sources, whereas unprotected dug wells, unprotected springs and surface water are regarded as unimproved water sources (Agbadi et al. 2019).

Sanitation is another essential basic need for human survival. Sanitation relates to the mechanisms of collection and disposal of human waste (Mara et al. 2010). These facilities differ inasmuch as societies differ. However, some basic requirements are applied to characterise a more dignified and improved sanitary system. The WHO and UNICEF (2017) define an improved sanitation facility as "an enhanced toilet or latrine which is not shared, from which excreta are safely disposed of in situ or treated off-site" (p. 1). The improved facilities include flush toilets, Blair toilets and pit latrines with a slab. A lack of proper sanitation conditions can lead to direct contamination of waterbodies, which can increase the spread of diseases such as cholera, typhoid and bilharzia (Antunes and Martins 2020; Mara et al. 2010). The mentioned negative health effects of poor sanitation access are mostly felt by poor people who constitute the bulk of the population in developing countries (Gomez et al. 2019). Hence, it is essential to examine access to water and sanitation by smallholder farmers in the study areas. As such, this study adopts the notion of identifying the water and sanitation sources in the smallholder farming sector. The factors impinging on access to water and sanitation are also identified.

2 Literature Review

2.1 Overview of Water and Sanitation Access in Zimbabwe

Since Zimbabwe's independence, tremendous strides have been made to enhance access to water and sanitation in rural areas, which house almost 70% of the country's total population (World Bank 2017). In addition, the World Bank (2017) recorded that around 60% of Zimbabwean households had access to safe water and 45% had access to sanitation services in 2016. It was noted that clean water supply in the communal areas of Zimbabwe had increased from 30% in the 1980s to over 60% by 2012 (ZimStats 2017). This can be attributed to the coordinated efforts of non-governmental organisations (NGOs) and the Government of Zimbabwe (GOZ), which resulted in boreholes being drilled in the communal areas of the country after independence. However, when the Fast Track Land Reform Programme (FTLRP) was introduced in 2000, there was a decline in access to clean water in the rural areas

(Madziyauswa 2017). As a social policy aimed at equitable distribution of land and access to land by poor people, the FTLRP resulted in decongesting communal areas when more indigenous farmers were resettled on former commercial farms under the A1 model (Moyo 2011). Studies done on agricultural productivity in the Shamva district revealed that smallholder farmers resettled under the A1 model were efficient in food crop production and considered as being food secure (Moyo 2011; Musemwa et al. 2013).

Despite the country's failure to meet its sustainability goal related to water and sanitation, the GOZ allocated only 0.4% of the national 2011 budget to water, sanitation and hygiene services (World Bank 2011). This illustrates how government resource allocation neglected some social responsibilities such as provision of access to water, health and education facilities while concentrating on the agricultural sector. As a result, the budget allocation for water and sanitation services in the rural areas was almost negligible. Since agricultural development precedes rural development, improving smallholder farmers' access to basic services is a prerequisite for sustainable development in Zimbabwe. It is imperative to study the level of access to water and sanitation services, as smallholder farmers are more vulnerable to health problems associated with poor sanitation and limited health facilities in rural areas (Sunanda and Masuka 2017). The study also seeks to understand the household socioeconomic factors influencing access to water and sanitation facilities in the smallholder farming community. This will help to provide empirical evidence that can be submitted to policymakers as recommendations on how to surge the use of clean and safe water sources as well as improved toilet facilities required for a healthy population.

2.2 Water Access Measures and Models Used in Related Studies

There are many studies that acknowledge the disparity in accessibility of water and sanitation services among low-, middle- and high-income countries (Gomez et al. 2019; Sommer et al. 2015). Gomez et al. (2019) conducted a global study on water access that showed water access challenges were common in lower-income countries and more pronounced in countries with agro-based economies. Furthermore, a study conducted by Tuyet-Hanh et al. (2016) focusing on Vietnam's urban households revealed that urban households were more likely to have water and sanitation access compared with farming households. Lastly, Sommer et al. (2015) conducted a multi-country cross-sectional survey that showed that there was limited use of improved water sources and toilet facilities in sub-Saharan countries, which has a negative impact on maternal and neonatal health in the region.

To understand the factors affecting household access to water and sanitation, the integrated behavioural model was applied partially. The model has been used in water, sanitation and hygiene (WASH) programmes (Morse et al. 2020). The theory asserts that the behaviour of households in using safe water and sanitation facilities is dependent on contextual, psychological and technological factors (Dreibelbis et al. 2013). The contextual factors relate to the socioeconomic characteristics of the study sample. Psychological factors include the cognitive abilities of individuals to change their behaviour, whereas technological factors take into consideration infrastructural issues and use of technology in water supply and waste disposal (Dreibelbis et al. 2013). Concerning the contextual factors, previous literature identified the socio-economic household-level determinants of accessibility to water and sanitation. Pahwaringira and Madobi (2013) conducted a case study in the Epworth informal settlement in Harare that indicated that the residents had limited access to essential water and sanitation services. Similarly, Chigonda and Chazireni (2018) used the case study approach and determined that about 26% of the households in smallholder resettlement areas in Zimbabwe did not have access to clean drinking water and about 58% did not have a toilet at their homesteads. Land tenure was recognised as a factor affecting accessibility to water resources, because the institutional arrangements around the management of land determine investment levels in terms of the infrastructural distribution and management of water. Where people own the land, they are likely to invest as they expect to reap the benefits of their investment in the long run. Sjöstedt (2011) reviewed land tenure in Botswana and Zambia and found that farmers are likely to invest in water infrastructure when they have secure land rights. However, based on the Land Acts in Zimbabwe, smallholder farmers rarely own land as they are settled under either communal or leasehold in resettlement areas (Sachikonye 2005).

Another study took place in the Lagos metropolis, Nigeria. Akoteyon (2019) used the analysis of variance (ANOVA) and concluded that income and costs were the most significant variables affecting access to water supply in the study area. Similarly, a study conducted in Harar, Ethiopia, by Gudeta (2016) using the bivariate analysis model revealed that residents with a higher socioeconomic status were associated with use of improved water and sanitation facilities in urban areas. In the Bomet municipality of Kenya, socioeconomic variables and the occupation and education level of the household head affected the use of improved water sources positively, while married household heads were more likely to use improved toilets than their unmarried counterparts (Koskei et al. 2013). Educated household heads are considered as having access to high-income jobs. They also have the ability to comprehend the dangers associated with the use of unsafe water and toilet facilities. Rhodes and McKenzie (2018) using data from the national living standards survey realised that the gender of the household head affected accessibility to water and sanitation facilities in South Africa. In that study, male-headed households were likely to use piped water and flush toilets, while the households headed by females did not.

Some studies that have been reviewed demonstrate how access to water and sanitation among the urban poor tends to be different from that in rural areas. In urban areas, piped water and flushing toilets comprise water and sanitation facilities, while that is not the case in rural areas. In this regard, the proxies used for some of the variables can differ slightly between rural and urban areas. For instance, in urban areas, wealth is based on variables such as home ownership, monthly salary and location, whereas livestock units, farm income and equipment can be used to measure wealth in rural areas (Muyanga et al. 2013). While comparative studies on access to sanitation between different municipalities in urban areas are common, studies focusing on rural areas are fewer.

3 Methodological Orientation

The study was carried out in the Shamva district of the Mashonaland central province of Zimbabwe (Fig. 6.1). The district is located 60 km north-west of the capital city Harare and boasts high agricultural productivity as it lies in an area classified as natural agro-ecologocal region II, which is suitable for intensive agricultural farming (Musemwa et al. 2013). The farmers in the area consist of commercial and smallholder farmers. The smallholder farmers comprise communal farmers and resettled farmers under the A1 model. Under the A1 model of resettlement, the land was acquired by the state, apportioned into plots and subsequently redistributed to the beneficiaries. These beneficiaries were given long-term tenure to settle and cultivate the land (Moyo 2011). Communal farmers are those settled on traditional farms where land is communally owned and managed by traditional leaders.

Data was collected from 187 smallholder farming households selected through a multistage random sampling procedure in the Shamva district. The sample was drawn from two wards consisting of 360 A1 and communal farmers in the area as determined by the Raosoft sample size calculator. Data was collected using a pretested, semi-structured questionnaire administered through face-to-face interviews.

A bivariate probit model was applied to examine access to improved water sources and sanitation facilities, which are jointly determined by

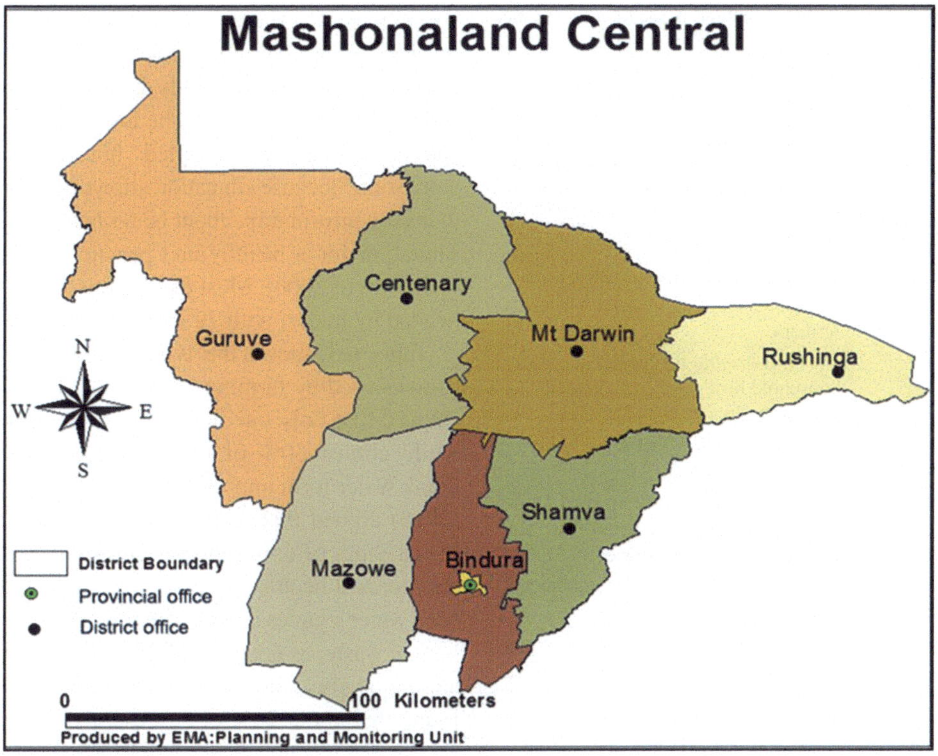

Fig. 6.1 Mashonaland central province map. (Source: Environmental Management Agency 2016)

the same explanatory variables that influence the accessibility of smallholder farmers of the Shamva district to the variable outcomes, after which the marginal effect for the joint probabilities was calculated using the estimated predicted probabilities for each of the possible outcomes that were achieved using STATA version 15. This model is classically used where there are two separate probit models with correlated disturbances, such as in a seemingly unrelated bivariate probit regression (SUR) that is comparable or like the basic bivariate probit model (Baslevent and El-hamidi 2009; Oduniyi and Tekana 2019). The model involves two separate dependent variables that are dichotomous in nature as well as two interrelated decisions by some determinants or factors.

The model is mathematically written as:

$$Y_1* = X_1\beta_1 + \varepsilon_1 \quad (1)$$

$$Y_2* = X_2\beta_2 + \varepsilon_2 \quad (2)$$

where y^* are unobservable and related to the binary dependent variables yj by the following rule:

$$yj = \begin{cases} 1 \text{if } yj* > 0 \\ 0 \text{if } yj* \le 0. \end{cases}$$

For $j = 1, 2$, that is, if the errors between the two probit models are independent of each other. In other words, the interest is in the joint probability of y_1 and y_2. This is better explained as follows: a farmer with access to improved water is coded 1 and 0 if otherwise (as shown in Table 6.1). Similarly, this applies to sanitation facilities. Thus, the two outcome variables are dichotomous, which were then regressed against the set of related explanatory variables such as age, land tenure, gender, marital status, household size, livestock units, level of education, farm equipment and household income. The choice of variables was influenced by the availability of data and reviewed literature.

Table 6.1 The variables included in the bivariate model

Variables	Description
Dependent variables Water source Sanitation facility	*Dummy variable* where 0 is no access to water e and 1 is access to water *Dummy variable* where 0 is no access to sanitation facility and 1 is access to an improved sanitation facility
Independent variables	
Land tenure	*Dummy variable* where 0 represents communal and 1 represents A1 farmers
Gender	*Dummy variable* for the gender of the household head where 0 is female and 1 is male
Age	*Continuous variable* of household head's age in years
Marital status	*Dummy variable* for the marital status of the household head where 1 is married and 0 is otherwise
Household size	*Continuous variable* for the number of people
Livestock units	*Continuous variable* for total livestock units using the universal livestock units calculator
Education	*Dummy variable* for the education level of the household head where 1 represents at least secondary level and 0 is otherwise
Farm equipment	*A proportion index* calculated from a summation of answers on availability of basic farm equipment
Household income	*A continuous variable of* annual household cash income in USD

Source: Author's computation, 2020

4 Presentation of Results and Discussion

Table 6.2 presents descriptive results of the study respondents. The average age of farming household heads was about 49 years, implying that youths are not much into farming. The low youth participation is a cause for concern because young farmers can use improved technology and increase productivity than the elderly. Generally, youths are more poised than older farmers to adopt sound agricultural practices in food safety, animal welfare, diversification, climate-smart agriculture and renewable energy production, which are key to sustainable development (Girei et al. 2017). On average, households consisted of six members. About 41% of the population received secondary school education. Thus, most heads of sampled households did not receive substantial formal education. The low levels of education among the sampled households are worrisome because education allows rural people to access information about better healthcare and sanitation for a healthy and productive population (Burchi 2006). Most of the households were headed by males, with 79% of them married.

Table 6.3 shows the typical drinking water sources used by farming households. The results showed that only one household used tap water, while almost 30% of the sampled households drank water form unprotected wells. Armah et al. (2018) argued that unprotected wells fall under the category of unimproved water sources. Thus, this poses a health risk to the respondents since such water sources can be contaminated easily by human waste, as noted by Mara et al. (2010). In a similar study, Chigonda and Chazireni (2018) found that about 26% of the resettled farmers in Zimbabwe did not have access to safe drinking water sources such as boreholes and protected wells.

Table 6.4 gives a summary of the sanitation facilities used by smallholders in the area. Following the WHO and UNICEF (2017) classification, pit latrines with slabs and Blair toilets are regarded as improved sanitation facilities. At least 59% of the sampled households had access to improved sanitation facilities. Despite the United Nations (2015) mandate to do away with the bush toilet system by 2030, it is alarming to note that almost 4% of the households used bush toilets. Besides health threats such as the spread of cholera, Saleem et al. (2019) further alluded that the dignity and psychological health of women and girls are compromised as they are vulnerable to other social ills such as rape where open defecation are practised.

Table 6.5 provides the bivariate model results of factors affecting access to water and sanitation by farming households. Land tenure, marital status of the household head, household size and farm equipment index were the variables identified as affecting farmers' access to improved water facilities. Land tenure was statistically

Table 6.2 Descriptive statistics of socio-economic variables

Continuous variables					
Variable		Minimum	Maximum	Mean	Standard deviation
Age of household head		24	93	49.35	14.29
Household size		1	20	5.84	2.60
Total livestock units		0	33.84	3.75	4.38
Household income		0	5 345	1 016.48	1 000.61
Farm equipment		0	1	0.70	0.46
Discrete variables					
Variable		Frequency ($n = 187$)		Percentage frequency ($n = 187$)	
Gender	Male	34		81.8	
	Female	153		18.2	
Education level	At least secondary	111		40.6	
	Otherwise	76		59.4	
Marital status	Married	148		79.1	
	Otherwise	39		20.9	
Land tenure	Communal	96		51.3	
	Resettled (A1)	91		48.7	

Source: Author's computation, 2020

Table 6.3 Water sources for smallholder farmers

Water source	Frequency ($n = 187$)	Percentage ($n = 187$)
Tap	1	0.5
Borehole	70	37.4
Protected well	60	32.1
Unprotected well	56	29.9

Source: Author's computation, 2020

Table 6.4 Sanitation facilities used by smallholder farming households

Sanitation facility	Frequency ($n = 187$)	Percentage frequency ($n = 187$)
Blair latrine	93	49.7
Pit latrine with slab	19	10.1
Pit latrine with no slab	68	36.4
Bush	7	3.7

Source: Author's computation, 2020

significant at 1%. Resettled farmers (A1) had more access to improved water sources compared with their communal counterparts. As alluded by Sachikonye (2005), land redistribution policies play a fundamental role in sustainable development. Land in communal areas is owned by the government and governed by traditional leaders. Therefore, farmers have limited land rights. As such, communal areas are characterised by low agricultural productivity and limited investments. In contrast, individual farmers in the resettlement schemes have long-term leases to farm on and invest in the land. Furthermore, the land reform beneficiaries under the A1 model inherited water infrastructure left behind by former white farmers (Matondi and Dekker 2011). This is consistent with findings by Sjöstedt (2011) who concluded that for water access to improve in Botswana and Zambia, the institutional arrangements regarding land should provide a form of certainty for citizens to invest in water infrastructure.

At a significance level of 1%, the variable marital status of household head influenced household access to water negatively. As shown by the negative coefficient, married household heads were less likely to have access to improved water facilities than their unmarried counterparts. This is in contrast with Agbadi et al. (2019) who found that married household heads had more access to clean, safe drinking water than unmarried household heads. However, the results of this study show that married household heads may invest much in farming activities and neglect other social needs such as water.

Household size affected the use of improved water facilities negatively at a 5% significance level, as families with a bigger number of people

Table 6.5 Bivariate model estimates

	Improved water facility		Improved toilet facility	
Variable	Coefficient	P>|z|	coefficient	P>|z|
Land tenure	0.595***	0.0078	−1.538***	0.000
Gender	0.503	0.239	−0.385	0.231
Age	0.010	0.197	−0.006	0.497
Marital status	−1.123***	0.003	0.493**	0.028
Household size	−0.129***	0.004	0.016	0.685
Livestock units	0.009	0.713	0.061**	0.020
Level of education	−0.123	0.581	0.442*	0.074
Farm equipment	−0.565**	0.024	0.674***	0.007
Household income	−0.000	0.234	−0.000	0.489
Constant	1.408	0.014	0.300	0.604
	Coefficient	Robust Std. Err.	Z	P>|z|
/athrho	−0.157	0.141	−1.11	0.266
Rho	−0.155	0.137		

Number of obs = 187
Wald $chi^2(18) = 97.74$
Log pseudo likelihood = −196.792
Prob > $chi^2 = 0.0000$
Wald test of rho = 0: $chi^2(1) = 1.23945$
Prob > $chi^2 = 0.2656$
Note: Significant at *$p < 0.1$; **$p < 0.05$ and ***$p < 0.01$ level, respectively
Source: Author's computation, 2020

residing together were less likely to use an improved water source. This agrees with a sub-Saharan multi-country longitudinal study showing that households with fewer people were likely to have access to improved water sources since the demand for water is lower in such households (Armah et al. 2018).

The variable farm equipment index was positive and statistically significant at 5%. Farm equipment as a productive asset can be used as a proxy for wealth in farming communities, and farmers with the higher index for farm equipment can boost agricultural productivity (Muyanga et al. 2013). With essential farm equipment, farmers may be able to drill protected wells, thus promoting access to water. The results from Table 6.5 indicate that land tenure, marital status, livestock units, level of education and farm equipment were impinging on access to improved sanitation services by the respondents.

The results in Table 6.5 indicate that land tenure significantly influenced household access to sanitation. The negative coefficient shows that resettled farmers had less access to sanitation compared with their communal counterparts. As mentioned in the literature review, communal households were assisted in building toilets after independence under the WASH projects coordinated by the GOZ and NGOs. However, most of the projects were not expanded to the resettlement areas, and therefore access to sanitation services was relatively low among A1 farmers (Madziyauswa 2017).

Married household heads were more likely to use improved sanitation facilities than unmarried heads. Koskei et al. (2013) had similar findings in Kenya. Evidence in rural Zimbabwe suggested poverty, food insecurity and a general lack of access to economic resources, including land, were concentrated in single-headed households (female-headed households in particular). As a result, single-headed households did not have enough financial and human assets to improve their toilet facilities (Kairiza and Kembo 2019).

Total livestock units for a household were found to be positive and statistically significant concerning the use of improved sanitation facilities ($p < 0.1$). In smallholder farming communities, livestock ownership can be considered a useful estimation of the socio-economic status of a household as farmers with a greater number of livestock units are considered wealthy (Bettencourt et al. 2015). Farmers with a better socio-economic status can be expected to build toilets, thus having better access to improved sanitation facilities (Angoua et al. 2018).

Household heads with at least secondary education were more likely to have access to improved toilet facilities ($p < 0.1$). According to Agboola et al. (2017), empowering household heads through education leads to financial stability. Education is a health factor in low-income countries as households with educated members are likely to consider the health outcomes of using improved toilet facilities and, as such, invest in sanitation services (Koskei et al. 2013). Furthermore, education allows households to have access to more financial resources, thus increasing the probability of improved sanitation facilities. This result is consistent with a survey conducted by Agbadi et al. (2019) in Ghana.

At a 5% significance level, the farm equipment variable influences access to sanitation positively. Farmers with sufficient equipment are able to generate more income due to improved productivity. Rhodes and McKenzie (2018) also indicated that, in South Africa, wealthy households are more likely to use flush toilets than their poor counterparts.

5 Conclusion

Smallholder farmers are pivotal to food security because of their role as food producers. To meet the SDGs of food security and economic development in rural areas, a healthy smallholder farmer population is required, since agricultural development precedes rural development. Moreover, farmers' health is a product of the water and sanitation facilities they use at household level. The study was thus aimed at assessing access to water and sanitation facilities and the determinants thereof in the smallholder farming sector of Zimbabwe. The results revealed that about 70% of sampled households had access to improved water sources as they sourced water from boreholes and protected wells. Also, Blair toilets and pit latrines constituted improved sanitation facilities used by smallholder farmers, with about 59% accessing such facilities. The results indicated a need for increased governmental and NGO coordinated efforts to improve sanitation facilities, especially in the resettled communities. Since the economic status of households, as indicated by level of education, farm equipment and livestock units, was positively associated with access to improved water and sanitation facilities, the GOZ should promote inclusive growth that targets development in the poverty-stricken farming households of Zimbabwe. The GOZ must promote sustainable agricultural livelihoods.

References

Agbadi, P., Darkwah, E. & Kenne P.L. (2019). A Multilevel Analysis of Regressors of Access to Improved Drinking Water and Sanitation Facilities in Ghana. *Journal of Environmental and Public Health*. https://doi.org/10.1155/2019/3983869

Agboola, P.T., Oyekale, A.S. & Oduniyi, O.S. (2017). Assessment of Welfare Shocks and Food Insecurity in Ephraim Mogale and Greater Tubatse Municipality of Sekhukhune Districts, Limpopo Province, South Africa. *Journal of Agriculture and Veterinary Science, 10*(4), 23–32. https://doi.org/10.9790/2380-1004022332

Akoteyon, I.S. (2019). Factors affecting household access to water supply in residential areas in parts of Lagos metropolis, Nigeria. *Bulletin of Geography: Socio-economic series, 43*(43), 7–24.

Angoua, E.L., Dongo, K., Templeton, M.R., Zinsstag, J. & Bonfoh, B. (2018). Barriers to access improved water and sanitation in poor peri-urban settlements of Abidjan, Côte D'Ivoire. *PLoS One, 13*(8). https://doi.org/10.1371/journal.pone.0202928

Antunes, M. & Martins, R. (2020). Determinants of access to improved water sources: Meeting the MDGs. *Utilities Policy, 63*(2020). https://doi.org/10.1016/j.jup.2020.101019.

Armah, F.A., Ekumah, B., Yawson, D.O., Odoi, J.O., Afitiri, A.R. & Nyieku, F.E. (2018). Access to improved water and sanitation in sub-Saharan Africa in a quarter-century. *Heliyon, 4*(11). https://doi.org/10.1016/j.heliyon.2018.e00931.

Baslevent, C. & El-hamidi, F. (2009). Preferences for Early Retirement among Older Government Employees in Egypt. *Economics Bulletin, 29*(2): 554–565.

Bettencourt, E.M.V., Tilman, M., Narciso, V., Carvalho, M.L.S. & Henriques, P.D.S. (2015). The Livestock Roles in the Wellbeing of Rural Communities of Timor-Leste. *Revista de Economia e Sociologia Rural*, *53*(Suppl.1), 63–80. https://doi.org/10.1590/1234-56781806-94790053s01005.

Burchi, F. (2006). Identifying the Role of Education in Socio-Economic Development. International Conference on Human and Economic Resources, Izmir, 2006.193–201. Izmir.

Chigonda, T. & Chazireni, E. (2018). Water supply and Sanitation in Zimbabwe's Resettlement areas: A case study approach. *European Journal of Social Sciences Studies*, *2*(11), 139–159.

Dinka, M.O. (2018). Safe Drinking Water: Concepts, Benefits, Principles and Standards. In: Glavan, M. (Ed.). *Water Challenges of an Urbanizing World*, IntechOpen, London, 163–181. https://doi.org/10.5772/intechopen.71352

Dreibelbis, R., Winch, P.J., Leontsini, E., Hulland, K.R., Ram, P.K., Unicomb L. & Luby, S.P. (2013). The Integrated Behavioural Model for Water, Sanitation, and Hygiene: a systematic review of behavioural models and a framework for designing and evaluating behaviour change interventions in infrastructure-restricted settings. *BMC Public Health 13*, 1015. https://doi.org/10.1186/1471-2458-13-1015

Environmental Management Agency. (2016). Provincial map for Mashonaland Central province. Retrieved from: https://www.ema.co.zw/about-us/provinces/provincial-information/item/24-mashonaland-central. (Accessed 10 August 2020).

Food and Agriculture Organization, FAO. (2014). The Water-Energy-Food Nexus: Understanding and managing the complex interactions between water, energy and food. Retrieved from: http://www.fao.org/3/a--bl496e.pdf. (Accessed 30 July 2020).

Girei, A.A., Saingbe, N.D., Ohen, S.B. & Gimba, E.A. (2017). Youth involvement in agricultural production in Obi Local Government Area, Nasarawa State, Nigeria. *International Journal of Environment, Agriculture and Biotechnology*, *1*(4), 1016–1023.

Gomez, M., Perdiguero, J. & Sanz, A. (2019). Socioeconomic Factors Affecting Water Access in Rural Areas of Low and Middle-Income Countries. *Water*, *11*(2), 202. https://doi.org/10.3390/w11020202

Gudeta, S.Y. (2016). The determinant of Access to Improved Latrine in Historical City of Harar, East Ethiopia. *International Journal of Advanced Research and Publications*, *1*(4), 126–129.

Kairiza, T. & Kembo, G.D. (2019). Coping with food and nutrition insecurity in Zimbabwe: does household head gender matter? *Agricultural and Food Economics*, *7*(24). https://doi.org/10.1186/s40100-019-0144-6

Koskei, E.C., Koskei, R.C., Koske, M.C. & Koech, H.K. (2013). Effect of Socioeconomic Factors on Access to Improved Water Sources and Basic Sanitation in Bomet Municipality. *Research Journal of Environmental and Earth Sciences*, *5*(12), 714–719.

Madziyauswa, V. (2017). Assessing sustainability of community managed NGOs' WASH interventions in rural Zimbabwe: the case of Chivi district in Masvingo province. *Journal of Water, Sanitation and Hygiene for Development*, *8*(4), 640–649. https://doi.org/10.2166/washdev.2017.049

Mara, D., Lane, J., Scott, B. & Trouba, D. (2010). Sanitation and Health. *PLoS Medicine, 7*(11). https://doi.org/10.1371/journal.pmed.1000363

Matondi, P. B. & Dekker, M. (2011).Land Rights and Tenure Security in Zimbabwe's Post Fast Track Land Reform Programme. A Synthesis Report for Land Ac. Project ID WS320005. Retrieved from: https://www.researchgate.net/publication/334375928_Plural_strategies_of_accessing_land_among_peri-urban_squatters. (Accessed 09 August 2020).

Morse, T., Tilley, E., Chidziwisano, K., Malolo, R. & Musaya, J. (2020). Health Outcomes of an Integrated Behaviour-Centred Water, Sanitation, Hygiene and Food Safety Intervention – A Randomised before and after Trial. *International Journal of Environmental Research and Public Health, 2648.* https://doi.org/10.3390/ijerph17082648

Moyo, S. (2011). Changing agrarian relations after redistributive land reform in Zimbabwe. *Journal of Peasant Studies*, *38*(5), 939–966.

Musemwa, L., Mushunje, A., Aghdasi, F., Muchenje, V. & Zhou, L. (2013). The efficiency of resettled farmers in Mashonaland Central Province of Zimbabwe in crop production: A DEA approach. *African Journal of Agricultural Research*, *8*(22), 2722–2729.

Muyanga, M., Jayne, T.S. & Burke, W.J. (2013). Pathways into and out of poverty: A study of determinants of rural household wealth dynamics in Kenya. *Journal of Development Studies*, *49*(10), 1358–1374.

Oduniyi, O.S. and Tekana, S.S. (2019). Adoption of agroforestry practices and climate change mitigation strategies in North West province of South Africa, *International Journal of Climate Change Strategies and Management*, 11 5, pp. 716-729. https://doi.org/10.1108/IJCCSM-02-2019-0009.

Pahwaringira, L. & Madobi, R. (2013). Water Supply and Sanitation Services for the Urban Poor: A Case Study of Epworth, Zimbabwe. *International Journal of Science and Research, 4*(10), 1804–1808.

Rhodes, B. & McKenzie, T. (2018). To what extent does socio-economic status still affect household access to water and sanitation services in South Africa*? Journal of Economic and Financial Sciences, 11*(1), a173. https://doi.org/10.4102/jef.v11i1.173

Sachikonye, L. (2005). The land is the economy: revisiting the land question. *African Security Review, 14*(3), 31–46. https://doi.org/10.1080/10246029.2005.9627368

Saleem, M., Burdett, T. & Heaslip, V. (2019). Health and social impacts of open defecation on women: a systematic review. *BMC Public Health*, *19*(158). https://doi.org/10.1186/s12889-019-6423-z

Sjöstedt, M. (2011). The impact of secure land tenure on water access levels in sub-Saharan Africa: The case of Botswana and Zambia. *Habitat international, 35*(1), 133–140. https://doi.org/10.1016/j.habitatint.2010.06.001

Sommer, J.M., Shandra, J.M., Restivo, M. & Coburn, C. (2015). Water, Sanitation, and Health in Sub-Saharan Africa: A Cross-national Analysis of Maternal and Neo-natal Mortality. *Human Ecology Review, 22*(1), 129–152.

Sunanda, R. & Masuka, N. (2017). Facilitators and barriers to effective primary health care in Zimbabwe. *African Journal of Primary Health Care & Family Medicine, 9*(1). https://doi.org/10.4102/phcfm.v9i1.1639

Tuyet-Hanh, T.T., Long, T.K., Van Minh, H. & Huong, L.T. (2016). Longitudinal Household Trends in Access to Improved Water Sources and Sanitation in Chi Linh Town, Hai Duong Province, Viet Nam and Associated Factors. *AIMS Public Health, 3*(4), 880–890.

United Nations. (2015). The Millennium Development Goals Report. UN, New York. https://www.un.org/millenniumgoals/2015_MDG_Report/pdf/MDG%202015%20rev%20(July%201).pdf. Accessed 21 January 2020.

United Nations General Assembly. (2015). Transforming our world: the 2030 Agenda for Sustainable Development, 21 October 2015, A/RES/70/1. Retrieved from: https://www.refworld.org/docid/57b6e3e44.html. (Accessed 06 August 2020).

World Bank Group. (2011). Water Supply and Sanitation in Zimbabwe: Turning Finance into Services for 2015 and Beyond. Retrieved from: https://www.wsp.org/sites/wsp.org/files/publications/CSO-Zimbabwe.pdf. (Accessed 15 January 2020).

World Bank Group. (2017). People using safely managed sanitation services (% of the population). https://data.worldbank.org/indicator/SH.STA.SMSS.ZS. (Accessed 02 March 2020).

World Health Organization and United Nations Children's Emergency Fund. (2017). Progress on drinking water, sanitation and hygiene: 2017 update and SDG baselines. Retrieved from: https://www.ircwash.org/resources/progress-drinking-water-sanitation-and-hygiene-2017-update-and-sdg-baselines. (Accessed 03 March 2020).

ZimStat. (2017). Zimbabwe National Statistics Agency Facts and Figures. Retrieved from: http://www.zimstat.co.zw/sites/default/files/img/publications/Facts%20and%20Figures/Fact_Figures_2017.pdf. (Accessed 05 January 2020).

7 Responsible Leadership and the Implementation of SDG 7: The Case of the UNDP Botswana Biogas Project

Newton Tawanda Runyowa and Willem Fourie

Abstract

Responsible leadership theory (RLT), an evolving and emergent leadership theory, is being recommended when developing and implementing projects across sectors. This is due to its elemental social and relational nature and its focus on social capital, ethical and responsible relationships. To meet their development objectives, countries especially in Africa need to, among other efforts, improve their population's access to clean energy. Invariably the successful implementation of the Sustainable Development Goals (SDGs) will require systems thinking and collaborative and effective leadership. Using an exploratory qualitative research approach and the UNDP Botswana Biogas project as a case study, this chapter focuses on identifying the leadership capacities demonstrated in the project and explores the extent to which these capacities conform to the constructs of RLT. The project is intended to promote the production and utilisation of biogas a clean energy source, contributing to the achievement of SDG 7. The key findings of the study are that the leaders interviewed understand RLT attributes and competencies and apply responsible leadership capacities in their development practice. This chapter recommends that for inclusive growth to be achieved in the African development context, governance and leadership need to be transformative and shared with collective problem-solving integrated in development practice.

Keywords

Responsible leadership · Biogas · Energy · Climate change · SDGs

1 Introduction

Botswana has been ranked highly as a stable democracy and has experienced high economic development since independence in 1966, attributable to the availability of strong institutions, judicious management of the economy and sustainable use of resources (Ministry of Finance and Economic Development 2016). Despite being a middle-income country since the late 1990s, and having made significant social sector investments, poverty, inequality and unemployment have remained high with the rural communities, women and youth disproportionately

N. T. Runyowa (✉) · W. Fourie
Faculty of Economic and Management Sciences, University of Pretoria, Pretoria, South Africa

G. Nhamo et al. (eds.), *Sustainable Development Goals for Society Vol. 2*, Sustainable Development Goals Series, https://doi.org/10.1007/978-3-030-70952-5_7

affected (United Nations 2016). This can even be reflected in energy consumption patterns.

In 2009 Botswana was only 46% energy self-sufficient, and by 2016 it had an electrification level of 60.7% (United Nations 2016). Current national data indicates that 71.5% of the population currently has access to electricity with 64.08% of them primarily relying on clean fuels and technology (Botswana Statistics 2018). However, for the rural population, the World Bank (2019) reports that access to electricity is still low at 37.5% but gradually increasing. Rural households in Botswana continue to use firewood as an energy source. This is mainly because poverty is more severe and widely spread in these areas. Due to poorly developed commercial energy infrastructure, electricity access for the rural population is limited.

The use of firewood as fuel has biodiversity, conservation and gender elements attached to it (Amigun et al. 2012). Women and children are mostly involved in the day-to-day collection of firewood. This might deprive them of time to do academic-, commercial- or governance-related activities. Amigun et al. (2012) indicate that the heavy use of firewood speeds up deforestation and increases soil erosion, desertification and the risk of flooding and loss of biodiversity. Additionally, the use of firewood is associated with health problems as a result of poor indoor air quality (Tumwesige et al. 2011; World Biogas Association [WBA] 2018).

Energy is a domestic necessity and is important in all development activities. Countries with abundant fossil fuel resources can achieve economic growth easily. However, low- and middle-income countries have little or no national energy infrastructure. They lack access to electricity, which is a chronic economic barrier on its own and another poverty trap. This results in poor productivity and less incomes and taxes to develop this crucial infrastructure. To meet their developmental objectives including the SDGs, developing countries will require an "improvement in the quality and magnitude of energy services" (Amigun et al. 2012: 35–36). Energy therefore becomes essential for all aspects of development: social, political, economic and environmental. It impacts on all services that have a bearing on people's quality of life.

Sachs (2018) describes how the fossil fuel-based modern economic growth model has propelled world economic growth. The author however notes that fossil fuels are now at the core of the sustainable development challenges due to their impact on the environment. Renewable energy sources remain practical solutions and are a huge potential for low- and middle-income countries to close the energy gap. Renewable or clean energy should ideally be locally available and be deployable in a decentralised approach. It can also be used to improve energy consumption in the rural areas and rejuvenate economies. Sources include solar, wind, hydropower, geothermal, tidal and wave energy and biomass energy. These sources and technologies can therefore be used to revitalise rural areas, decrease energy poverty, improve energy consumption levels and contribute to gender equality, poverty alleviation and rural development.

Given that energy plays a critical role in development, clean energy sources like biogas present a huge opportunity to Africa and Botswana to positively contribute directly to SDG 7 and indirectly to numerous other SDGs. This chapter presents a biogas energy-based case study positioned in the theoretical leadership literature and SDG context. The nexus between achieving the SDGs and the roles leaders play in the process will be established further in this chapter with a special lens on Africa. The concepts of sustainable development and leadership will be given some space in this chapter starting with global, regional (Africa) and local (Botswana) implementation of the SDGs in the next section.

2 Literature Review

2.1 Contextualising Sustainable Development: Globally, in Africa and in Botswana

At the United Nations General Assembly in 2015, a resolution, *Transforming our world: the 2030 Agenda for Sustainable Development*, was adopted. This resolution expounded the interaction between the world economy and the Earth's physical environment. It addressed the issues of environmentally sustainable and socially inclusive development. This resolution anchored on 17 SDGs and 169 targets recognises the interlinkages of the economy, society, environment, politics and governance (United Nations 2015). The inclusive SDG development process involved multiple stakeholders. Among them are individuals, governments, civil society organisations, business, academia, parliaments and agencies of the United Nations. They delivered an overarching transformative international development agenda without prescribing what would be universal or national goals.

Bhattacharya, Khan and Salma (2014: 165–177) report that the SDGs were developed through a mandated "balance of economic, social, and environmental dimensions of sustainable development in an integrated and coherent fashion". The authors identified that a total of 40.2% of the targets belonged under the social pillar, 38.5% were associated with the economy and 21.3% represented the environmental agenda. It should be noted though that due to multifunctionality it might be difficult to place one goal under an exclusive development pillar. The SDGs on the other hand have been criticised for being too many, ambitious and sidelining the communities in their own development (Pogge and Sengupta 2016). While acknowledging this, the SDGs remain the best deliberate action to change things for the better. This is because of their universal scope and often targeting eradication and not reduction.

The 2030 Agenda for Sustainable Development was fully adopted on the African continent. To this end the SDGs have been aligned to the goals and key priority areas of the African Union Agenda 2063. The African Union Commission (AUC), the Economic Commission for Africa (ECA) of the United Nations, the African Development Bank (AfDB) and the United Nations Development Programme – Regional Bureau for Africa (UNDP-RBA) since 2017 annually publish the Africa Sustainable Development Report on Agenda 2063 and the SDGs.

This is done to track the domestication and implementation of the two, though the report is deficient in scope and depth of analysis. Findings from their inaugural 2017 Africa Sustainability Report indicate that in Africa close to 60% of the SDG indicators cannot be tracked mainly due to lack of acute data availability. This gap points to the need for stronger institutional collaborations and partnerships, both within and between governments, civil society organisations, business, academia and other stakeholders. Africa needs to successfully implement the 2030 Agenda for Sustainable Development and the Africa Agenda 2063 developmental policies to meet the expectations and needs of its people. To facilitate this, the capability of the state and a transformed public administration that is "effective, coherent, representative, competent and democratic" are prerequisites (Tshiyoyo 2017: 173). Consequently, the need for strong political and government leadership in development can not be over emphasised.

Botswana subjected herself to the SDG's Voluntary National Review process in 2017. They have also developed and launched the SDG roadmap. The country has domesticated the SDGs in alignment with the National Development Plan (NDP) running up to 2023 and the National Vision 2036 which is the national development vision. Botswana plans a phased implementation approach for the SDGs. A baseline assessment was conducted, and a preliminary indicator report is available (Botswana Statistics 2018). The Government of Botswana and the United Nations Agencies are currently working together under the United Nations Sustainable Development Framework for the years 2017–2021. They have managed to mainstream the Agenda 2030 into national priorities and leadership of the Vision 2036 (UNDP 2017). While the implementation level of SDG-related activities at the global, regional and national level (in Botswana) is welcome and encouraging, the role of leaders and the leadership process needs to be illuminated. This is the focus of the next section.

2.2 Leadership, Sustainable Leadership and Responsible Leadership Theory

2.2.1 Leadership

Leadership is a concept discussed and characterised by many experts and scholars with the majority of theoretical concepts, propositions and constructs being developed in the Northern Hemisphere and diffused to the South (Day and Antonakis 2012; Fourie et al. 2017; Northouse 2016). Leadership is presumed to hold the key to the successful implementation of the Sustainable Development Goals (SDGs), and leaders are expected to effectively and judiciously formulate and implement public policy and deliver services to fulfil the needs and goals of the people (Bahauddin 2018; Dartey-Baah 2014). This relationship between achieving sustainable development (in the era of the SDGs) and the role of leadership and/or leaders play(s) in that process has to be established. Sustainable development is complex and multifaceted in nature. To this end Dartey-Baah (2014: 207) observes that a "more balanced relationship among the environment, society and economy in pursuit of development and improved quality of life" is required. This balance is reliant on leaders and the process of leadership across the entire development spectrum.

The concept of leadership and leadership theory grounds this case study, focusing on leadership capacities. Leadership, after more than a hundred years of scholarship, remains complex and diverse, with multiple dimensions, many theories and an even wider variety of theoretical approaches (Day and Antonakis 2012; Northouse 2016). To this end, leadership remains without a universal definition and has no grand theory on the subject. This leadership literature review is done from the point of understanding leadership either as a person, process, results, position or purpose (Kempster et al. 2011). This approach is relevant because leadership manifests itself in either one or more of these forms.

A brief discussion of leadership in Africa is presented below to link African Leadership Theory to its developmental challenges and the SDGs. The gaps in African Leadership Theory are also highlighted. This is done through the critical appraisal of what has been published on the topic and what is attributed to cause Africa's development challenges. Such the Afrocentric voice in leadership theory was sought in the literature and explored in the implementation of the UNDP Botswana Biogas project.

Tshiyoyo (2017) observes that Africa is in a state of anxiety to correct its course from the current trajectory of underdevelopment and stagnation towards a more sustainable development resulting outcome. This observation was made while recognising that political systems in Africa remain critical drivers in social and economic reforms. Fourie et al. (2017), in their seminal paper that reviewed 60 years (1950–2009) of research on leadership in Africa, acknowledge that political leadership has been and continues to be momentous and a key area of research interest. They report varied opinions on Africa's developmental challenges, including the lack of responsible leadership and a proposal of leadership as the panacea for the continent's developmental inertia.

Fourie et al. (2017) provide evidence that political leadership at the local and national level in post-colonial Africa receives most attention from researchers. Additionally, no leadership theory is described as being African in its origin. In their synthesis and conclusion, it is "not possible to speak of a distinctly African theory of leadership" yet (p. 239), and they recommend African leadership scholars to develop theories based on "African experiences and socio-cultural resources…" (p. 245) to interrogate dominant leadership theories.

To this list, Gumede (2017) adds that inclusive development led by the people, on issues that they themselves consider important be the focus of leadership. As this can be achieved through appropriate polices, institutions and political systems. This is critical since leadership or challenges thereof originate and manifest at local levels in society, including in development project and practice settings. For instance, Amigun et al. (2012) and Roopnarain and Adeleke (2017) all agree that biogas technology is a viable alternative to biomass-based fuel, with a huge potential to significantly contribute to sustainable development in Africa. However they identify factors including political, social, cultural, economic, informational, financial, technical and training as negatively impacting on its absorption. These leadership factors can be personal, social and/or institutional and are pivotal in achieving positive change in biogas technology adoption and utilisation.

Gumede (2017) and Fourie et al. (2017) agree on the notion that leadership theory in Africa be centred on Afrocentric histories, experiences and philosophies. They posit this on Africa's high level of diversity and cultural dynamism. Dartey-Baah (2014) and Kagema (2018) mention related narratives especially when relating Africa's leadership with sustainable development. Although no Afrocentric leadership theory currently exists, literature has demonstrated that political leadership is a key determinant to sustainable development. This can be recognised through the amount of research on it and its major role in socio-economic development. Consequently political leadership in Africa is fundamental to its sustainable development. In this context, the next two sections narrow down the leadership discussion to sustainable leadership and the responsible leadership theory which was assessed in the case study.

2.2.2 Sustainable Leadership

The UNDP Botswana Biogas project was explored using the single case study method to better understand the phenomenon of leadership in a sustainable development practice setting which involves multiple stakeholders. The case study approach was used to engage with leadership theories in this context. To help develop the baseline to the investigation of leadership and sustainability, Bendell, Sutherland and Little (2017: 434) posit that:

> Sustainable leadership is any ethical behaviour that has the intention and effect of helping groups of people address shared dilemmas in significant ways not otherwise achieved.

The authors identify the presiding models of leadership as having causal links to the developmental crisis that the world is facing. Change is a social process; hence leadership in sustainable development has to be collaborative, collective, distributed and relational (Bendell et al. 2017). The assertion is that leadership is not about a position or a preserve of special individuals because the developmental dilemmas the world faces are shared. To successfully implement the SDGs, the leadership at all levels have to adopt leadership styles that promote shared responsibility and a collective approach focused on the long term. The need for leaders and leadership in the context of sustainable development to be ethically grounded and think beyond self-interest has been recognised by many other researchers (Bahauddin 2018; Dartey-Baah 2014; Gumede 2017; Kagema 2018). When leaders are ethically balanced and decide for posterity, they exhibit responsibility. This is the focus of the next section.

2.2.3 Responsible Leadership Theory

Responsibility is self-evident in all leadership theories (Kempster and Carroll 2016). Leadership is focused on responsibility, relational and ethical, and leaders act responsibly: responsible relationships are key in stakeholder engagement, leaders respect all stakeholders and leadership is shared (Bendell et al. 2017; Fourie et al. 2017; Frangieh and Yaacoub 2017; Voegtlin 2016). To this list, Maak and Pless (2006) add that leadership occurs in relationships in which leaders take the centre stage and ethics are the key building block. Responsible leaders build mutually beneficial and sustainable relationships. With proactive engagement and using inclusive and ethical approaches, these relationships build into social capital.

Leadership moves past relationships between a leader and follower to the calibre of a leader's relationships with internal and external stakeholders and become a socio-relational and moral phenomenon (Antunes and Franco 2016; Maak and Pless 2006). This approach is important especially when looking at community or national development projects. As is explained by Frangieh and Yaacoub (2017), at the centre of leadership theory are responsible relationships, ethical and multidisciplinary in nature. Several researchers share the sentiment that responsible leadership creates networks within the stakeholder universe to obtain capital and have deeper engagements. Responsible leadership therefore is capable of translating an agreed vision into life and contributing to sustainability and authenticity (Maak 2007).

Change is a social phenomenon and development occurs in a social context. The responsible leadership theory (RLT) can be viewed with a social connection. When this is done, it acquires dimensions that view the interaction between a leader and stakeholders as focused on deploying solutions in areas with a wider influence in social and environmental consequences (Voegtlin 2016). As a result, the UNDP Botswana Biogas project intends to promote the development and utilisation of biogas a clean energy source and at the same time sustainably manage agro-waste. This fully fits the tag of a socially and environmentally impactful solution to energy access. The next section focuses on biogas technology.

2.3 Overview of Biogas Technology in Africa and Botswana

2.3.1 Biogas Technology in Africa

Biomass in the forms of firewood, agriculture residue, charcoal and animal waste contributes significantly to primary energy supply and the total final energy consumption in most rural households in Southern African countries (Amigun et al. 2012; Tumwesige et al. 2011 & Zhou et al. 2009). Biogas is a type of biomass-derived fuel. The word "biogas" refers to a clean energy gas. This is produced as a result of anaerobic fermentation of organic matter that can be stored and used when needed. When integrated, the biogas system is used to generate multiple products. Some by-products like fertiliser can become inputs in food production, whose waste becomes part of inputs of the same system as shown in Fig. 7.1.

Biogas technology adoption is still suboptimal in Africa. This is in spite of the continent having vast biomass resources and large volumes of waste which present huge opportunities to expand access to energy using renewable energy resources (Kemausuor et al. 2018; McCabe and Schmidt 2018; Scarlat et al. 2018). However efforts are underway to promote its production and utilisation in Africa. This is being done as a way to contribute modestly to sustainable development, overcome energy poverty, integrate waste management and energy generation, preserve the environment, substitute for traditional cooking fuels, provide an alternative source of lighting, empower women, develop rural areas and significantly contribute to the development of low-carbon economies (Amigun et al. 2012; Mshandete and Parawira 2009; Roopnarain and Adeleke 2017; Scarlat et al. 2018; Tumwesige et al. 2011). The interconnectedness and interdependencies inherent in sustainable development are evident in these expected outcomes.

Amigun et al. (2012: 42) maintain that assimilation of a technology like biogas requires the "proper social, cultural, political and economic institutions to support adoption, dissemination and appropriate contextual use". Consequently the presence of appropriate institutions and their leadership becomes key. Botswana is listed among countries with biogas technology installations in the African region (Amigun et al. 2012; Khatibu 1986; Mshandete and Parawira 2009; Roopnarain and Adeleke 2017). The Botswana government is reported to have started looking at reliable sources of energy as far back as 1976 (Kebadiretse 2010). The Rural Industries Innovation Centre (RIIC), an appropriate technology centre, set up in 1975 is credited with the initial work done on biogas in the country (Chanda 1999). They built and tested three types of digesters for use mostly as a diesel fuel substitute for powering pump engines at boreholes (Khatibu 1986; Kebadiretse 2010; Roopnarain and Adeleke 2017).

Most of the earlier biogas installations in Botswana used cow dung as feedstock and were used to pump water. Later household biogas plants were also developed, and a total of 12 plants were installed for cooking and demonstration between 1982 and 1987 (Government of

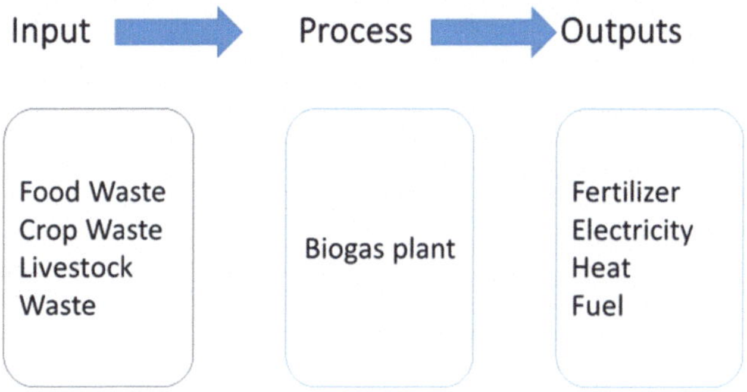

Fig. 7.1 Logic framework model of a biogas system. (Source: Authors based on McCabe and Schmidt (2018: 6))

Botswana and UNDP 2004: 23). Oladiran (1995: 193) observed that the country had a potential to generate 60% of its national electricity needs by using biogas alone. This was due to the large herds of cattle mostly kept using the free range farming method in rural areas. Despite the identified potential to generate heat and cooking fuel from biogas, the country did not witness a significant uptake of the technology. This was due to challenges with the collection of cow dung, open grazing and very high investment costs (Mangole et al. 2011; Oladiran 1995). Seanama Conservation Consultancy (n.d.) identifies similar opportunities in the abundant cow dung resources which together with coal and solar they presume could address the country's energy insecurities and increase access to energy services.

The Biomass Strategy Final Report of a 2007 Department of Energy Affairs commissioned biogas utilisation assessment provided country context recommendations focusing on social, cultural, political and economic institutions (Zhou et al. 2009). The report highlights opportunities to improve, optimise and close gaps on biogas technology uptake. Additional findings from the report recommend the development of a strategy to disseminate biogas technology and create demand for household plants (requires cheap finance), collaboration with zero-grazing farmers to generate electricity and schools hosting community plants.

Notwithstanding the identified opportunities and potential, lack of knowledge was also identified as a reason for poor acceptance. This was because locals were not accepting alternative forms of energy like biogas regardless of it being cheap and easy to use (Sunday Standard 2016). The Botswana government undertook an initiative to encourage the use of biogas as a strategy to preserve the environment, reduce deforestation and improve waste management practices in cases where food waste can be used as feedstock (Kebadiretse 2010).

2.3.2 UNDP Botswana Biogas Project

By 2015, biogas was still not well established in Botswana. The country had an installation base of 15 biogas plants most of which were nonfunctional and on the other hand had an objective to reach a renewable energy mix target of 25% by the year 2030 (UNDP 2015). Although biogas is only one of the renewable energy sources, given its long deployment history, one would expect a higher biogas installation base. The Botswana National Development Plan 11 aims to reduce the number of households that use wood as the main source of fuel by half in 6 years (Ministry of Finance and Economic Development 2016).

Given government's commitment towards increasing renewable energy sources and reducing the use of wood as the main fuel source, the biogas project was timely. The project is intended to install a total of 1000 small-scale biogas plants in rural communities in South Eastern Botswana covering 14 subdistrict councils. Additionally three medium-sized biogas plants with a 1MW electricity generation capacity each would also be installed. This would grow the biogas market in the country. The project was supposed to be implemented in phases. According to the project's theory of change and the authors' review of the project work plans, budgets and other documents, the project implementation milestones were reconstructed as shown in Table 7.1.

Table 7.1 UNDP Botswana Biogas project implementation timelines (planned and actual)

Year	Planned activity	Actual activity
2014		Pre-feasibility Feasibility (-roject documentation)
2015	Project preparation (May) Project 5mplementation (August 2015–August 2019)	GEF funding approval (June) Letters of commitments from stakeholders (June)
2017	Midterm evaluation	Benchmarking trip to Uganda and Ethiopia Establishment of multi-stakeholder platform (MSP)
2018	28 demonstration plants (at two per subdistrict) 300 small-scale plants across PPP framework developed Guidelines and regulations approved	Project officially starts on 20 April 2018 18 demonstration small-scale biogas plants constructed
2019	1 medium scale biogas plant operational, 1 PPP signed	Midterm review, feasibility for BMC medium-scale biogas plant conducted
2020	3 medium-scale biogas plants operational, 3 PPP signed and 1000 small-scale biogas plants operational	

Source: Authors based on UNDP (n.d.); UNDP (2015: 42–43)

The next section will describe the research approach in this case study including the selection of participants, the design of the data collection tools and data analysis approach.

3 Research Design

3.1 Research Approach

The research approach used was the exploratory qualitative research. This was focused on capturing the participants' realities and perceptions (ontological). In addition it sought to understand the epistemological context of leadership capacities in the development and implementation of SDG-related projects. The case study method has been used in similar research about individuals, programmes, implementation process and organisational change (Antunes and Franco 2016; Maak and Stoetter 2012; Pless 2007). It represented the best method to understand leadership in this context.

The UNDP Botswana Biogas project was selected because of its focus on improving access to clean energy and the sustainable management of agro-waste. In addition was the critical role renewable energy plays as mitigation for climate change. Given this focus on SDG implementation and the absence of a grand theory on leadership described earlier, only a unique critical case could practically be deployed to extend the existing theories on leadership (Xiao 2012). For this reason a single case study design approach was used. The rationale of studying the leadership capacities in project development and implementation was purposive. It was premised on both their uniqueness and suitability to highlight or replicate the constructs of emergent leadership theories.

The single case study method has a unique and revelatory nature with the ability to offer both depth and insight (Farquhar 2013; Yin 2003). This motivated its selection over other research methods. The method is also recommended because of its capability to deal with multiple sources of evidence that include documents, artefacts, interviews and observations (Yin 2003). The data collection methods used were face-to-face interviews, observation of events and review of documents. This was aided by the availability of the project documents including work plans and budgets in the public domain as per UNDP standards.

3.2 Sampling

The study initially targeted a "judgement sample" of individuals in the project implementation team and the project steering committee based on their direct leaderships in this project. When access to this cohort was not successful, the same purposive sampling technique was used to identify participants among informed outsiders who knew about the project or had previously worked on the project in similar or related roles. As a method with no randomisation, purposive sampling is not free from bias. Selection can result in a change of behaviour, and inherently the methodology is also vulnerable to judgement errors or selection bias which potentially can impact the generalisation and reliability of the results (Tongco 2007). In spite of these weaknesses, the availability and willingness of the informed outsiders to participate were the key determining factors. They were able and willing to share their reflective and observed experiences about the project.

The criteria for participant selection was based on the attendance lists of institutional stakeholders and individuals who represented their organisations during the UNDP Botswana Biogas project consultative meetings and follow-up face-to-face meetings (Wright 2014). A total of 10 participants were expected from the 29 potential interviewees who were identified and contacted via email and follow-up telephone calls. Only a total of three participants were finally interviewed.

3.3 Data Collection and Analysis Approach

An interview schedule was developed to inform and help answer the overall research questions using semistructured interview questions. These facilitated the comparison of the participant responses while at the same time understanding their individual experiences (Barlow 2012). The Responsible Leadership Theory (RLT) described earlier was preselected as the theoretical construct for the case. It is focused on the inspired role(s) of leaders and the importance of leadership in the achievement of positive change. This is demonstrated through their behaviours, capacities, perceptions and attitudes. Selected responsible leadership theoretical constructs were used to develop the data collection tools and form part of the interview questions. The rationale for their selection was based on the evolution of the social and relational nature of RLT identified through an extensive review of literature on attributes, capacities and performance competencies as shown in Table 7.2.

Muff, Liechti and Dyllick (2017:1–37) use the Globally Responsible Leadership Initiative (GRLI) model to propose a RLT definition that incorporates key competencies areas. The competencies focus on:

> creating, managing and securing good relations with multiple stakeholders, ethically correct and values-based behavior, highly developed self-awareness, good understanding of the interdependencies with a larger system, and the ability to lead change and innovation towards sustainable development. (Muff et al. 2017: 8)

Our case study adapted these competency areas to categorise the key attributes and frame the interview questions. The unit of observation used were leaders, and the unit of analysis were their respective opinions, capacities or perceptions. The data analysis focused on content, thematic and narrative analysis. The use of more than one analysis method was intended to improve accountability, depth and breadth of the research findings. The aim was to identify from the participants' constructs of the RLT that are important, together with the challenges that impact their application. The data obtained represented the participants' "lived experiences, perceptions, opinions and beliefs" (Roulston 2013: 297–312). Coding and data analysis was conducted manually by the researchers.

4 Presentation and Discussion of Results

Interviews were the core production of data for the case. Semistructured questions were used to collect data that can be compared and at the same time allow the interviewees to share their individual experiences with the researchers. The data is based on three in-depth interviews with two males and one female, all over 35 years of age. Their names were anonymised to interviewee 1, 2 and 3, respectively. Though only three, the interviews were very rich as the interviewees are

Table 7.2 Emergent RLT constructs and their relationship to interview questions

Responsible leadership construct	Key attributes	Relevant literature	Related interview questions
Leadership influence Leadership development Motivation	Leadership focused on responsibility Leaders act responsibility Responsible relationship key in stakeholder engagement Leadership is shared Leaders respect all Sustainable relationships	Antunes and Franco (2016) Frangieh and Yaacoub (2017:293) Maak and Pless (2006:99–112) Pless (2007:438–439) Kagema (2018:9)	Q1 Q2 Q3
Creating, managing and securing good relations	Identify and integrate stakeholders Long-term relations	Maak (2007:331)	Q2 Q7 Q12 Q13
Ethically correct and values-based behaviour	Know own values Have a value system Role model Responsibility towards society	Pless (2007:438)	Q3 Q5 Q9 Q10
Self-awareness	Weakness Strengths Learning from mistakes Reflective behaviour Handle conflict	Muff, Liechti and Dyllick (2017:1–37) Maak and Pless (2006:99–112) Pless (2007:450)	Q3 Q4 Q11 Q13 Q14
Good understanding of interdependencies with larger system	Interconnectedness Able to deal with complexity Transgenerational See big picture	Voegtlin (2016:591) Pless and Maak (2009:59)	Q3 Q4 Q12
Ability to change and innovation towards sustainable development	Drivers/enablers of innovation Visionary, translate ideas into action Flexible and adaptable Responsive	Kagema (2018:9) Pless (2007:450) Maak and Pless (2006:99–103) Muff, Liechti and Dyllick (2017:1–37)	Q4 Q8 Q10 Q14
Environment Situational Relational	Leadership practice in context Situational characteristics Context in which the influencing process takes place	Day and Antonakis (2012:5) Maak and Stoetter (2012:413–420) Pless and Maak (2009:59)	Q4 Q12 Q13 Q14

Source: Authors based on relevant literature

subject-matter experts and have worked in multiple settings, development programmes and projects. The participants are development practitioners within the resource management, environment, energy and development consulting sectors and were to share their opinions and experiences on the case. The following key findings are presented to demonstrate the interviewees' perceptions, practices and understanding of RLT.

4.1 Leadership Influence on Development Projects

On a leader influencing attitudes and perceptions about a project, all interviewees agreed with the influence of leadership. They highlighted the need for leaders to be exemplary in word and deed and for the leader not to influence uptake or implementation but rather use outcomes of consultation or feasibility studies. Responsible leaders positively influence both the process and followers. They identify and utilise the capacity to showcase extraordinary effort and rationalise resource use. All interviewees revealed that responsible leadership is participatory and inclusive. The competency of leaders in observing their environments and acting on their observations was reported. This relates to the leader being a visionary translating idea into action. Responsible leaders must understand the sustainability challenge and demonstrate their responsibility towards society.

On leaders influencing others, all interviewees agreed that leadership is an influencing process. Such leaders do influence the perceptions, actions and processes in projects. Responsible leaders should inspire positive attitudes and encourage followers and other stakeholders to work across disciplines and sectoral boundaries. This is because development is complex and multidimensional. These views point to the capability to work across disciplines and boundaries as a key competency under the systems thinking theoretical construct of RLT. Maak and Pless (2006) provide evidence that leaders are visionaries with a sense of purpose, who can mobilise and lead cross-sectoral teams and encourage collaboration that enhances responsiveness to their stakeholders. To this list, Kagema (2018:9) adds that a responsible leader "must be responsive to the needs, concerns and interests of those he leads". This competency demonstrates the ability to change and adapt which is critical in problem-solving.

4.2 Influence of Political Systems on Development

On the influence of political systems on development, interviewee 1 indicated that governance is subjective; African countries though having resources and following western types of democracy were developmentally "poor", while their counterparts from the Middle East, with similar resources but not following western democracy were not "poor". Interviewee 3 identified the influence of political systems as the "driving force" in leadership influence as political considerations take precedence over development. Within development practice, leadership is practised in a social context, and responsible leaders must understand the context within which development is practised. All development should be people led and reflect the people's interests, needs and environment. Pless and Maak (2009:64) succinctly capture this concept with their statement "Who determines if, and which direction, a local community should develop".

4.3 Stakeholder Relations

In response to the question of how a responsible leader resolved potential and actual conflicts, all the interviewees responded and captured the main theme of discussion that of working with multiple stakeholders. Conflict or contradictory interests in relationships might occur. When they do they are to be effectively dealt with proactively. Using a bottom-up approach, fairness and mediation if conflict still occur despite prevention. This will be a demonstration of knowledge in managing stakeholder relations. Responsible leadership according to Maak (2007:334) is a "relational and ethical phenomenon" which takes place in social interaction among stakeholders.

This competency becomes critical for responsible leaders. There is never-too-much stakeholder engagement in development practice as expectations have to be continuously managed and multiple stakeholders have conflicting interests. Responsible leaders have to be able to identify and integrate stakeholders and adequately consult while appreciating the positive in diversity. Antunes and Franco (2016:139) report that responsible leadership is understood well beyond the immediate leader-follower setting to involve social networks that "are important to create opportunities and transfer knowledge". The study of leadership in complex contexts in which there are multiple stakeholders and levels of action therefore provides a look in to explore the relational nature and collective dimensions of leadership (Ospina 2016).

4.4 Leadership Qualities

The interviewees mentioned that a responsible leader should demonstrate the following values and ethics: talent recognition and acknowledgement, open mindedness, willingness to learn, build capacity, transfer skills, being humble and honest. Responsible leaders are objective, transparent, fair and accountable in their approach to development issues. They are values driven and act as role models to followers, team members and their wider stakeholder universe. Additionally, responsible leaders are self-aware and can reflect on their leadership style. They also understand the interdependencies involved in development practice and appreciate the interconnectedness of development issues. This allows them to appreciate their weaknesses and be adaptive in their behaviour.

The competency requirement, for a leader to have a value system, is well documented in literature (Maak and Pless 2006; Muff et al. 2017; Pless 2007). Responsible leaders are objective, transparent, fair and accountable in their approach to development practice. A leader uses their relational influence, through assuming specific roles that enables them to access further resources in the form of social capita. Muff et al. (2017: 7) report that the exigency of self-awareness and the capacity to reflect on one's leadership style are the "inner dimension of RL". This competency demonstrates the ability of a leader to understand the the possible impact of one's action on the organisation and its stakeholders. Maak and Pless (2006:106) argue that when leaders critically review themselves, the organisation and stated interests from their stakeholders, it will be a demonstration of their "reflection skills and critical thinking capability". Once a leader is self-aware, they appreciate their weaknesses. They can adapt their behaviour recognising that for them to achieve their personal and organisational goals they require other role players and skills that they might not personally have.

4.5 Systems Thinking

Responding to questions on how the perception of followers impacted the actions of leaders in project implementation, and how leaders put together teams to develop solutions to complex developmental challenges, all interviewees identified that a responsible leader is one who is able to deal with complexities associated with development practice innovatively by being able and willing to learn from practice and from others. This leader is also flexible and adaptable to change. Responsible leaders adopt a long-term perspective. They are transgenerational and translate ideas into action.

Muff et al. (2017) discuss how important having a vision is to the change process. More especially with regard to driving or enabling innovation, responsible leaders adapt their communication to reach all audiences. According to Voegtlin (2016:591), responsible leadership emphasises "collective problem solving through communication". Additionally Maak and Pless (2006:111) in their role models of a leader describe responsible leaders as "creators of shared systems of meaning through sense making and dialogue". Communication therefore becomes critical in developing and maintaining trustful relationships and most importantly in facilitating the translation of ideas into action.

5 Conclusion

It is important to report that the data is missing the voice of the community as no interviews were sought or conducted with a community leader nor a member of the community in the areas intended to benefit from the Biogas technology. In human development, issues are never clear cut. Development projects tend to impact beyond their initial implementation framework. Therefore leadership practice has to be aware of these interdependencies. The UNDP Botswana Biogas project, although primarily mooted as an environmental project, has potential to impact multiple other sectors including energy, health, women empowerment, the economy and partnerships among others.

All this is possible in a multiple stakeholder context. The interviewees mentioned the following as key enablers to successful development projects; the role of tradition in establishing relationships, engaging local or traditional leadership in times of conflict, the need for local leadership led projects to enhance their footprint, giving people a platform to speak, and contextual leadership knowledge. Tradition remains important in establishing and maintaining relationships and there is need for an Afrocentric approach to leadership given the diversity and dynamism in Africa (Fourie et al. 2017; Gumede 2017).

As the research findings discussed earlier imply, leadership depends on context and competencies. When these are applied there is more consensus, understanding and collaborative actions. These are are predictors of success. The study found that, most of the RLT constructs assessed for, were identified and presented by the interviewees. Furthermore the case demonstrated, that when qualitative research methodology is applied to development practice it is liberating. This is because it allows a reflective approach in understanding, explaining theory and study findings, respectively.

References

Amigun, B., Parawira, W., Musango, J.K., Aboyade, A.O. & Badmos, A.S.(2012). Anaerobic biogas generation for rural area energy provision in Africa. Retrieved from: http://cdn.intechopen.com/pdfs/31319/InTechAnaerobic_biogas_generation_for_rural_area_energy_provision_in_africa.pdf. (Accessed 26 September 2019).

Antunes, A., & Franco, M. (2016). How people in organizations make sense of responsible leadership practices. Multiple case studies. *Leadership & Organization Development Journal*, 37(1), 126-152.

Bahauddin, K. MD. (2018). The essence of leadership for achieving the Sustainable Development Goals. Retrieved from: http://sdg.iisd.org/commentary/generation-2030/the-essence-of-leadership-for-achieving-the-sustainable-development-goals/. (Accessed 12 August 2019).

Barlow, C.A. (2012). Interviews. In A.J. Mills, G. Durepos & E Wiebe (Eds.), *Encylopedia of case study research* (pp. 496-499). Thousand Oaks, CA: SAGE Publications Inc.

Bendell, J., Sutherland, N. & Little, R. (2017). Beyond unsustainable leadership: Critical social theory for sustainable leadership. *Sustainability Accounting, Management and Policy Journal*, 8(4), 418-444.

Bhattacharya, D., Khan, T.I. & Salma, U. (2014). A commentary on the Final Outcome Document of the Open Working Group on SDGs. *SAIS Review*, XXXIV(2), 165-177.

Botswana Statistics. (2018). Botswana Domesticated Sustainable Development Goals Indicators Baseline Report. Retrieved from: http://www.statsbots.org.bw/sites/default/files/special_documents/Botswana%20Domesticated%20Sustainable%20Development%20Goals%20Indicators%20%20Baseline%20Stats%20Brief_0.pdf. (Accessed 26 July 2019).

Chanda, W. (1999). Technology Design and Manufacture - The Technology Transfer Programme in Botswana. Retrieved from: http://library.fes.de/fulltext/bueros/botswana/00554002.htm. (Accessed 07 August 2019).

Day, D.V. & Antonakis, J. (2012). Leadership: Past, Present, and Future. In D.V. Day & J. Antonakis (Eds.), *The nature of leadership*, 2nd (3-25). Thousand Oaks, CA: SAGE Publications Inc.

Dartey-Baah, K. (2014). Effective leadership and sustainable development in Africa: Is there "really" a link?. *Journal of Global Responsibility*, 5(2), 203-218.

Fourie, W., van der Merwe, S.C. & van der Merwe, B. (2017). Sixty years of research on leadership in Africa: A review of literature. *Leadership*, 13(2), 222-251.

Farquhar, J.D. (2013). *Case study research for business*, London: SAGE Publications Ltd.

Frangieh, C.G. & Yaacoub, H.K. (2017). A systematic literature review of responsible leadership: Challenges, outcomes and practices. *Journal of Global Responsibility*, 8(2), 281-299.

Government of Botswana & UNDP. (2004). Botswana technology needs assessment on climate change: Final report. Retrieved from: http://www.ungsp.org/sites/default/files/documents/botswana_tna.pdf. (Accessed 04 July 2019).

Gumede, V. (2017). Leadership for Africa's development: Revisiting indigenous African leadership and setting the agenda for political leadership. *Journal of Black Studies*, 48(1), 74-90.

Kagema, D.N. (2018). Responsible leadership and sustainable development in post –independent Africa: a Kenyan experience. *The Journal of Values-Based Leadership*, 11(1), 1-13.

Kebadiretse, F. (2010). Government encourages biogas use as alternative energy. Retrieved from: http://www.mmegi.bw/index.php?sid=4&aid=929&dir=2010/March/Monday15. (Accessed 22 September 2019).

Kemausuor, F., Adaramola, M.S. & Morken, J. (2018). A Review of Commercial Biogas Systems and Lessons for Africa. *Energies*, 11(2984): 1-21.

Kempster, S. & Carroll, B. (Eds.). (2016). *Responsible Leadership: Realism and romanticism*. Abingdon-on-Thames: Routledge.

Kempster, S., Jackson, B., & Convoy, M. (2011). Leadership as a purpose: Exploring the role of purpose in leadership practice. *Leadership*, 7(3), 317-334.

Khatibu M. (1986). Biogas Technology for Water Pumping in Botswana. In M.M El-Halwagi (Ed.), *Biogas Technology, Transfer and Diffusion*, pp. 604. Dordrecht: Springer.

Maak, T. (2007). Responsible leadership, stakeholder engagement, and the emergence of social capital. *Journal of Business Ethics*, (74), 329-343.

Maak, T. & Pless, N.M. (2006). Responsible leadership in a stakeholder society- A relational perspective. *Journal of Business Ethics*, 66, 99-115.

Maak, T. & Stoetter, N. (2012). Social entrepreneurs as responsible leaders: 'Fundación Paraguaya' and the case of Martin Burt. *Journal of Business Ethics*, 111(3), 413-430.

Mangole, L., Atlhopheng, J.R. & Chanda. (2011). Botswana – Boteti Biogass Installation. Retrieved from: http://www.desire-his.eu/en/download-documents/doc_download/301-wp43-boteti-biogas-summary. (Accessed 21 March 2019).

McCabe, B.K. & Schmidt, T. (2018). Integrated biogas systems: Local applications of anaerobic digestion towards integrated sustainable solutions, IEA Bioenergy Task 37. Retrieved from: https://www.ieabioenergy.com/wp-content/uploads/2018/06/Integrated-biogas-systems_WEB.pdf. (Accessed September 2019).

Ministry of Finance & Economic Development. (2016). National Development Plan 11 Volume 1 April 2017 – March 2023. Retrieved from: https://www.finance.gov.bw/images/NDP_11_2017-2023.pdf. (Accessed 26 July 2019).

Mshandete, A.M., & Parawira, W. (2009). Biogas technology research in sub Saharan African countries – A review. *African Journal of Biotechnology*, 8(2), 116-125.

Muff, K., Liechti, A. & Dyllick, T. (2017). The Competency Assessment for Responsible Leadership (CARL). Consolidating the responsible leadership discourse into an operationalized definition and an online tool for practice and education. Retrieved from: https://carl2030.org/wp-content/uploads/2017/11/Muff-Liechti-Dyllick-CARL-consolidating-the-RL-discourse-into-an-online-tool-210817-002.pdf. (Accessed 24 April 2019).

Northouse, P.G. (2016). *Leadership: Theory and practice*, 7th edn. Thousand Oaks, CA: SAGE Publications Inc.

Oladiran, M. T. (1995). Energy applications in Botswana. *Energy Conversion and Management*, 36(3), 191-195.

Ospina, S.M. (2016). Collective leadership and context in public administration: Bridging public leadership research and leadership studies. *Public Administration Review*, 77(2), 275-287.

Pless, N.M. (2007). Understanding responsible leadership: Role identity and motivational drivers. The case of Dame Anita Roddick, Founder of The Body Shop. *Journal of Business Ethics*, 74, 437–456.

Pless, N. & Maak, T. (2009). Responsible leaders as agents of world benefit: Learnings from "Project Ulysses". *Journal of Business Ethics*, 85, 59-71.

Pogge, T. & Sengupta, M. (2016). Assessing the sustainable development goals from a human rights perspective. *Journal of International and Comparative Social Policy*, 32(2), 83-97.

Roopnarain, A. & Adeleke, R. (2017). Current status, hurdles and future prospects of biogas digestion technology in Africa. *Renewable and Sustainable Energy Reviews*, 67(2017), 1162-1179.

Roulston, K. (2013). Analyzing interviews. In U. Flick (Ed.), *The SAGE Handbook of Qualitative Data Analysis* (pp. 297-312). London: SAGE Publications Ltd.

Sachs, J.D. (2018). 'Age of Sustainable Development'. Retrieved from: https://www.edx.org/course/age-of-sustainable-development. (Accessed 10 March 2019).

Scarlat, N., Dallemand, J. & Fahl, F. (2018). Biogas: Developments and perspectives in Europe. *Renewable Energy*, 129 (2018), 457-472.

Seanama Conservation Consultancy. n.d. Energy Policy Brief Reflecting on the Challenges of Attaining a Green Economy for Botswana. Retrieved from: https://sustainabledevelopment.un.org/content/documents/1009National%20Report%20(Energy)%20-%20Botswana.pdf. (Accessed 8 August 2019).

Sunday Standard. (2016). 'Locals consumers shun biogas'. Retrieved from: http://www.sundaystandard.info/local-consumers-shun-biogas. (Accessed 11 February 2019).

Tongco, M.D.C. (2007). Purposive sampling as a tool for informant selection. *Ethnobotany Research & Applications*, 5, 147-158.

Tshiyoyo, M. (2017). Assessing the catalytic role of the African Charter and the African Peer Review Mechanism in the realization of the objectives set out in Africa Agenda 2063 and the SDGs. *African Journal of Public Affairs*, 9(8), 172-184.

Tumwesige, V., Avery, L., Austin, G., Balana, B., Bechtel, K., Casson, E. et al. (2011). Small scale biogas digester for sustainable energy production in Sub Saharan Africa. Paper presented at 1st World Sustainability Forum, MDPI, Basel, 1-30 November.

United Nations. (2015). Transforming our world: the 2030 Agenda for Sustainable Development. Retrieved from: http://www.un.org/ga/search/view_doc.asp?symbol=A/RES/70/1&Lang=E. (Accessed 10 February 2019).

United Nations. (2016). UNDP country programme document for Botswana (2017-2021). Retrieved from: https://www.undp.org/content/dam/botswana/docs/Legal%20Documents/CPD-Botswana.pdf. (Accessed 28 February 2019).

United Nations Development Program. (2015). PRODOC Biogas BW. Retrieved from: https://open.undp.org/projects/00098758. (Accessed 12 April 2019).

United Nations Development Program. (2017). Government of Botswana and United Nations Sustainable Development Framework (UNSDF) 2017-2021. Retrieved from: https://www.bw.undp.org/content/dam/botswana/docs/Publications/BW_UNSDF%202017.pdf. (Accessed 29 July 2019).

United Nations Development Program. (n.d.). Biogas Project Workplan 2018. https://open.undp.org/projects/00098758. (Accessed 18 April 2019).

Voegtlin, C. (2016). What does it mean to be responsible? Addressing the missing responsibility dimension in ethical leadership research. *Leadership*, 12(5): 581-608.

World Bank. (2019). Access to electricity, rural (% of rural population). Retrieved from: https://data.worldbank.org/indicator/EG.ELC.ACCS.RU.ZS. (Accessed 08 April 2019).

World Biogas Association. (2018). World Biogas Association Fact Sheet 3: How to achieve the sustainable development goals through biogas. Retrieved from: http://www.worldbiogasassociation.org/wp-content/uploads/2018/07/WBA-SDGs-Biogas-factsheet-3.pdf. (Accessed 13 September 2019).

Wright, N. (2014). Promoting Production and Utilization of Bio-methane from Agro Waste in South-Eastern Botswana, Stakeholder Consultation Final Report. BebeZip (Pty) Ltd.

Xiao, H. (2012). Single case design. In A.J. Mills, G. Durepos & E. Wiebe (Eds.), *Encylopedia of case study research* (pp. 868-870), Thousand Oaks, CA: SAGE Publications Inc.

Yin, R. (2003). *Case study research: Design and methods*, 2nd ed. Thousand Oaks, CA: SAGE Publications.

Zhou, P.P., Batidzirai, B., Simbini, T., Odireng, M., Wright, N. & Tadzimirwa, T. (2009). Botswana Biomass energy strategy: Final report. Retrieved from: http://www.lse.ac.uk/GranthamInstitute/wp-content/uploads/laws/1090.pdf. (Accessed 25 May 2019).

Elements of Responsible Leadership in Driving Climate Action (SDG 13)

8

Karien Erasmus and Yolande Steenkamp

Abstract

The 2015 Paris Agreement on climate change calls for ambitious greenhouse gas (GHG) mitigation efforts to limit global warming to a maximum 2 °C to curb climate change. Both the Paris Agreement and the Initiative for Climate Action Transparency (ICAT) emphasise the role of non-state actors, such as private sector companies, in setting ambitious mitigation targets. Leaders in the private sector are seen as critical in developing and implementing GHG emission targets. Thus, this study aimed to identify three elements of responsible leadership (RL) (the building of stakeholder relationships and RL drivers and roles) in climate action, specifically GHG target setting. Methodologically the study followed a grounded theory, inductive approach, analysing a single case study and the Equity Method for GHG emission target setting developed by Promethium Carbon. The findings showed that elements of RL did contribute to driving climate action in the private sector and are, therefore, the appropriate approach to address climate change. However, there is room for expansion in the RL definition and elements of the theory to develop an appropriate description of responsible climate leadership for developing countries. Based on the findings and analysis, a conceptual framework for responsible climate leadership in a developing country context was proposed.

Keywords

Responsible leadership · GHG targets · Roles · Drivers

1 Introduction

This study addressed the business leadership approaches required to enable climate action, exploring whether aspects of RL were essential factors in developing and implementing the Equity Method – a procedure for setting GHG emission targets in developing country contexts. The RL approach was selected because it recognises the need for urgent action, within the complexity of the current economic environment, for long-term gains. RL represents a move away from past leadership styles, specifically within the last two centuries, where leaders have traditionally focused on maximising short- to medium-term shareholder profits, often at the expense of strategic resources such as the environment.

K. Erasmus (✉) · Y. Steenkamp
Department of Business Management, University of Pretoria, Hatfield, South Africa
e-mail: karien@promethium.co.za; yolande.steenkamp@up.ac.za

G. Nhamo et al. (eds.), *Sustainable Development Goals for Society Vol. 2*, Sustainable Development Goals Series, https://doi.org/10.1007/978-3-030-70952-5_8

Climate change is one of the most significant challenges currently facing the world. This is reflected in the World Economic Forum's (WEF) Global Risks Report, which ranks extreme weather events and the failure of climate change mitigation and adaption as the two highest risks in terms of likelihood and impact (WEF 2019). The 2015 Paris Agreement culminated in a commitment from the world's leaders to take collective action to curb climate change through emission reductions. To achieve these objectives, countries have submitted nationally determined contributions (NDCs) outlining their respective emission trajectories to mitigate the release of GHG emissions. The 2015 Paris Agreement is very clear that equity should form the basis of long-term mitigation efforts (Stern 2015). Stern (2015:271) notes that mitigation efforts should be "consistent with the notion of equitable access to sustainable development", implying that although all counties commit to decarbonisation, developed countries should decarbonise at a faster pace than developing countries.

Increasingly, leaders in the private sector are seen as critical contributors to develop and implement plans to mitigate climate change (Falkner 2016). Both the Paris Agreement and the recent Initiative for Climate Action Transparency (ICAT) call for the increasing role of non-state actors, such as private sector companies (ICAT 2018), to not only contribute to NDCs but to also take the lead in setting ambitious mitigation targets. The Science Based Targets initiative (SBTi) encourages companies to set mitigation targets in line to limit global temperature to a 2 °C increase by using defined SBTi methodologies (SBTi 2018). However, whereas the Paris Agreement clearly articulates the principle of common but differentiated responsibilities, the methodologies developed by the SBTi do not consider the socio-economic context of developing countries that have not enjoyed the same benefits of industrialisation as most developed countries (Promethium Carbon 2018). Since the Equity Method was developed to address the need to take the context of developing countries into account when setting emission targets, it was selected as a case for this study to determine which, if any, elements of RL played a role in actions to address climate change. Three elements of responsible leadership (RL) received attention, namely, the building of stakeholder relationships, RL drivers and RL roles in climate action, specifically GHG (GHG) target setting.

2 Literature Review

2.1 Responsible Leadership Overview

RL was deemed an appropriate theoretical lens for the study due to its emphasis on urgent action in the interest of long-term gains while engaging all stakeholders of an organisation, including the environment. This section provides a brief literature review of the RL approach, with emphasis on the elements highlighted in the research objectives, namely, stakeholder engagement and the roles and drivers of RL.

Maak (2007:334) defines RL as "…the art and ability involved in building, cultivating and sustaining trustful relationships to different stakeholders, both inside and outside the organisation, and in coordinating responsible action to achieve a meaningful, commonly shared business vision". He further highlights that RL includes the ability to develop and implement a particular vision, with the support of stakeholders, which can then contribute to long-term business viability (Maak 2007:334). Pless (2007:438), in turn, defines RL as:

> A values-based and through ethical principles driven relationship between leaders and stakeholders who are connected through a shared sense of meaning and purpose through which they raise one another to higher levels of motivation and commitment for achieving sustainable values creation and social change.

In addition to this definition, Mirvis (2010:11) notes that RL is a function of the interrelationships between businesses and organisations and society in economic, socio-political, ecological and moral spheres. These spheres are defined as the investment by shareholders (economic sphere); the responsibility and accountability to several different stakeholders (socio-political sphere); the use, intrusion or impact on natural systems and resources (ecological sphere); and the responsibility for the impact of business, its services or products as acknowledged by businesses (moral sphere) (Mirvis 2010). This dynamic interrelationship is an appropriate approach to the interrelated and complex nature of climate change.

Voegtlin, Patzer and Scherer (2012:14) add to the concept of RL by considering additional interactions and arguing that relations have moved beyond mere employee or shareholder engagements and now include a much broader stakeholder context which relates to various perspectives on sustainable corporations (Voegtling, Patzer and Scherer 2012). Voegtlin's work adds to this concept with the idea of interaction as a means of moderating between different stakeholders in order to obtain organisational legitimacy (Voegtlin 2016). Frangieh and Yaacoub (2017:282) add morality to RL theory, describing RL as a function of ethical reasoning and moral imagination that enables a responsible leader to make decisions and understand its impacts on others.

2.2 The Role of Stakeholder Relations in Responsible Leadership

Maak (2007) emphasises the value of positive stakeholder relationships by qualifying them as key to organisational viability and business success. There is a direct link between an organisation's ability to maintain positive stakeholder relations and the reputation of an organisation (Frangieh and Yaacoub 2017). This is a sensitive relationship that is nurtured by appreciating and integrating the concerns of various stakeholders in organisational activities, structures and strategies (Frangieh and Yaacoub 2017). RL thus demands a balanced approach to the diverse needs of stakeholders and interest groups in decision-making and actions (Waldman and Galvin 2008).

However, managing such diverse needs and priorities is no easy task (Maak 2007), and leadership should incorporate proactive engagement practices and ensure stakeholder inclusivity (Maak 2007). By including and acknowledging stakeholders, RL enables better understanding, builds trust and cooperation and ultimately contributes to social capital development (Maak 2007).

In current climate change narratives, the role of stakeholders is becoming more valuable, and an inability to find a balance between significantly diverse views may hamper and delay climate action. This is why the stakeholder component of RL is so pertinent to the discussion on climate action and why the Equity Method aimed to find a balance between global GHG emission target setting methodologies and the context of developing countries to achieve climate action in a just manner.

2.3 Responsible Leadership Roles and Drivers

Pless' (2007) description of the role of identity and drivers of RL was a key reference text used in this study, informing to a large extent our evaluation of the presence of RL elements in the development of the Equity Method.

2.3.1 Responsible Leader Roles

The role model of RL is based on the fact that RL needs to address and manage the needs of a variety of stakeholders (Pless 2007). Describing the different characters of a responsible leader, the model comprises both values-based roles and operational roles (Pless 2007) and can be summarised as follows:

- The responsible leader, as a steward, understands the global context of business and has an appreciation for the social, environmental and cultural context in which they operate (Maak and Pless 2006).
- The responsible leader is a guardian of values (Maak and Pless 2006).
- The responsible leader is an active and reflective citizen and promotes active citizenship both within and outside the organisation (Maak and Pless 2006).
- Responsible leaders, as visionaries, can work with diverse stakeholders to develop a vision that appeals to the various followers and pro-

vides a sense of direction (Maak and Pless 2006).

- The responsible leader is a servant and requires attentiveness, humility and modesty as well as an aspiration to care for others and their unique needs (Maak and Pless 2006).
- Responsible leaders play an important role as storytellers, possessing a critical ability to develop a coherent dialogue that makes sense to the various followers (Maak and Pless 2006).
- Responsible leaders are essential change agents (Maak and Pless 2006).
- The responsible leader as a networker refers to a female concept of leadership which includes the idea of a web of inclusion (Pless 2007). This refers to the leader being able to build and sustain strong relationships.

2.3.2 Responsible Leader Drivers

In addition to these roles, Pless (2007) identifies underlying driving forces that support and develop RL and also distinguishes between intrapsychic (motivated by personal needs) and moral drivers (embedded in values and norms) (Table 8.1).

The roles and drivers of RL are discussed in the following sections concerning the findings from the case study analysis.

Table 8.1 Discussion on responsible leadership motivational drivers

Drivers	Definition
Intrapsychic drivers	
Need for exploration and assertion	This need is related to cognitive functions that affect our ability to learn, experiment and work and that provide a sense of competency
Need attachment and affiliation	Fundamentally we need connectedness and meaningful relationships, whether on a personal level or through association with an organisation
Sense of enjoyment	Experiencing joy and having fun are critical to individual and organisational mental health and well-being
Normative drivers	
Need for justice	The need for justice refers to a deep-seated need for fairness and a moral framework for human interaction
Need for recognition	This refers to the individual need to be both recognised and valued as well as recognise others
Sense of care	This refers to the reciprocal "seeing and responding to need" that takes place in a relationship through "sustaining the web of connection" (Pless 2007:441)

Source: Authors, based on Pless (2007:439–441)

3 Methodology

The research methodology applied to this study was qualitative and followed a grounded theory approach, which allowed the researcher to best identify elements of RL that drive climate action (specifically GHG emission target setting) in the private sector. A single case study design allowed for critical case analysis that adopted as particular approach the Equity Method – a tool developed by Promethium Carbon to set GHG emission reduction targets in the private sector – and semi-structured interviews were conducted with purposefully selected participants.

The Equity Method aims to provide a just approach to GHG emission reduction target setting for companies operating in developing countries (Promethium Carbon 2018:1). The study analysed the elements of RL portrayed by the project leader in developing and implementing the Equity Method along the three study objectives, namely, (i) key aspects that contribute to overcoming stakeholder inclusiveness challenges in the realisation of climate actions, (ii) the RL roles specifically relevant to the development of the Equity Method as it relates to climate action and (iii) the most prominent RL drivers in developing the Equity Method and setting GHG targets.

3.1 Research Approach: Selecting the Equity Method as a Single Case Study

This study followed an inductive approach to reach "testable theoretical propositions" based on various data sources (Antonakis et al. 2004:52), specifically utilising the descriptive nature (Meuser et al. 2016) of the inductive approach. Grounded theory was particularly relevant to this study as it allowed for the use of different data sources, each of which can be similarly coded (Corbin and Strauss 1990), recurrently considered and systematically refined to inform research propositions (Antonakis et al. 2004). The development and implementation of the Equity Method was selected as a single case study for the following reasons:

1. Aiming to enable companies to set science-based GHG emission targets that support economic growth within the context of developing economies, the Equity Method addresses target 13.2 of Sustainable Development Goal (SDG) 13: Climate Action. The Equity Method is distinct from other SBTi methods because it focuses on setting GHG emission targets within developing country contexts, as opposed to developed country contexts.
2. The process of the development and implementation of the Equity Method resonated with RL in that it had to bring together various stakeholders, develop a desired end state and balance a diversity of needs related to emissions management and economic growth.
3. As a national initiative, the Equity Method involved several companies in the development and implementation phases, allowing for potentially diverse views around stakeholder engagement and the roles and drivers of RL in the different phases of the project. At the time of conducting this study, the Equity Method had already been developed and applied, enabling the study to identify key role players during these phases.

Although structured according to the research objectives, the semi-structured interviews were conversational and evolved through the interview (Barlow 2012). This allowed the participants to provide their interpretations and insights, enabling the data to contribute to the expansion of the elements of RL within the context of climate change. Interview participants were selected to represent stakeholders from within Promethium Carbon who were part of the development of the Equity Method, as well as stakeholders outside of the company who actively participated in both the development and implementation of the Equity Method. Set criteria ensured that participants would have both exposure to the Equity Method and an established relationship with the project leader, enabling them to reflect on his leadership style.

The unit of observation selected for this study was the project leader responsible for the development of the Equity Method, while the unit of analysis was RL, more specifically those elements of RL that contributed to the facilitation of climate action. Limitations of the study included its sample size of only seven participants, a limited timeframe and research participants' limited knowledge of the RL approach. While the latter was addressed by availing essential literature before the interviews, not all participants perused this, limiting some feedback in the interviews.

3.2 Context and Case Overview

The Equity Method, the case selected for this study, is linked to SDG 13: Climate Action, specifically target 13.3 which is to "improve education, awareness-raising and human and institutional capacity on climate change mitigation, adaptation, impact reduction and early warning" (UN 2019:14).

3.2.1 Climate Change: The Wicked Problem of Our Time

Due to its numerous and interdependent components and a lack of proven theoretical approaches towards a solution, "human interaction with the natural environment is a highly complex and therefore a wicked problem" (Metcalf and Benn 2012:371). Also, the time pressure involved, complications involving developing and developed nations, weak centralised decision-making systems and a consistent overlooking of the gravity of the problem resulted in Levin et al. (2012:124) to define climate change as a "super wicked problem".

To prevent the loss of ecosystems, the drastic change of human systems and widespread global pressure (Levin et al. 2012), and in response to calls by scientists and key role players such as the Intergovernmental Panel on Climate Change (IPCC) (IPCC 2018), the Paris Agreement (2015) succeeded in setting an international target to reduce GHG emissions limit global temperatures to 2 °C above pre-industrial levels. Because short-term targets will not provide the required levels and rates of decarbonisation (Ge and Ross 2019), the Paris Agreement has accordingly recognised the need for the development of long-term, low GHG emission strategies in efforts to limit global warming. This contributes to SDG indicator 13.2, namely, to "Integrate climate change measures into [national] policies, strategies and planning" (UN 2019). For this case study, this indicator will be used to show change, where "change" will refer to the integration of climate change mitigation targets in organisational policies, strategies and planning.

3.2.2 Overview of the Equity Method

Corporate sector involvement in GHG emission target setting is becoming more relevant and necessary as part of the transition to a global low-carbon economy (Immink 2019:56; Krabbe et al. 2015:1). Andrade and De Oliviera (2015:376) eloquently echo this:

> Environmental institutions need to recognise that sovereign states no longer tackle challenges alone; globalisation requires strong cooperation among nation-states and non-state actors, such as private companies and private environmental standards organisations, environmental groups, and indigenous people, as well as subnational governments such as municipalities and provinces.

For the corporate sector to actively participate in lowering global GHG emissions, methodologies for setting targets should be translated appropriately from a national perspective to a corporate level (Krabbe et al. 2015). The development of the SBTi (developed by the CDP, WWF, World Resources Institute and UN Global Compact) aimed to motivate urgent and ambitious climate mitigation actions (SBTi 2018). Its agenda is to align the emission reduction targets of companies to scientifically established global carbon budgets. However, although the partner countries of the Paris Agreement agreed on the principles of equity and common but differentiated responsibilities and respective capabilities in the light of different national circumstances, the Paris Agreement does not define particular emissions allocation processes for developed, developing and least-developed parties to the agreement (Promethium Carbon 2018).

Promethium Carbon's Equity Method has been developed using the principles of the SBTi while furthermore taking cognisance of the common but differentiated responsibility principles as set out by the Paris Agreement, which is specifically relevant to developing countries that should be afforded an opportunity to grow their economies to meet basic developmental objectives, while the associated GHG emissions could be tapered down over the longer term. Failing to recognise and consider these aspects presents an ethical challenge for the SBTi and its target setting methodologies (Promethium Carbon 2018). The targets generated from the methodologies may result in an unfair curbing of output as a result of restrictive methodologies related to GHG mitigation action while not encouraging commitments at higher levels of ambition from other companies (Promethium Carbon 2018:1).

The Equity Method makes a distinction between the scientific principle of setting GHG targets and the ethical principle related to common but differentiated responsibilities (Promethium Carbon 2018). It also takes into consideration various aspects related to setting a corporate GHG emission reduction target. These include that the management of the targets set by a company should form part of the target setting method. That the credibility and integrity of the target should be supported by the standard on which the method is built and that the method guiding this should address issues such as responsibility for the targets in a company, the practical implementation thereof as well as the monitoring, reporting and verification of the target. This approach is illustrated in Fig. 8.1.

Although well-received by the South African National Business Initiative (NBI) and the com-

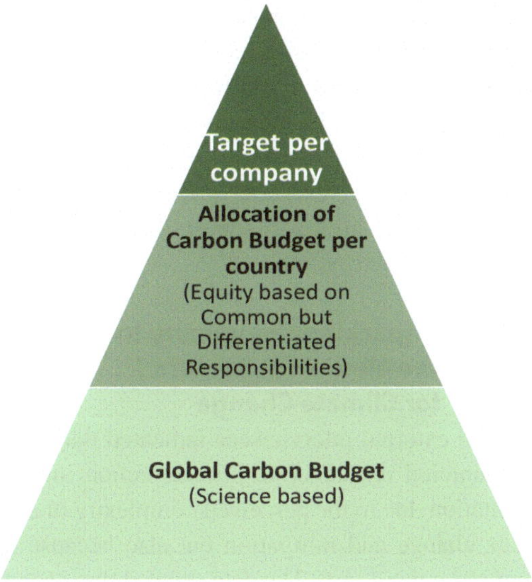

Fig. 8.1 Approach to the development of the Equity Method. (Source: Promethium Carbon 2018 (reproduced with permission from author))

panies that participated in its development, the SBTi has not agreed to consider the Equity Method as an alternative to current SBTi methodologies that do not consider geographic differentiation. The SBTi believes that in a global market, all target setting methodologies must be equal in their approaches and that all companies have similar challenges and opportunities, regardless of geographical differences (Promethium Carbon 2018:3). This does not imply that the project failed; the Equity Method has opened the door to more robust discussions on the importance of GHG emission target setting and the means with which this is done, as envisioned by the Paris Agreement.

4 Findings and Discussion of Results

4.1 The Equity Method: Recognising the Importance of Stakeholder Engagement as Depicted in Responsible Leadership Theory

Stakeholder engagement is a key component of RL. Voegtlin indicates that RL is the task of interacting with and moderating between different stakeholders (Voegtlin 2016:586). However, the varied nature of stakeholders and their respective priorities potentially complicates leadership (Maak 2007:330). The development of the Equity Method included a diverse number of stakeholders, each with particular objectives and ideas regarding GHG emission target setting.

4.1.1 Long-Term Target Setting Versus Short-Term Decisions

One of the key challenges raised during the interviews was that GHG emission target setting requires a long-term perspective. Indeed, of Allen's (2018) four identified worldview shifts that should guide leaders, specifically with regard to environmental issues such as climate change, the third recognises that leaders should move from short-term to long-term thinking (Allen 2018). This applies especially in a developing context, due to the pressures of immediate, short-term and often reactive decisions. Levin et al. (2012) reference the tragedy of policies focusing on short-term timeframes, which often have a negative long-term impact on our ability to reduce GHG emissions. Responsible climate action often necessitates impeding short-term choices (Stern 2015).

The tension between immediate, reactive decisions and long-term target setting is even more pronounced in view of factors resulting in continuous delay, especially in terms of climate change mitigation. South Africa, although a fossil fuel-dependent country, has committed to lowering GHG emissions aligned to the Paris Agreement (De Jongh and Möllmann 2014). However, the country is also faced with extreme poverty, high unemployment levels and inequality (De Jongh and Möllmann 2014). Therefore, climate action competes with short-term political and social priorities.

Setting GHG emission targets requires a transformation from strategic and operational perspectives to long-term perspectives, and it is uncertain how this transformation enables companies to participate in a low-carbon economy while still being able to meet immediate business objectives (Immink 2019). As a result, there is resistance to the changes required in setting these targets. In this regard, finding the balance between short- and long-term business and growth objectives becomes all the more critical in ensuring buy-in from stakeholders.

In developing and implementing the Equity Method, the project leader was able to obtain collaboration between stakeholders by (i) making a sound argument for the need for climate mitigation based both on scientific principles and the value thereof to business objectives, (ii) understanding and appreciating the context of the companies he worked with within the context of climate change and (iii) finding a middle ground based on the data and the context of companies.

4.1.2 Building Trust and Relationships Through Respect

During an interview, the project leader stressed the importance of respect, saying that "…it [leadership] starts with respect and with acceptance of the person across the table from you". The literature supports the notion of respect as fundamental to leadership. Frangieh and Yaacoub (2017:293) mention respect as one of the themes that prompt followers to see their leaders as responsible, and Maak and Pless (2006) further note that it forms part of a suite of attributes related to the character of a responsible leader.

Another attribute that Maak and Pless highlight and that was mentioned in most of the interviews is trust (Maak and Pless 2006). Trust was mentioned as important in relation to the project leader's efforts to overcome stakeholder engagement challenges and build strong relationships. Internal Interviewee 2 mentioned that "…time [in terms of Promethium Carbon's operational existence] builds trust" and that this was key to the development of the Equity Method. She stressed that to obtain buy-in from the various stakeholders, it was imperative that these long-standing relationships were in place and that the stakeholders had trust in the project leader. Leadership theory also indicates a shift from understanding leadership in terms of single relationships to a social process which includes many and varied complex relationships (Metcalf and Benn 2012).

Through constant engagement, the project leader managed to build long-standing relationships that provided a platform to discuss the Equity Method. The concept of respect as a means of building trust is a potential area of expansion for RL research, as such relationships are critical in allowing robust discussion followed by climate action.

4.1.3 Unpacking Complexity to Ease the Mind Shift Required for Climate Change

Several external interviewees indicated that they encountered resistance to climate action implementation due to the perceived complexity of climate change and mitigation but also because it requires a *mind shift*. The data showed that one of the key ways in which the project leader managed to engage stakeholders and obtain buy-in for climate change mitigation was through his ability to unpack the complexity of climate change science. More specifically, he was able to translate not only the impacts of climate change but also the benefits of climate action with regard to various sectors and business operations.

Maak and Pless (2006) stress that responsible leaders should reduce complexity. The data showed that responsible climate leaders require an understanding of the holistic context of climate change. This enables them to appreciate the interconnectedness of climate change and its impacts and, as a result, empowers them to engage sectors and organisations in an appropriate relational approach to climate change.

When communicating climate issues, leaders must take note of the nature of complex systems, a realistic view of human behaviour and wellbeing, the critical role of natural and social capital and the irreducible uncertainty surrounding these issues (MacTaggart and Lynham 2018; Metcalf and Benn 2012). Allen (2018) reiterates this as the first world shift required for sustainability leaders, a shift which implies leaders should see systems as open and dynamic rather than isolated and closed and that climate resilience requires taking note of social, environmental and economic systems (Allen 2018). Allowing stakeholders to understand their sector or company within an open system enables them to understand both its contribution to and role in mitigating climate change. Since understanding the broader environmental science context and developing systems thinking skills are principally highlighted for sustainability leadership (Rogers 2011), this is a potential area of expansion in RL research pertaining to climate action.

4.1.4 Finding a Common Middle Ground

Internal Interviewee 1 discussed the project leader's ability to compromise in terms of short- and long-term perspectives to ensure the buy-in and uptake of the Equity Method:

> ...but then also a willingness to ... compromise for progress in the right direction. Because at the end, the method was only possible when we compromised a bit. We compromised on how much the existing emissions would weigh, how much new growth would weigh...so there was a lot of compromise in getting a method going. (Internal Interviewee 1)

Finding the middle ground is mainly related to understanding and appreciating the context of the company needing to set the GHG emission target. Given Maak and Pless' concept that RL must balance short-term gain with long-term sustainability (2006), in the context of climate action, responsible leaders must be able to illustrate the imperative of immediate action to achieve long-term sustainability. The data suggested that responsible climate leaders focus on addressing long-term gain from a global perspective (as per Allen's suggested worldview shift (Allen 2018:215) through decisive short-term action.

4.1.5 Communicating the Vision as a Journey Instead of a Fixed Point

In discussing the critical role of the vision as part of the definition of RL, the project leader proposed that the vision, within the context of GHG emission target setting and the Equity Method, takes cognisance of three aspects:

1. The vision must be cognisant of the context of where the company is.
2. The leader must understand that change does not happen overnight.
3. The leader must be able to articulate the vision in terms of three components – where the company is, where they want to be and how they are going to get there.

This is specifically relevant in terms of understanding climate change as a wicked problem where there is both a need for immediate action and a continued delay in taking action. Maak and Pless (2006) stress that responsible leaders must be able to guide, using a vision as an essential tool in overcoming challenges to stakeholder engagement. The project leader managed to develop an end vision, appreciating the current context of the companies involved in the development of the Equity Method, and developed a methodology that would provide stepping stones to get to the desired end state. GHG emission target setting is not only about the end goal. It is also about determining a possible way in which to reach the end goal and communicating this "journey" to facilitate buy-in. Climate action requires long-term thinking, as illustrated by Allen's worldview shifts, where the third shift relates to long-term thinking (2018).

The case study confirmed the idea that the vision to drive climate action should be a process rather than a stoic end point. This is further supported by the notion that greater attention must be given to the generation of path-dependent policy interventions that can "constrain our future collective selves" (Levin et al. 2012:124). Further research is needed concerning the vision component of RL, which could be useful in integrating long-term and systems-related perspectives to develop achievable and practical guidance within the context of climate change.

4.2 Key Roles to Drive Climate Action

The following provides an overview of the key roles emerging from the interview as relevant to driving climate action, with a view to expanding the roles described by Maak and

Pless (2006:109) and Pless (2007:438–448) for a climate context.

4.2.1 The Leader as a Climate Action Visionary

The development of a vision and the subsequent role of the leader as a visionary is an important part of RL. The data revealed the significance of the project leader's role as a visionary in driving the development of the Equity Method and also in garnering support for the principle that GHG target setting in developing country contexts should be based on common but differentiated responsibilities. This is aligned to Maak and Pless' (2006) role description of the leader as a visionary. They note that a vision supported by stakeholders gives direction and advice that it should be built on the principle of shared value creation.

Within the context of climate change, three key areas emerged from the data that could expand the current understanding of the visionary role of the leader in RL:

1. Recognising the complexity of climate change and contextualising this in the vision.
2. Recognising that the vision and the process of achieving it must allow for flexibility in order to evaluate or re-evaluate progress.
3. Integrating the priority of long-term gains in relation to short-term needs in the vision.

The first relates to the project leader's ability to appreciate the complexity of the problem. Climate change exemplifies the fact that nature is an open system with a myriad of interrelated processes (Murphy 2009). Unpacking the technical aspects includes recognising the integrated nature of climate change.

Secondly, the project leader was able to re-evaluate progress in terms of the final vision. This speaks to the urgency of immediate action while re-directing the process to reach the desired end state. The project leader noted that, within the context of the IPCC's latest report and the need to achieve a 1.5 °C rather than 2 °C temperature increase, the Equity Method might not be sufficient to meet greater climate ambition:

> If we want to go to net neutral by 2050, then we have to have … disruptions. So it means that when we started to work on the Equity Method, it was the right thing to do. But we have got new insights, we have got new perspectives and we have to adjust our end goal. Our end goal has gone from a 2 degrees to net neutral world. So it's a fundamental shift in the end goal. (Project leader)

Finally, the vision must speak to the long-term benefits of climate action in relation to the short-term and immediate needs of developing countries. Recognising the interrelationships between social, economic and environmental spheres and linking these to a long-term vision which speaks to a sustainable end state will become increasingly important (Kempster and Carroll 2016:8). The vision must create a shared meaning to facilitate buy-in and ultimately drive action (Bennis and Thomas 2002).

4.2.2 The Leader as a Storyteller: Moving Away from Doomsday Prophecies

Good communication skills emerged as crucial for engaging stakeholders and implementing GHG emission reduction targets. Communication also played an important role in deciphering the complexity of climate change in relation to the various sectors in which the companies operate. The notion of "connecting the dots" pointed to a need to move away from doomsday prophecies around climate change. Moser and Dilling (2011:164) argue that, while the use of fear in climate change communication manages to grab attention, it is not a behavioural change driver. Redekop and Thomas (2018) similarly noted that fear is disempowering and switches people off. Aligned with RL literature, the ability to be a change agent within the context of driving climate action, therefore, relies on the ability to communicate in a way that results in buy-in and changing behaviours, not in switching off or isolating the intended audience. Research is needed to expand the role of responsible climate leaders as storytellers who focus on solution-based alternatives and adaptive capacity. Responsible leaders must anchor communication in real-world practices rather than theoretical tales (Redekop and Thomas 2018). This speaks directly to the need to expand RL theory with regard to climate change in terms of creating a practical and appropriate vision. Redekop and Thomas hold that the focus of climate communication must be on *pragmatic arguments*, linking the impacts of climate change to issues such as energy security or health and related localised solutions.

4.2.3 From Change Agent to Climate Change Agent

Maak and Pless (2006:112) state that leaders as change agents are "responsible for mobilising stakeholders, building and sustaining commitment among followers through ongoing sense-making activities, reducing complexity and anxiety, and ultimately, keeping momentum in times when change causes insecurity and disorientation". This role may be expanded in the climate context given Voegtlin's (2016) description of RL as the best practice that exceeds legal requirements and compliance guidelines. Climate change, as a *wicked problem*, requires urgent action, with leaders moving into the realm of decisive, practicable commitments. The Equity Method was developed to move outside of the bounds of compliance and made a case for the best practice, practicable GHG target setting methodologies in the context of developing countries.

4.3 Key Drivers Required for Climate Action

The following section considers the most prominent drivers for responsible climate leadership that emerged from the interviews.

4.3.1 Sense of Care: Finding Solutions

Kempster and Carroll relate RL to the duty of care (Kempster and Carroll 2016), and Pless (2007) similarly identifies a sense of care as a normative driver for RL. The interviewees perceived the project leader as motivated to develop and implement the Equity Method due to his sense of care. The project leader confirmed this, indicating the recognition of his duty of care towards his fellow (wo)man as a driver to effect climate action. Bennis and Thomas (2002) state that leaders can develop meaning out of devastating events, typically such as the hardships that result from climate change. Leaders are further able to identify solutions or a plan of action to address these issues (Bennis and Thomas 2002). Similarly, climate action is not just about caring but about identifying and developing appropriate solutions to difficult challenges, as the Equity Method demonstrates.

4.3.2 The Need for Climate and Social Justice

Pless (2007) identifies the need for justice as a normative driver of RL. This refers to a fundamental human need for fairness and a moral framework as a basis for human interaction (Pless 2007). Several interviewees indicated that the need for justice is a critical driver in the development of the Equity Method, while the project leader specified the need for justice as his motivation for developing the Equity Method. A difference emerged, however, between Pless' description and the interviewees' understanding of justice.

Usually referring to the human sphere, justice must now be expanded in the context of climate change to include the biosphere (Satterwhite 2018). Satterwhite emphasises that human justice cannot be separated from justice in the environmental context and vice versa, and as such justice impacts on the social as much as on the environmental dimension, with obvious implications for leadership (Satterwhite 2018).

Given the context of the case study, the data suggested that the need for justice with regard to climate action relates to the principle of common but differentiated responsibilities under the Paris Agreement. Developing countries face both developmental challenges and the impacts of climate change, even though they can be argued to bear the least responsibility for climate change, having typically not benefited from centuries of fossil fuel use. The combustion of fossil fuels is widely recognised as one of the key drivers of both socio-economic development in developed countries and the global greenhouse effect, resulting in climate change.

Climate justice and social justice are, therefore, inseparable. Responsible leaders within the climate change context should understand that enabling developing countries to achieve the SDGs will require balancing GHG emissions with developing the necessary infrastructure to achieve the goals. As Uddin (2017:106) states in his exploration of this problem, the argument of whether developed or developing nations should accept more responsibility for climate change is often called the North–South debate in global environmental politics. This is a complex debate that is far from settled, but what is at least clear is that, however relevant Pless' driver may be, in a climate context the need for justice needs elaboration to capture the intricacies related to climate action within developing countries.

4.3.3 Courage to Address Climate Change

One theme emerging from the interviews that differs from Pless' drivers for RL is courage, although Voegtlin (2016) mentions courage as part of the ethical literacy of RL. Several interviewees noted the project leader's courage in publicly raising concerns regarding the fairness of the SBTi methodologies and proceeding to develop a methodology (the Equity Method) that aims to be just in setting GHG emission targets in developing countries.

Murphy (2009) notes that proactive future-orientated decisions in such a context require extraordinarily courageous and visionary leadership. Mirvis (2010) similarly describes several stakeholders that worked on a smarter city concept. Facing a lack of guidance on how to go about this, the team had to call on conviction, courage and a willingness to invent as things progressed, which resulted in innovation as the teams built internal capacity to manage the project (Mirvis 2010). In the climate context, the data pointed to the need for innovation due to the uncertain and urgent nature of climate change.

4.4 Towards a Conceptual Framework for Responsible Leadership in the Context of Climate Change

Figure 8.2 provides a proposed conceptual framework for responsible climate leadership in the context of developing countries, as developed by the authors based on this study. The framework starts with the components of leadership as defined by Northouse (2016:6) and links these to a summary of Maak and Pless' definition of RL while integrating the key findings and its analysis in terms of climate action in the context of the case study. The framework shows how responsible climate leadership for developing countries could potentially address the challenges of climate change as a wicked problem.

5 Conclusion

The study found that elements of responsible leadership contributed to driving climate action in the private sector through the development and application of the Equity Method. This method aims to enable companies, as critical non-state actors, to address climate change by setting GHG emission targets that do not sacrifice the developmental priorities in developing countries. It relies on active stakeholder engagement to achieve its objectives and ensure buy-in from a wide range of stakeholders. The Equity Method illustrates the most appropriate roles and drivers of RL theory that could serve as a framework for climate action.

The *key challenge* to RL in terms of climate change relates to balancing long-term gain over short-term and immediate needs. Key aspects that could contribute in overcoming this challenge include building and maintaining relationships of trust, unpacking the complexity of climate change and highlighting the interrelated

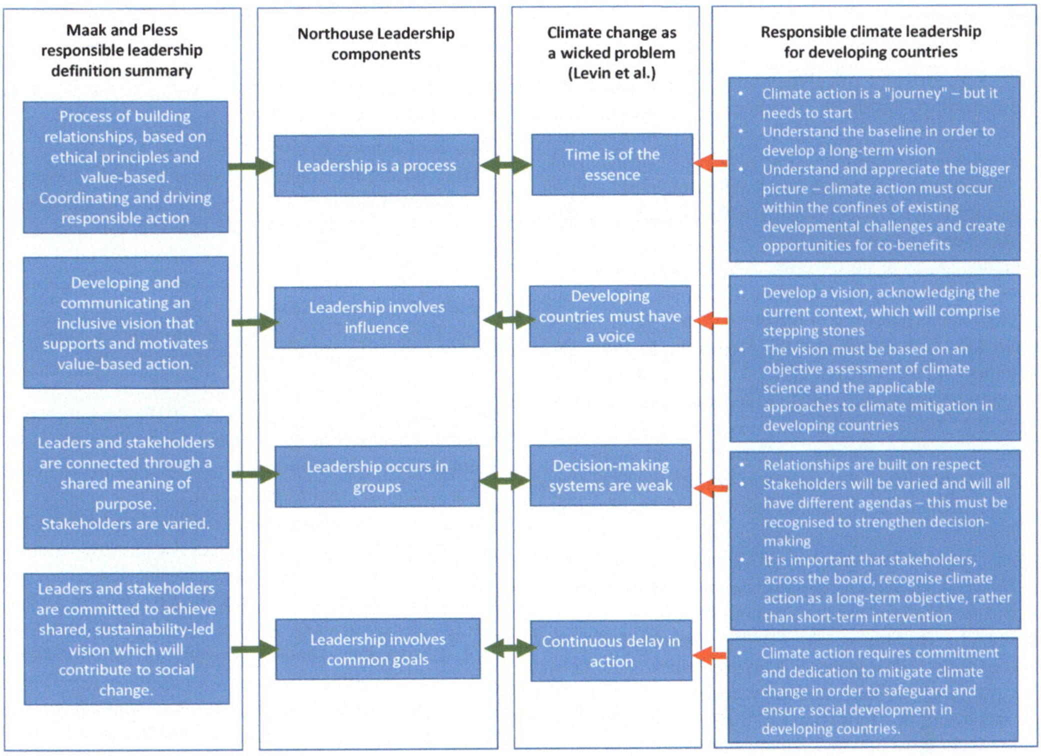

Fig. 8.2 Concept framework to inform responsible climate leadership in developing countries. (Source: Developed by the Authors (2019))

nature of climate change to various sectors while being able to find common middle ground.

Critical roles of responsible leaders in climate action emerged as (i) the leader as a climate action visionary, (ii) the leader as a solution-driven storyteller and (iii) the leader as a climate change agent moving into the best practice with regard to climate action.

In terms of *drivers*, a sense of care emerged as most relevant to responsible leadership and climate action, both in terms of climate change awareness and in terms of finding appropriate solutions. Another driver is the need to recognise the interface between social and climate justice and, finally, have the courage to proactively and consistently create awareness and affect climate action.

Due to its focus on stakeholder engagement on the one hand and its described roles and the drivers on the other, RL is suitable to addressing *wicked* climate challenges. RL practices can encourage a multi-objective assessment of development challenges and as such deliver effectively on the envisioned partnerships under the SDGs. This is specifically relevant to SDG target 13.2: to integrate climate change mitigation as part of national policies, strategies and planning. The private sector, as a non-state actor, has a significant role to play in mitigating climate change.

The Equity Method relies on active stakeholder engagement to achieve its objectives and ensure buy-in from a wide range of stakeholders. The development of the Equity Method could, therefore, potentially illustrate the most appropriate roles and drivers of RL theory that could serve as a framework for climate action.

This research suggested a conceptual framework for climate action (SDG 13) in the developing country context as it pertains to responsible leadership. While RL is an appropriate lens through which to drive climate action, there is room for expansion to develop a robust description of responsible climate leadership for developing countries.

References

Allen, K.E. (2018). Critical internal shifts for Responsible Leadership. In: B.W Redekop, D.R. Gallagher & R. Satterwhite (Eds.), *Innovation in Environmental Leadership: Critical Perspectives*. New York: Routledge. https://bookshelf.vitalsource.com/#/books/9781351795388/cfi/6/46!/4/2/2/2/4@0:0. Accessed 25 October 2019.

Andrade, J.C.S. & de Oliveira, J.A.P. (2015). The Role of the Private Sector in Global Climate and Energy Governance, *Journal of Business Ethics*, 130.

Antonakis, J., Schriesheim, C., Donovan, J., Pillai, K., Pellegrini, E. & Rossomme, J. (2004). Methods for studying leadership. In: J. Antonakis, A.T. Cianciolo & R.J. Sternberg (Eds.), *The Nature of Leadership, Second Edition*. California: Sage Publications Inc.

Barlow, C.A. (2012). *Interviews in Encyclopedia of Case Study Research*. Thousand Oaks: SAGE Publications, Inc.

Bennis, W.G., & Thomas, R.J. (2002). *Geeks and geezers: How era, values, and defining moments shape leaders*. Boston: Harvard Business School Press.

Corbin, J., & Strauss, A. (1990). Grounded Theory Research: Procedures, Canons, and Evaluative Criteria. *Qualitative Sociology, 13*(1).

De Jongh, D., & Möllmann, C. (2014). Market Barriers for Voluntary Climate Change Mitigation in the South African Private Sector. *South African Journal of Economic and Management Sciences, 17*(5).

Falkner, R. (2016). The Paris Agreement and the new logic of international climate politics. *International Affairs, 92*(5),1107-1125.

Frangieh, C.G., & Yaacoub, H.K. (2017). A systematic literature review of responsible leadership. *Journal of Global Responsibility, 8*(2),281-299.

Ge, M., & Ross, K. (2019). Which Countries Have Long-term Strategies to Reduce Emissions?. Resource document. https://www.wri.org/blog/2019/09/which-countries-have-long-term-strategies-reduce-emissions. Accessed 25 October 2019.

ICAT. (2018). *Initiative for Climate Action Transparency: Non-State and Subnational Action Guidance*. Washington DC: World Resources Institute

Immink, H. (2019). *GHG forecasting and target setting using an ex-post analysis*. Dissertation. Stellenbosch: Stellenbosch University.

IPCC. (2018). *Global Warming of 1.5 Degree Celsius: An IPCC special report o the impacts of global warming of 1.5 degree Celsius above pre-industrial levels and related global GHG pathways, in the context of strengthening the global response to the threat of climate change, sustainable development, and efforts to eradicate poverty (Summary for Policy Makers)*. Geneva: Intergovernmental Panel on Climate Change.

Kempster, S., & Carroll, B. (2016). Introduction: Responsible leadership – realism and romanticism, In: S. Kempster & B. Carroll (Eds.), *Responsible Leadership: Realism and romanticism*. New York: Routledge.

Krabbe, O., Linthorst, G., Blok, K., Crijns-Graus, W., van Vuuren, Detlef P., Höhne, N., Faria, P., Aden, N., & Pineda, A. C. (2015). Aligning corporate greenhouse-gas emissions targets with climate goals. *Nature Climate Change, 5*(12),1057-1060.

Levin, K., Cashore, B., Bernstein, S., & Auld, G. (2012). Overcoming the tragedy of super wicked problems: Constraining our future selves to ameliorate global climate change. *Policy Sci, 45*,123-152.

Maak, T. (2007). Responsible Leadership, Stakeholder Engagement and the Emergence of Social Capital. *Journal of Business Ethics,74*,329-343.

Maak, T., & Pless, N. (2006). Responsible Leadership in a Stakeholder Society: A Relational Perspective. *Journal of Business Ethics,66*,99-115.

MacTaggart, R.W., & Lynham, S.A. (2018). An integrative literature review of Responsible Leadership: Knowns, unknowns, and implications. *Journal of Leadership, Accountability and Ethics, 15*(3).

Metcalf, L., & Benn, S. (2012). Leadership for Sustainability: An Evolution of Leadership Ability. *Journal of Business Ethics, 112*(3),369-384.

Meuser, J.D., Gardner, W.L., Dinh, J.E., Hu, J., Liden, R.C., & Lord, R.G. (2016). A Network Analysis of Leadership Theory. *Journal of Management, 42*(5),1374-1403.

Mirvis, P.H. (2010). *Responsible Leadership Emerging: Individual, Organizational, and Collective Frontiers*. Boston: Boston College Center for Corporate Citizenship.

Moser, S.C. & Dilling, L. (2011). Communicating Climate Change: Closing the science-action gap, In: Dryzek, J.S., Norgaard, R.B. & Schlosberg, D. (eds.), The Oxford Handbook of Climate Change and Society, Oxford, Oxford University Press.

Murphy, R. (2009). *Leadership in disaster: Learning for a future with global climate change*. Quebec: McGill-Queen's University Press.

Northouse, P.G. (2016). Leadership : Theory and Practice. Seventh edition. International student edition. ed, Los Angeles, SAGE.

Pless, N.M. (2007). Understanding Responsible Leadership: Role Identity and Motivational Drivers. *Journal of Business Ethics, 74*(4),437-456.

Promethiume Carbon. (2018). *Equity Method for GHG Target Setting*. Johannesburg: Promethium Carbon.

Redekop, B.W., & Thomas, M. (2018) Climate Change Leadership: From tragic to comic discourse. In: B.W. Redekop, D.R. Gallagher, R. Satterwhite (Eds.), *Innovation in Environmental Leadership: Critical perspectives*. New York: Routledge. https://bookshelf.vitalsource.com/#/books/9781351795388/cfi/6/38!/4/2/2/2/4@0:0. Accessed 25 October 2019.

Rogers, K.S. (2011). Leading sustainabililty. In: W.H Mobley, M. Li, Y. Wang (Eds.), *Advances in Global Leadership*. Bingley: Emerald Books.

Satterwhite, R. (2018). A Case for Universal Context: Intersections of the biosphere, systems, and justice using a critical constructionist lens. In: B.W. Redekop, D.R. Gallagher, R. Satterwhite (Eds.), *Innovation in Environmental Leadership: Critical perspectives*. New York: Routledge.

SBTi. (2018). Science-Based Targets. https://science-basedtargets.org/about-the-science-based-targets-initiative/. Accessed 28 September.

Stern, N. (2015). *Why are we waiting: The logic, urgency, and promise of tackling climate change.* Cambridge: The MIT Press.

Uddin, Md. K. (2017). Climate Change and Global Environmental Politics: North-South Divide. *Environmental Policy and Law 47*(3-4), 106-114.

UN. (2019) *Global indicator framework for the Sustainable Development Goals and targets of the 2030 Agenda for Sustainable Development.* New York: United Nations Press.

Voegtlin, C., Patzer, M. & Scherer, A.G. (2012). Responsible Leadership In Global Business: A New Approach to Leadership and its Multi-Level Outcomes, *Journal of Business Ethics*, 105.

Voegtlin, C. (2016). What does it mean to be responsible? Addressing the missing responsibility dimension in ethical leadership research. *Leadership, 12*(5),581-608.

Waldman, D.A. & Galvin, B.M. (2008). Alternative Perspectives of Responsible Leadership, *Organizational Dynamics, 37*(4),327-341.

WEF. (2019). *The Global Risks Report 2019.* Geneva: World Economic Forum.

9 Leadership Capabilities for Successful Implementation of SDG 7 Targets at Energy Company X

Tatenda Mangondo and Andani Thakhathi

Abstract

This chapter sought to identify initiatives that led to the realisation of several Sustainable Development Goal (SDG) targets by a company referred to as "Energy Company X", a renewable energy company operating in Southern Africa. The study explored phenomenological perceptions of leadership capabilities that enabled Energy Company X to successfully implement a renewal energy project to facilitate the transition towards clean energy in the region. The insights offered herein were generated through an exploratory qualitative case study research design drawing on semi-structured interviews with key expert participants who played critical leadership roles which contributed to the project's success. The case study was augmented by documentation including policy briefs, website pages, research publications and company-specific documents. Several SDG targets were simultaneously realised even though the project's primary pursuit was SDG 7 – Clean Energy. This demonstrates how the pursuit of one SDG results in the concurrent realisation of other related SDGs, thus confirming their interdependence. The findings revealed that transformational attributes, creating shared value (CSV) and bottom-up collaboration (BUC) were central leadership capabilities that enabled the project's success. This reveals that collective – shared and collaborative – leadership is best suited for the Southern African context, thereby echoing the efficacy of Ubuntu-based collective leadership approaches that are bottom-up in nature.

Keywords

SDGs · Renewable energy · Transformational leadership · Shared value

1 Introduction and Background

According to the United Nations Department of Economic and Social Affairs (UNDESA) Population Division (2019), as of mid-2019, the world's population was about 7.7 billion. However, the global population growth is estimated to reach 8.5 billion in 2030, 9.7 billion in 2050 and 10.9 billion in 2100. Eastern and Southeastern Asia were the two most populous regions in 2019, but in the coming decades, most of the global population will be concentrated in

T. Mangondo
Dainfern Golf Estate, Midrand, South Africa

A. Thakhathi (✉)
University of Pretoria, Pretoria, South Africa
e-mail: andani.thakhathi@up.ac.za

G. Nhamo et al. (eds.), *Sustainable Development Goals for Society Vol. 2*, Sustainable Development Goals Series, https://doi.org/10.1007/978-3-030-70952-5_9

sub-Saharan Africa. A growing population equates to an increased demand for and utilisation of natural resources that, eventually, will be depleted, leaving the demands of billions of people unmet (Schincariol 2017).

This current era has been termed "Anthropocene" (Sachs 2012), the era in which humans have pushed the earth's planetary boundaries through climate change, stratospheric ozone depletion, flooding, ocean acidification, biochemical flows, atmospheric aerosol loading, freshwater use, land system change, chemical pollution and loss of biodiversity (Rockström et al. 2009; Sachs 2012). The growing population's activities and resource use have drastically increased, posing challenges, compromising the environment and giving rise to the possibility of depleting resources to the extent of risking future generations (Sachs 2012).

Consumption and production patterns are unsustainable, calling for urgent remedying to ensure sustainability of the planet, to combat the damage done to the environment and to ensure that future generations that can also enjoy the earth and all it has to offer (Akani 2017). This requires re-evaluation to ensure sustainable development in the form of Sustainable Development Goals (SDGs). The emergence of SDGs resulted from concerns around the current rate of economic growth and human development, seemingly coming at the cost of the environment to the detriment of future generations (Akani 2017).

A/RES/41/128 Article 1 of the United Nations Declaration on the Right of Development (1986:186) states that every human being has the right to development which "is an inalienable human right by virtue of which every human person and all peoples are entitled to participate in, contribute to, and enjoy economic, social, cultural and political development, in which all human rights and fundamental freedoms can be fully realised". For this right to become attainable to all, the stress humanity is currently placing on planet earth through rising standards of living, pollution, overconsumption and overproduction must reduce or change considerably. We need to adopt mindfulness through lifestyle and behavioural change towards sustainable consumption and production (Akani 2017; Dhandra 2019).

The most used definition of sustainable development defines it as "development that meets the needs of the present without compromising the ability of future generations to meet their own needs" (Brundtland 1987). Should we continue the sustainable development trajectory of business as usual, the UN's SDGs may be the best solution. The current era requires a shift away from business as usual to one of sustainable consumption, production and development (Sachs 2012).

The transformation required by the SDGs is a probable justification for a relationship between leadership and the role of leaders as "transformational and inclusive" leadership is implied as necessary to deliver on the SDGs (Hajer et al. 2015). This type of leadership is an under-researched and under-theorised area with a "gap" that needs to be filled. The body of knowledge on leadership is a debated field (Carroll and Nicholson 2014). There is no grand or "all-inclusive theory' on leadership; there are constructs of leadership such as transformational leadership and methodical classifications of how leaders should behave (Riggio and Harvey 2011).

Furthermore, there are multiple theories of leadership because none is able to fully define what leadership is and establish how one becomes a leader; the available literature does not agree upon a single definition of leadership and has different perspectives on the role of a leader (Grint 2005). The existing theories of leadership are suited to Euro-American and Western countries and are oftentimes duplicated by African scholars (Chime and Enor 2016). The case study presented in this chapter hopefully contributes to shedding light on the "gaps" in the current leadership thinking and possibly lays the foundation for new theories of leadership which are more suited to the African context.

The SDGs were established by the United Nations General Assembly (UNGA). The goals originated from the United Nations Conference on Sustainable Development in Rio de Janeiro in 2012 through a convergence of various agencies and conferences such as the Millennium Declaration, 2005 World Summit Outcome, Rio Declaration on Environment and Development and the United Nations Development Programme (UNDP) and were adopted by UN member states in 2015.

These universal goals form part of an agenda to reach sustainable development for all countries, developed and developing, by 2030 based on the three levels of sustainable development: environmental, social and economic. Eradicating poverty in all its forms and dimensions is the purpose of the agenda as it is the greatest global challenge and hindrance to achieving sustainable development (Sachs 2012; UNDESA Population Division 2019). The 2030 Agenda is also guided by principles and other instruments of the UN Charter, Universal Declaration of Human Rights (Akani 2017; UNDESA Population Division 2019).

The SDGs are 17 goals with 169 targets and 230 indicators, which are used to monitor and assess successful implementation of the goals (Shula et al. 2019). The SDGs took over and built on from where the Millennium Development Goals (MDGs) were unable to achieve set targets. The MDGs were established in 2000 under the UN Millennium Declaration to fight poverty in all its forms and have 8 goals and 68 indicators that were to be achieved by 2015 (UN 2015a). The purpose was to eradicate extreme poverty and hunger, improve maternal health, reduce child mortality, achieve universal primary education, prevent and combat diseases, global partnership for development, promote gender equality, empower women and ensure environmental sustainability (Solberg 2015; UNDESA Population Division 2019; World Health Organization 2015).

The purpose of the SDGs is to move beyond the MDGs target of eradicating poverty and achieving development for all by 2015, achieving universal sustainable development inclusive of all countries by 2030 and providing solutions and guidelines to address shared global challenges (Sachs et al. 2019; Stevens and Kanie 2016; UN 2015a, b). According to Sachs (2012), sustainable development has a triple bottom line approach to the wellbeing of people. Globally, all nations strive for social inclusion, economic development, environmental sustainability and good governance, albeit goals and objectives differ internationally, between and within societies (Sachs 2012).

Being established and guided by a universally inclusive and transformational vision (Hajer et al. 2015), the SDGs require a change in our thoughts and habits so that the choices we make today do not infringe upon future generations (UN 2015b), calling for prosperity for all, a healthy thriving planet, global partnership and a peaceful world, one that is all-inclusive, with equal opportunities for all, thus leaving no one behind (UNDESA Population Division 2019).

The implementation of the SDGs calls for a revitalised global partnership which facilitates engagement on a global scale in support of the implementation of the SDG targets. This partnership implies multi-stakeholder engagement of civil society, businesses and the private sector, governments, the UN, intergovernmental organisations and other actors, therefore encouraging public-private partnerships (Emerson et al. 2012; Shula et al. 2019; Stibbe et al. 2018). Leaving no one behind therefore requires a global effort to ensure the inclusion of each country, government and individual.

The 2030 Agenda also calls for implementation at national and regional levels. Governments are encouraged to adopt cohesive nationally owned sustainable development strategies substantiated by integrated national financing frameworks, reiterating the responsibility of governments for their own national, social and economic development, reinforced through national policies (National Planning Commission 2011, 2012; Shula et al. 2019). The agenda respects each country's policy space and leadership for the implementation of policies for sustainable development and poverty eradication, policies which adhere to international standards and commitments. Implementation at the national level would feed into implementation at the regional level through lessons learnt, sharing of best practices and peer learning and reviews. The regional level would use national-level reviews to contribute at a global level (Shula et al. 2019; Stibbe et al. 2018; UNDESA Population Division 2019).

This study focuses on Energy Company X in South Africa, one of the 112 Independent Power Producers (IPPs) of the Renewable Energy Independent Power Producer Procurement Programme (REIPPPP), which was established as a public-private partnership (PPP) between the then Department of Energy (DoE) and Eskom, Development Bank of Southern Africa (DBSA) and the National Treasury (NT) in 2010. Subsequently, the Independent Power Producer Office (IPPO) was formed to procure energy and monitor projects for the 20-year contracts in the form of Implementation Agreements (IA) signed by the IPPs, IPPO, DoE, DBSA and NT (DoE, NT and DBSA 2015).

In late 2007–2008, energy security challenges emerged in South Africa, and the country was unable to meet its electricity demand with a 10% capacity shortfall. Eskom, the country's power utility, had to introduce load shedding with hopes of stabilising the national power grid (DoE, NT and DBSA 2015; Jain and Jain 2017; Sebitosi 2008). Load shedding led to intervention by government through the establishment in 2010 of an Inter-Ministerial Committee (IMC) to address the electricity supply constraints experienced because of severe financial constraints in Eskom during 2008–2010.

The increase in demand for electricity supply called for energy efficiencies and various government interventions through the adaptation of policies to enable policy frameworks which would add renewable energy to the national electricity generation energy mix. The renewable energy sources that were found to be feasible in South Africa are solar, wind, geothermal, hydropower, biomass, waste to energy and tidal wave energy (Jain and Jain 2017).

The REIPPPP was primarily designed to respond to the energy security challenges which resulted in load shedding for the period 2008–2010 as this was recognised as the quickest way to bring much-needed additional electricity generation onto the grid to ensure security of electricity supply while ensuring a diversified national electricity generation energy mix (Jain and Jain 2017). The REIPPPP would also respond to the call from the NDP to meet the nation's development imperative through the Economic Development Framework, developed for the REIPPPP, focusing on issues such as employment creation, Social Economic Development (SED) and Enterprise Development (ED) by promoting Small, Medium and Micro Enterprises (SMMEs) and broadening economic ownership (DoE 2015).

Energy Company X is the focus of this case study as one of the IPPs contributing to the energy mix and adding capacity to the national grid. With energy production in over 29 countries worldwide, a focus on development of renewable energy has, over the years, become more focused on green and sustainable development through an energy transition by producing clean and cheaper energy and including ED and SED as a part of the project's agenda (Energy Company X 2016, Energy Company X 2017a). The company is also committed to social development of communities in which its independent power projects are located. Though its primary focus is on clean affordable energy production, Energy Company X is an avid supporter of building and uplifting local communities by addressing their priority needs in key areas of education, skills development, youth development, health care, welfare provision for the aged and infrastructure building, among other initiatives (Energy Company X 2017b, Energy Company X 2019a).

It is through the ED and SED initiatives that Energy Company X has realised some SDG targets. Key to successful project execution is the role of leadership through utilised leadership capabilities. From this background, the research focused on those who played a leadership role in project delivery in the numerous initiatives. For the purposes of this study, a leader is defined as an individual placed in a position of power to implement an initiative and is able to get buy-in from and encourage community engagement; leaders are, therefore, the individuals who had decision-making powers in the project and were able to communicate the vision effectively and motivate others to adopt the vision (Energy

Company X 2016, Energy Company X 2017b, Energy Company X 2019b).

2 Literature Review

2.1 SDGs and Leadership Theories

The assumption is that, as a result of the interconnectedness of the SDGs, together with the various stakeholders and actors at local and international levels, the successful implementation of the 2030 Agenda necessitates leadership to endorse and rally actors, both big and small, in implementing and achieving these goals. The complexity of achieving the goals and the challenges we face in the world today require practical solutions. There is therefore a need to change the way we act and think, which calls for transformational and collaborative leadership as the SDGs were created for the purposes of transforming our world through collaboration among and between communities, governments and nations at local, regional and global levels (Leal Filho et al. 2019). Indisputably, the current era calls for leadership, leaders and leadership styles that consider and incorporate the purpose and objectives of SDGs through coordination, transformation and collaboration of multi-stakeholders (Shula et al. 2019).

2.2 SDGs and Transformational Leadership

The SDGs call for transformational leadership by requiring introspection and re-evaluation of self-interest for the benefit of society, particularly those in developing regions and the less advantaged (Akani 2017). For the SDGs to succeed, those championing change must genuinely believe in sustainable development as being for the greater good. Foresight, courage and the ability to inspire others with a moral conviction for the improvement of society are needed by those championing for the SDGs to lead others in the vision and purpose, thereby implementing the SDGs. Leaders therefore have an important role to play here (Leal Filho et al. 2019).

The UN has thus far demonstrated transformational leadership and the need for it globally by endorsing the SDGs (Akani 2017:182). This argues for transformational leaders being able to communicate strategic goals and give direction to achieve sustainability (Berson et al. 2004). This suggests the UN recognises the importance of partnership and multi-stakeholder engagement of people and nations, considering the world's resources – which entail technological advances and a global economy, all impacting on the planet and development – being expendable, limited and unable to meet the demands of 7.4 billion people. These resources require global optimisation in use to avoid further pressure on the earth or depletion of available resources (Akani 2017; Rockström et al. 2009; Sachs 2012).

This calls for focus on the five Ps, people, planet, prosperity, peace and partnership, meaning prosperity for all people, a healthy thriving planet and a peaceful and just world that calls for global partnership and collaboration (UNDESA Population Division 2019). This will facilitate global engagement in support of the implementation of the SDGs' 169 targets through multi-stakeholder engagement of civil society, businesses and the private sector, governments, the UN, intergovernmental organisations and other actors, therefore encouraging public-private partnerships (Stibbe et al. 2018).

2.3 SDGs and Collaborative Leadership

Collaborative leadership is relevant to the design and implementation of the SDGs. Different actors partnered to form the SDGs to ensure sustainable development for all (Leal Filho et al. 2019). These actors have, so far, included the UN, governments, civil society, multi-stakeholders, nations and other non-governmental organisations (NGOs), providing and encouraging collaboration across sectors to respond to sustainable development and successfully achieve the 2030 Agenda globally (Stibbe et al. 2018; UN et al. 2019).

Collaborative or shared leadership is based on a premise of bringing the right people together to work towards a common goal. It has been argued to be an alternative to the traditional top-down leadership approach. In contrast to transformational or visionary leadership, it allows for open communication and empowers those working towards effective change, as their perspectives and insights are considered and incorporated into the decision-making process. It largely involves the collaboration of different stakeholders to address concerns that affect a community or organisation. The coming together of different stakeholders, working across functions or sectors, enables the addressing of complex issues, promotes community building and allows for social capital building by creating opportunities for dialogue, knowledge transfer, collective responsibility and action (Kramer and Crespy 2011).

By different stakeholders working together, hierarchies are not as important as actors share responsibility and control; leadership functions are shared among peers or group members. Collaborative leaders foster collaborative relationships with positive outcomes, increased productivity and satisfaction because of collective leadership, stakeholder engagement and shared decision-making. This type of leadership brings diverse groups and different competencies together and can be used or implemented at a community, organisational, governmental or international level, allowing for strategic decisions to be made and implemented to deal with universal societal concerns (Chrislip 2002).

Collaborative leaders enable collaboration by specifically defining public concerns. They steward individuals into action rather than merely delegating tasks or giving directives. It is a process not limited to certain individuals but all group members (Emerson et al. 2012). This fosters relationships among peers by creating open dialogue on issues, which replaces hierarchy (Kramer and Crespy 2011). Collaborative leaders empower followers by giving them authority to make decisions and placing responsibility on them to do so, creating value and a sense of purpose. Collaborative leadership encourages member-to-member exchange and not just leader-to-group member exchange like other leadership styles, resulting in group effectiveness, encouraging participation and more commitment to produce favourable outcomes (Zhang et al. 2018).

This study identified initiatives that led to the realisation of several SDG targets by Energy Company X. Furthermore, the study identified those that played critical roles in the initiatives and explored their perceptions of leadership capabilities that enabled successful project implementation. The findings of the case study were compared to contemporary leadership theories to confirm or expand existing knowledge of leadership and change. This study confirmed the probable justification for a relationship between leadership and the role of leaders in successful implementation of SDG targets

3 Methodology

The researchers conducted exploratory qualitative research to identify an initiative implemented by Energy Company X that led to the realisation of several SDG targets. The data collected from this case study guided the researchers in identifying leaders perceived to have played important roles in successful project delivery and the values, behaviours, competencies and traits that empowered them to play leadership roles in Energy Company X. This research approach was appropriate for this study because it allowed for the exploration of participant perceptions of leadership capacities utilised for successful project delivery, with a focus on understanding the views and experiences of practising leaders within Energy Company X, using theory to guide data analysis (Caelli and Mill 2003:2–3). The main research question answered by this chapter is: What leadership capabilities enabled successful project implementation in Energy Company X?

Qualitative research was carried out to gather data that is well suited for studies that occur within a context or an event, studies concerning organisations and work environments (Creswell 2017; O'Neil and Koekemoer 2016). Specifically, for this case study, research was conducted to identify leaders who played critical roles in successful project delivery. The focal point of the research was their perceptions of the identified behaviours, values, traits and capabilities that enabled them to play leadership roles contributing to successfully implementing the initiative.

Four participants were identified as having played critical leadership roles in Energy Company X, using purposive selection, with the main selection criteria being the possession of expertise and specialised knowledge relevant to the two research questions (Daniel 2012; Flick 2018; Guest et al. 2012). Expert/purposive selection was chosen to target a specific type of practitioner from within Energy Company X, namely, those that played leadership roles (the "experts"), in the SDG-oriented project's delivery. By targeting these role players, the research was able to yield knowledge based on their experience and involvement as leaders in Energy Company X's SDG-oriented project (Daniel 2012; Guest 2012). The identified leaders were two participants at the senior management level – the head of the Sustainability Department and a senior sustainability analyst – and two participants at community engagement level – a community liaison officer and a community engagement officer.

In-depth semi-structured interviews were the primary method of data collection. Telephonic, computer and internet-based interviews via Skype were the chosen interviewing methods because of geographical divides and convenience in terms of access to participants, time and cost constraints (Cole 2017; Parcell and Rafferty 2017; Stephens 2010). A mobile audio recording device was used to record interviews, and notes were used to recall what was discussed during the interviews (Parcell and Rafferty 2017).

Participants were asked the same standardised questions to obtain their perceptions on adopted leadership capabilities. Follow-up questions were asked to obtain as much data as possible and probing for further clarification on responses (Morris 2015; Rubin and Rubin 2005). The audio recordings were converted into interview transcripts in text form using NVivo (QSR International), a computer-assisted qualitative analysis software (CAQDAS) (Guest et al. 2012; Hilal and Alabri 2013; Silver and Lewins 2014). The transcripts went through extensive data analysis to arrive at findings (Kowal and O'Connell 2014).

As stated above, qualitative research in the form of an embedded single-case design was selected, specifically looking at Energy Company X as a whole, which, as the main unit of analysis, focused on the leaders' perceptions that led to successful project delivery (Scholz and Tietje 2002). This was to delimit the study to the overall success of a single project out of 112 projects in the entire REIPPPP and the company's realisation of SDG targets. Case studies allow for the focus on an organisation within a specific context during a period while gathering in-depth information using different approaches to data collection (Creswell 2017). Energy Company X was selected because of its successful alignment to SDG targets through its education, nutrition and renewable energy projects, to name a few.

To ensure rigour, the researcher represented participants' views, experiences and perceptions without imposing personal views on those of the participants. The researcher followed the criteria of credibility, transparency, transferability and volume and depth of data (Morgan and Ravitch 2018). The researcher ensured credibility by making transcripts and findings readily available to participants after the interview for review upon request. This allows participants to address any ambiguity or make changes to preserve credibility as the researcher seeks to ensure a true representation of participants' experiences in the study (Rheinhardt et al. 2018).

The researcher detailed the steps and choices made throughout the research process. Notes were made regarding any changes in the process, issues arising within the research design and any other key data. This ensures transparency and increases the credibility of the study (Rheinhardt et al. 2018).

Transferability ensures that the results of the study are applicable to other contexts. Thick descriptions representative of participants' perceptions of leadership capacities for successful project execution were used to ensure findings can be applied to other contexts (Rheinhardt et al. 2018). In line with ethical considerations, the researcher obtained institutional permission to conduct academic research on Energy Company X (Bell and Kothiyal 2018). On receipt of the signed informed consent forms from participants, the researcher adhered to confidentiality rules of keeping the identities of participants anonymous in the research report by assigning pseudonyms (Bell and Kothiyal 2018). The same was applied to interview audio recordings and transcripts.

The next section presents the emerging key findings.

4 Findings

The participants' responses show that leadership capabilities are important for successful project delivery and are summarised below. For the purposes of this study, leadership capabilities are defined as behaviours, skills, values, traits and competencies that enabled the study participants to play leadership roles (Coates et al. 2013), enabling them to further the goals and objectives of Energy Company X. Two central themes emerged as being pertinent for successful project implementation: creating shared value (CSV) and bottom-up collaboration. Subthemes also emerged and have been classed under the central themes as listed in Table 9.1.

Table 9.1 Leadership Capabilities

Creating shared value	Bottom-up collaboration
Accountability and ownership	Ability to listen
Adaptability and flexibility	Relating
Engage others	Teamwork
Passion	Self-awareness
Trustworthiness – openness and honesty	

Source: Authors' own design

4.1 Creating Shared Value (CSV)

Participants discussed the notion of CSV in detail and fervently mentioned it at any opportunity, demonstrating that it is not only a capability or value that Energy Company X subscribes to but is essentially the vision and a leadership style employed by the organisation. Participants described CSV as part of the organisation's "genetic make-up" and a driving force behind what they do, such as giving back to communities through corporate social investment. CSV was described as pertinent to successful project delivery by creating social and economic value and addressing the needs and challenges of society.

This shows the relationship between the success of business and social development or progress. Through the implementation of renewable energy projects, Energy Company X not only gained financially but also added value to the communities where power plants are located. Energy Company X's main goal was to build power plants and generate energy over a 20-year period. This entailed investing in infrastructure and ports, building roads and connecting communities to electricity grids. Communities benefitted owing to access to affordable and clean energy and job creation.

By advancing and CSV, financial success was achieved while providing solutions to the pressing needs and challenges of the community. This was possible through consultation with the municipality and the community at large by conducting community needs assessments to determine how best to address community conditions and pressing needs that could otherwise pose limitations and barriers to business.

Participants highlighted CSV in excerpts below:

> Making sure that whenever you get to the community, you leave them better off than when you found them. And how we do that, it's listening to our communities. We listen, we listen, we listen, we listen. And then we develop whatever projects that we come up with based on what, the communities said they want. (Participant 1)
>
> Understanding the community dynamics. Issuing studies, like a community needs assessment based on objectives to understand where the community

> is at that point in time, and aspirational objectives they want to reach. (Participant 3)
> So, what that means is whatever projects we identify, we first need to do what's called the needs analysis, meaning a sister study to understand what are the needs in that area because we try and not apply, a top-down approach, but instead of try a bottom-up approach. So, we rely on our community to tell us what it is they need, based on our very transparent budget. (Participant 4)

Consultations during community needs assessments fostered engagement between Energy Company X and the community. Engaging others as a subtheme under CSV allowed for constant community engagement that encouraged dialogue between the different stakeholders and further contextualised the understanding of community needs, concerns and priorities. This dialogue promoted accountability and shared ownership by both the community and Energy Company X. The community was encouraged to accept responsibility and take ownership for initiatives to put in place for their benefit as they were involved from the onset in identifying, developing, implementing and participating in the operations of the initiatives. Engagement allowed for understanding the existing conditions in communities, which legitimised Energy Company X and its business intentions by pursuing the shared value initiatives that promoted social progress. Participants emphasised the need to focus on the objectives of the company, the reason they were doing what they do and why they made the decisions they did to be accountable to the community.

The extracts below highlight community engagement, accountability and ownership:

> community ownership of the project, which starts with community. I think for any community development project to succeed; it must have a support of its community right from the beginning. (Participant 1)
> Communities all need to own their projects. They need to see and work towards the goal being the implementation of that project. So, we engage from negotiation stage, to, implementation stage, as far as even identifying the service provider who is going to implement the project. We engage at that level with local communities so that they can have a sense of ownership. (Participant 4)

The findings also revealed the importance of adaptability and flexibility as imperative skills for community engagement. These skills facilitated the ability to handle change, adopt new ways of doing things, compromise, navigate conflict and address challenges from increased community involvement due to the difference of opinions in discussions about which projects to implement. This aided in adapting to new situations, group and community dynamics, embracing innovative thinking and fresh ideas where SED initiatives are concerned:

> …adapt to new ways of doing things with different people and communities disagree in the middle of a project. So, it's important to be able to adapt to whatever situation that comes up which will take us to the next step. (Participant 1)
> …sit and evaluate, to critically evaluate, you know what can go wrong, what can go right so that you only have a contingency plan. And I think other capabilities also very important because I think most of us know that very few projects go as planned so you might be able to be fluid and flexible. (Participant 1)

Attention was drawn to having passion as a requisite and driving force for contributing to the uplifting of communities, working towards problem-solving and decision-making. The desire to make a difference drove that passion and encouraged commitment and enthusiasm from start to finish to ensure successful implementation. Participants underlined this as below:

> I think one of the things, that has shaped what we do, in my role here is the kind of leadership that I've put forward here, which is, which I want to call passionate leadership … it was the passion that drives us to step up to the plate and help make the bigger step forward in terms of assisting the communities in their development. (Participant 1)

4.2 Bottom-Up Collaboration

The participants advocated for a bottom-up approach and collaboration to achieve the desired outcomes for projects to benefit all stakeholders. Participants stated the importance of mobilising different stakeholders: municipalities, NGOs, managers and even schools to work together. It allowed sharing of information, innovative ways to identify and solve problems, promoting a bottom-up approach which encouraged diversity in new ideas, problem-solving and opinions. This approach encouraged more engagement and teamwork from all stakeholders and willingness to take accountability and ownership of work and initiatives. This cultivated more trust between Energy Company X, the communities and other stakeholders and advanced community buy-in for programs and initiatives. This finding is demonstrated in the excerpts below by Participant 4:

> We apply a bottom up approach in that we try and frame our discussions with our community members in such a way that it links to the, the imaging and the framework that we ascribe to of creating shared value.
> any stakeholder is welcome, and we then engage from there. So, in my view, it's a, it's a democratic decision where, we, we, we try and engage in seeing what is needed. We don't, um, assume we understand what is needed in that community.

In all the engagements and bottom-up collaboration with the different stakeholders, participants highlighted listening as an important attribute for successful project delivery. Listening attentively and actively allows others to voice their opinions and helps one to understand what it is they are communicating. This is the best way to encourage community buy-in and ownership. The excerpt below underlines this:

> Learning to listen, learning to empathise, but also being proactive in yielding to their request, and understanding again their lived experience and how we can assist and find a synergy, in developing community. (Participant 3)

On a personal level, participants referred to self-awareness as an essential trait for project delivery and for any leader to have. Knowing one's own personality, emotions, needs and values and possessing emotional and social intelligence in interaction with others play a huge part in better decision-making, building relationships and strengthening communication. It helps one realise the impact of their decisions on others. Below, participants explained the importance of self-awareness:

> experience has really forced me to focus on the world around me, again, you know, not just in the bubble that I live in with my loved ones in and the people that I know. I think yeah. Um, um, I am more focused on the world around me, much better than ever. (Participant 1)
> It's one thing for you to be aware and in the best action that you can attach to it. And I've never been one to be a passive spectator. And Energy Company X has afforded me that opportunity to, and I don't mean outside of my consciousness, but be proactive in society. (Participant 3)
> I'm very cognizant of my surroundings ... I think it has changed my personality. (Participant 4)

5 Discussion

The first key finding of the case study is that as an energy production company, Energy Company X produces clean and affordable energy, adds electricity to the national grid and contributes to the South African energy mix, thereby promoting sustainable development in the region (Rockström et al. 2009; Sachs 2012). In line with adhering to its 20-year contractual obligations, Energy Company X is also required to contribute to meeting the nation's development imperatives through its various ED and SED initiatives (Solberg 2015). The organisation contributes to industrialisation and infrastructure development, promotes skills development, broad-based black economic empowerment (BBBEE) objectives through enterprise development and ownership, thus promoting the inclusivity required and emphasised by scholars such as Hajer et al. (2015) and Stevens and Kanie (2016). In line with its energy power projects and ED and SED initiatives, Energy Company X has successfully managed to realise the targets under SDG7, thus contributing to a better future for the coming generations as called for by Akani (2017) and Schincariol (2017). A further conclusion to draw on, is the cross-cutting factor that the energy power projects have on other SDG targets such as SDGs 1, 8, 9 and 10 (Dhandra 2019; Fischer et al. 2017). These synergistic gains are imperative and should continue to be documented and disseminated for continuous monitoring of progress towards the 2030 Agenda for Sustainable Development (Sachs et al. 2019).

Purposive selection was used to identify critical players in successful project implementation (Creswell 2017; O'Neil and Koekemoer 2016). Due to the nature of the study, the selection criteria were based on expertise and specialised knowledge to identify participants from Energy Company X (Daniel 2012). Applying this method of selection aided in targeting specific role players, the experts provided knowledge based on their experience and involvement in the initiatives implemented by Energy Company X (Guest et al. 2012). The researchers identified four key role players: the head of the Sustainability Department, a senior sustainability analyst, a community liaison officer and a community engagement officer. This shows how, based on the selection criteria and the nature of the study and organisation, the role players were indeed key in the successful implementation of projects and provided data to address the study's research objectives (Flick 2018).

Participants identified several leadership capabilities that they perceived as imperative for successful project delivery. Participants indicated leadership capabilities were instrumental in enabling successful project implementation (Shula et al. 2019). The central themes emerging from the findings were creating shared value and bottom-up collaboration with cross-cutting subthemes (Kramer and Crespy 2011).

The perceptions of participants drew that engaging others, accountability and shared ownership, adaptability and flexibility are key to meeting intended outcomes under the central theme of CSV (Morse 2010). The ability to listen attentively and actively and being self-aware were considered great traits in a leader when dealing with stakeholders as this impacts on trustworthiness and acceptance by the intended beneficiaries. These were grouped under the bottom-up collaboration theme (Chrislip 2002; VanVactor 2012).

These findings demonstrate that the perceived leadership capabilities discussed by all four participants are key in enabling successful project implementation. Dialogue is key in all perceived leadership capabilities and for project delivery. It is best facilitated through a community needs assessment as a framework to understand the community needs while simultaneously meeting business objectives (Thakhathi 2019). Dialogue promoted and facilitated through a project steering committee provides opportunity for promoting accountability and shared ownership by both the community and Energy Company X. This communication via community engagement involves intended beneficiaries from negotiation to implementation stage (Martiskainen 2017). Furthermore, discourse allows for adopting new ways of doing things, fresh ideas, addressing challenges, adapting to new situations, group and community dynamics and innovative thinking where problem-solving is concerned (Thakhathi et al. 2019).

The findings were compared to transformational and collaborative leadership theories which the researchers preliminarily chose to inform the choice of leadership style possibly adopted by Energy Company X (Stibbe et al. 2018). The findings revealed that transformational leadership is less aligned with the leadership style adopted by Energy Company X. Collaborative leadership theory is more compatible and suited to the organisation as it is closely related to the leadership capabilities that double as leadership styles of CSV and bottom-up collaboration (Berson et al. 2004). It is apparent that although the SDGs are transformational in nature, transformational leadership theory is not necessarily the best theory to implement them. Thus, there is a relationship between leadership and the role of leaders, as one cannot exist without the other (Leal Filho et al. 2019). The implementation of SDG targets requires leadership, and there are several appropriate leadership styles (Thakhathi et al. 2019). It is best to adopt leadership styles that are suited to organisational aspirations, rather than clinging to a single prescription of leadership and placing a limit on how many styles are adopted (Cleveland and Cleveland 2018).

Transformational and collaborative leadership theories are not enough to implement SDG targets, and this demonstrates the importance of also being constructively critical (Nhamo and

Nhamo 2016). When adopting a leadership style to suit organisational aspirations, there are prescriptions to choose from and no limit to how many styles may be adopted (Coates et al. 2013). Finding an equilibrium, a synergy of leadership theories that speak to each other is important (Zhang et al. 2018). Three types of leadership emerged within Energy Company X: passionate, democratic and CSV leadership (Jia et al. 2018). Participant responses illustrated that CSV has been adopted as a leadership style, evident in the use of the model to achieve business priorities and community aspirations (Emerson et al. 2012). Together with bottom-up collaboration, these are the leadership styles adopted and practised within the organisation and across engagement with other stakeholders, thus echoing the importance of Ubuntu-based approaches when leading in the African context (Thakhathi and Netshitangani 2020).

6 Conclusion

The study confirmed a gap in the body of knowledge on leadership. It is still an under-researched and under-theorised field. As a complex, debatable and evolving concept, there is a need for new ways of thinking and new ways of bringing about successful change. When implementing the SDGs, leadership is required but the type of leadership cannot be prescribed. A leadership style adopted in one organisation or context may be incompatible with the next. This is because of varying group and organisational dynamics within all organisations, settings and contexts such as the nature of business, the size of the workforce, workforce requirements, organisational aspirations and priorities. As the CSV model adopted by Energy Company X has demonstrated, in order to achieve buy-in and ownership, there is a need for collaboration through a bottom-up approach. To achieve this, needs assessments and materiality matrixes need to be conducted with increased engagement and involvement by all stakeholders to achieve the transformational vision of the SDGs. The traditional top-down approach of prescribing what executives think is best for the organisation will no longer work in a society rife with global societal challenges. It is advisable, rather, to unite a diversity of opinions, fresh ideas, innovative thinking and problem-solving to agree on needs and therefore develop the right tools to successfully implement goals and ensure a collective impact.

The following are recommended for the successful implementation of the SDG targets: partnerships and collaborations between various IPPs and other stakeholders, bottom-up collaboration between project companies and communities, encouraging more PPP initiatives for collective impact and commitment to SDG targets, community buy-in and ownership and knowledge sharing of lessons learnt and what works for successful project delivery.

The study focused on a single organisation and context which posed as a limitation considering the wide scope of implementing the SDG targets by other IPPs from the REIPPPP. Data on their successful initiatives, specific to targets under SDG7, would have enabled more in-depth findings of leadership capabilities utilised and leadership styles adopted for successful project delivery. Further research should therefore build on and advance this work by testing the transferability of the findings to other contexts in pursuit of scalable and impactful analytically and statistically generalisable findings.

References

Akani, E. (2017). Sustainable Development Goals (SDGS) and the Politics of Development., *Journal of Emerging Trends in Educational Research and Policy Studies*. 8(2): 136–142. https://UnivofPretoria.on.worldcat.org/oclc/7358644064.

Bell, E. & Kothiyal, N. (2018). Ethics creep from the core to the periphery. In Sage Handbook of Qualitative Business and Management Research Methods: History and Traditions. (pp. 546-561). London: SAGE Publications Ltd. https://doi.org/10.4135/9781526430212.

Berson, Y. & Avolio, B. J. (2004). 'Transformational Leadership and the Dissemination of Organizational Goals: A Case Study of a Telecommunication Firm', The Leadership Quarterly, 15(5), 625–646. https://doi.org/10.1016/j.leaqua.2004.07.003.

Brundtland, G.H. (1987). 'Our Common Future—Call for Action', Environmental Conservation. Cambridge University Press, 14(4), 291–294. https://doi.org/10.1017/S0376892900016805.

Caelli, K.R. & Mill, J. (2003). 'Clear as Mud', Toward Greater Clarity in Generic Qualitative Research'. *International Journal of Qualitative Methods*, 1–13. https://doi.org/10.1177/160940690300200201.

Carroll, B. & Nicholson, H. (2014). 'Resistance and struggle in leadership development', Human Relations, 67(11), 1413–1436. https://doi.org/10.1177%2F0018726714521644.

Chime, J. & Enor, F.N. (2016). The Post-Colonial State and the Nexus between Leadership and Poverty. *Developing Country Studies*, 6 (4), 105–111. https://pdfs.semanticscholar.org/5a8c/04d324a426238976f155e1f5250a41480410.pd

Chrislip, D.D. (2002). The collaborative leadership field book (Vol. 255). John Wiley & Sons.

Cleveland, M., & Cleveland, S. (2018). Building engaged communities—A collaborative leadership approach. *Smart Cities*, *1*(1), 155-162.

Coates, H., Meek, L., Brown, J., Friedman, T., Noonan, P., & Mitchell, J. (2013). VET leadership for the future–characteristics, contexts and capabilities. *Journal of Further and Higher Education*, 37(6), 819-843. https://doi-org.uplib.idm.oclc.org/10.1080/0309877X.2012.684042

Cole, A.W. (2017). Online Interviews. In *The Sage Encyclopedia of Communication Research Methods*, 4, 1144–1145. California: SAGE Publications. https://doi.org/10.4135/9781483381411.

Creswell, J.W. (2017). Research design. In Qualitative, quantitative, and mixed methods approaches. Sage publications.

Daniel, J. (2012). Sampling essentials: Practical guidelines for making sampling choices (pp1–22). California: SAGE Publications.: https://doi.org/10.4135/9781452272047

Department of Energy. (2015). State of Renewable Energy in South Africa. Livro. https://doi: 9781920435080. Accessed 06 March 2019.

Department of Energy, Department of National Treasury and Development Bank of Southern Africa. (2015). Independent Power Producers Procurement Programme (IPPPP), An Overview. https://doi.org/10.1002/ejoc.201200111. Accessed 03 April 2019.

Dhandra, T.K. (2019). Achieving triple dividend through mindfulness: More sustainable consumption, less unsustainable consumption and more life satisfaction. *Ecological Economics*, 161, 83–90. https://doi-org.uplib.idm.oclc.org/10.1016/j.ecolecon.2019.03.021.

Emerson, K., Nabatchi, T., Balogh, S. (2012). An integrative framework for collaborative governance. *J. Public Adm. Res. Theory* 22 (1). https://doi.org/10.1093/jopart/mur011.

Energy Company X (2016). https://www.energycompanyx.com/media/press/d/2016/05/energycompanyx-starts-production-at-its-largest-solar- power-plant-in-South-Africa. Accessed 05 March 2019.

Energy Company X (2017a). https://www.energycompanyx.com/stories/a/2017/02/our-energy-in-the-heart-of-South-Africa. Accessed 05 March 2019.

Energy Company X (2017b). https://www.energycompanyx.com/about-us/a/2017/10/the-company. Accessed 05 March 2019.

Energy Company X (2019a). https://www.energypowerx.com/media/news/d/2019/06/renewafrica-initiative-sustainable-future-africa. Accessed 05 March 2019.

Energy Company X (2019b). South Africa and Zambia Project Cards.

Flick, U. (2018). The Sage handbook of qualitative data collection. London: SAGE Publications Ltd.: https://doi.org/10.4135/9781526416070.

Grint K. (2005). Problems, problems, problems: The social construction of 'leadership'. Human Relations 58 (11):1467–1494.

Guest, G., MacQueen, K.M., & Namey, E.E. (2012). Choosing qualitative data analysis software. In Applied thematic analysis (pp. 217–240). California: SAGE Publications Inc. https://dx-doi-org.uplib.idm.oclc.org/10.4135/9781483384436.n9.

Hajer, M., Nilsson, M., Raworth, K., Bakker, P., Berkhout, F., De Boer, Y., Rockström, J., Ludwig, K. & Kok, M. (2015). Beyond cockpit-ism: Four insights to enhance the transformative potential of the sustainable development goals. *Sustainability*, 7(2),1651–1660. https://doi.org/10.3390/su7021651.

Hilal, A.H. & Alabri, S.S. (2013). Using NVivo for data analysis in qualitative research. *International Interdisciplinary Journal of Education*, 2(2),181–186. https://doi.org/10.12816/0002914.

Jain, S. & Jain, P.K. (2017). The rise of renewable energy implementation in South Africa. *Energy Procedia*, 143,721–726. https://doi.org/10.1016/j.egypro.2017.12.752.

Jia, J., Liu, H., Chin, T., & Hu, D. (2018). The continuous mediating effects of GHRM on employees' green passion via transformational leadership and green creativity. *Sustainability*, *10*(9), 3237.

Kowal, S. & O'Connell, D. (2014). Transcription as a crucial step of data analysis. In The Sage Handbook of Qualitative Data Analysis (pp. 64–78). London: SAGE Publications Ltd. https://doi.org/10.4135/9781446282243.

Kramer, M.W. & Crespy, D.A. (2011). Communicating collaborative leadership. *The Leadership Quarterly*, 22(5),1024–1037. https://doi-org.uplib.idm.oclc.org/10.1016/j.leaqua.2011.07.021.

Leal Filho, W., et al. (2019), "Assessing the impacts of climate change in cities and their adaptive capacity: towards transformative approaches to climate change adaptation and poverty reduction in urban areas in a set of developing countries", The Science of the Total Environment, Vol. 692, pp. 1175–1190, https://doi.org/10.1016/j.scitotenv.2019.07.227.

Martiskainen, M. (2017). The role of community leadership in the development of grassroots innovations. *Environmental Innovation and Societal Transitions*, *22*, 78–89.

Morgan, D. L., & Ravitch, S. M. (2018). Trustworthiness, .in Frey, B. (ed.), The sage encyclopedia of educational research, measurement, and evaluation, 4, 1729–1730, SAGE Publications, Inc., Thousand Oaks, CA. https://dx-doi-org.uplib.idm.oclc.org/10.4135/9781506326139.n716.

Morse, R. S. (2010). Integrative public leadership: Catalyzing collaboration to create public value. *The Leadership Quarterly*, *21*(2), 231–245.

Morris, A. (2015). The what and why of in-depth interviewing. In A practical introduction to in-depth interviewing, (pp. 1–16). London: SAGE Publications Ltd. https://doi.org/10.4135/9781473921344.

Nhamo, G., & Nhamo, S. (2016). Paris (COP21) Agreement: Loss and damage, adaptation and climate finance issues. *International Journal of African Renaissance Studies-Multi-, Inter-and Transdisciplinarity*, *11*(2), 118–138.

O'Neil, S. & Koekemoer, E. (2016). Two Decades of Qualitative Research in Psychology, Industrial and Organisational Psychology and Human Resource Management within South Africa: A Critical Review. *SA Journal of Industrial Psychology*, 42(1), 1–16. https://doiorg.uplib.idm.oclc.org/10.4102/sajip.v42i1.1350.

Parcell, E.S. & Rafferty, KA. (2017). Interviews, Recording and Transcribing. In The Sage Encyclopaedia of Communication Research Methods, vol. 4, (pp. 801–803). California: SAGE Publications, Inc. https://doi.org/10.4135/9781483381411.

QSR International (n.d.). https://www.qsrinternational.com/nvivo/what-is-nvivo. Accessed 30 April 2019.

Riggio, R.E. & Harvey, M., (eds.). (2011). Leadership studies: The dialogue of disciplines. Edward Elgar Publishing. https://books.google.co.za/books?id=IhNumf9m7OIC&pg=PA12&lpg=PA12&dq=no+grand+theory+of+leadership&source.

Rheinhardt, A, Kreiner, G, Gioia, D & Corley, K. (2018). 'Conducting and publishing rigorous qualitative research', in The Sage Handbook of Qualitative Business and Management Research Methods: history and traditions, SAGE Publications Ltd, 55 City Road, London, 515–531. https://doi.org/10.4135/9781526430212.

Rockström, J., W. Steffen, K., Noone, A., Persson, F. S., Chapin, III, E., Lambin, T. M., Lenton, M., Scheffer, C., Folke, H., Schellnhuber, B., Nykvist, C. A., De Wit, T., Hughes, S.,van der Leeuw, H., Rodhe, S. Sorlin, P. K., Snyder, R., Costanza, U., Svedin, M., Falkenmark, L., Karlberg, R. W., Corell, V. J., Fabry, J., Hansen, B., Walker, D., Liverman, K., Richardson, P., Crutzen, P. & Foley, J. (2009). Planetary boundaries: exploring the safe operating space for humanity. *Ecology and Society* 14(2):32. http://www.ecologyandsociety.org/vol14/iss2/art32/.

Rubin, H. J., & Rubin, I. S. (2005) Qualitative interviewing, The art of hearing data, 2nd ed, Thousand Oaks, CA: SAGE Publications, Inc. https://doi.org/10.4135/9781452226651.

Sachs, J.D. (2012). From Millennium Development Goals to Sustainable Development Goals. Lancet 379(9832), 2206–11. https://doi.org/10.1016/S0140-6736(12)60685-0.

Sachs, J.D., Schmidt-Traub, G., Mazzucato, M., Messner, D., Nakicenovic, N. & Rockström, J. (2019). Six Transformations to achieve the Sustainable Development Goals. *Nature Sustainability*, (pp. 1–10). https://www.nature.com/articles/s41893-019-0352-9.

Schincariol, V.E. (2017). Joan Robinson on Population Growth. *Review of Political Economy*, 29(2), 267-281. https://doiorg.uplib.idm.oclc.org/10.1080/09538259.2016.1257026.

Scholz, R. W., & Tietje, O. (2002). Embedded case study methods (pp. 9–15). California: SAGE Publications, Inc. https://doi.org/10.4135/9781412984027.

Sebitosi, A.B. (2008). Energy efficiency, security of supply and the environment in South Africa: Moving beyond the strategy documents. *Energy*, 33(11),1591–1596. https://doi.org/10.1016/j.energy.2008.08.003.

Shula, K., Filho, W.L., Lardjane, S., Sommer, J.H., Salvia, A.L. & Borgemeister, C. (2019). The contribution of Regional Centers of Expertise for the implementation of the 2030 Agenda for Sustainable Development. *Journal of Cleaner Production,* Volume 237. https://doi-org.uplib.idm.oclc.org/10.1016/j.jclepro.2019.117809.

Silver, C. & Lewins, A. (2014). Using software in qualitative research. London: SAGE Publications Ltd. https://doi.org/10.4135/9781473906907.

Solberg, E. (2015). From MDGs to SDGs The Political Value of Common Global Goals. *Harvard International Review*, 37(1), 58. http://search.proquest.com/openview/7ad34f170994c1ebeb6267861bce385a/1?pq-origsite=gscholar&cbl=32013.

Stevens, C. & Kanie, N. (2016). The transformative potential of the sustainable development goals (SDGs). *International Environmental Agreements: Politics, Law and Economics*, 16(3), 393–396. https://doi.org/10.1007/s10784-016-9324-y.

Stephens, N. (2010). Collecting data from elites and ultra-elites: telephone and face- to-face interviews with macroeconomists. In Atkinson, P. & Delamont, S. (eds), *Sage qualitative research methods* (pp. 204–216). California: SAGE Publications, Inc. https://doi.org/10.4135/9780857028211.

Stibbe, D.T., Reid, S. & Gilbert, J. (2018). Maximising the Impact of Partnerships for the SDGs. The Partnering Initiative and UN DESA: New York. https://sustainabledevelopment.un.org/content/documents/2564Maximising_the_impact_of_partnerships_for_the_SDGs.pdf.

Thakhathi, A., le Roux, C., & Davis, A. (2019). Sustainability Leaders' Influencing Strategies for Institutionalising Organisational Change towards Corporate Sustainability: A Strategy-as-Practice Perspective. *Journal of Change Management*, *19*(4), 246–265.

Thakhathi, A. (2019). Creative start-up capital raising for inclusive sustainable development: A case study of Boswa ba Rona Development Corporation's self-reliance. *Journal of Cleaner Production*, *241*, 118161.

Thakhathi, A., & Netshitangani, T. G. (2020). Ubuntu-as-Unity: Indigenous African proverbs as a 're-educating' tool for embodied social cohesion and sustainable development. *African Identities*, 1–14.

United Nations, Department of Economic and Social Affairs, Population Division. (2019). World Population Prospects 2019: Highlights (ST/ESA/SER.A/423). https://population.un.org/wpp/Publications/Files/WPP2019Highlights.pdf. Accessed 03 September 2019.

UN General Assembly. (1986). Declaration on the Right to Development: resolution /adopted by the General Assembly, 4 December 1986, A/RES/41/128, 186–187. https://www.refworld.org/docid/3b00f22544.html.

United Nations. (2015a). Transforming our world: The 2030 agenda for sustainable development. General Assembly 70 session. A/RES/70/1. https://sustainabledevelopment.un.org/index.php?page=view&type=400&nr=2125&m enu=1515.

United Nations. (2015b). The millennium development goals report 2015. https://www.un.org/millennium-goals/2015_MDG_Report/pdf/MDG%202015%20rev%20(July%201).pdf. Accessed 02 September 2019.

VanVactor, J. D. (2012). Collaborative leadership model in the management of health care. *Journal of Business Research*, *65*(4), 555–561.

World Health Organization. (2015). Health in 2015: from MDGs, millennium development goals to SDGs, sustainable development goals (pp. 3–11). https://apps.who.int/iris/bitstream/handle/10665/200009/9789241565110_eng.pdf. Accessed 04 September 2019.

Zhang, L., Cao, T., & Wang, Y. (2018). The mediation role of leadership styles in integrated project collaboration: An emotional intelligence perspective. *International Journal of Project Management*, 36(2), 317–330. https://doi-org.uplib.idm.oclc.org/10.1016/j.ijproman.2017.08.014.

10 Designing Effective Social Protection for Food and Nutrition Security Among Farm Workers: Lessons from Masvingo, Zimbabwe

Joseph Tinarwo

Abstract

Implementing effective social protection programmes can be a strategy to address food and nutrition insecurity and is important for the achievement of the Sustainable Development Goals (SDGs). Yet, the implementation of social protection systems to end hunger and malnutrition among farm workers remains unexplored. Farm workers are confronted with a horde of challenges that make them perpetually poor, and social protection instruments are often used as a better approach to cushion them from the related shocks and stresses. This article analyses the efficacy of social protection interventions in enhancing food and nutrition security among farm workers in Masvingo, Zimbabwe. A qualitative research design was adopted, and data were gathered using focus group discussions (FGDs) and key informant interviews (KIIs) from purposively sampled key participants from government, development partners, academia as well as the farm workers in Masvingo. The study found that farm workers in Masvingo are enduring the burden of hunger and malnutrition. Hence, social protection systems, if well designed and implemented, reinforced by strong legal and capacitated institutional frameworks, adequate budgetary support and good governance, has a potential to buffer this poor and vulnerable group against food and nutrition insecurity, address lifecycle vulnerability and achieve SDGs.

Keywords

Social protection · Food and nutrition security · Farm workers · Masvingo · Zimbabwe

1 Introduction

Farm workers are a key pillar for ensuring food and nutrition security and, ultimately, economic growth in many developing countries (Hurst et al. 2005). Nevertheless, literature concerning the food and nutrition security status of farm workers is scant (Devereux and Tavener-Smith 2019). Hurst et al. (2005) define farm workers as the people who are employed in the primary agriculture industry, particularly on farms and primary manufacturing facilities, to ensure food security and the availability of other agricultural raw materials. In most cases, farm labourers work on smallholder farms and also big industrialised

J. Tinarwo (✉)
Julius Nyerere School of Social Sciences, Great Zimbabwe University, Masvingo, Zimbabwe

School of Public Management and Governance, University of Johannesburg, South Africa

G. Nhamo et al. (eds.), *Sustainable Development Goals for Society Vol. 2*, Sustainable Development Goals Series, https://doi.org/10.1007/978-3-030-70952-5_10

agricultural estates and farms (Hurst et al. 2005). In most developing countries, farm workers are confronted with a horde of challenges which include hunger and malnutrition, being marginalised and vulnerable to exploitation, earning low and unsustainable incomes and being perpetually poor (Sachikonye 2003; Devereux and Tavener-Smith 2019). The government of Zimbabwe initiated its land reforms in the 1980s to correct the land imbalances that existed in the country before independence in 1980. In year 2000, the Zimbabwean government implemented the Fast Track Land Reform Programme (FTLRP) to expedite the land redistribution process (Chilunjika and Uwizeyimana 2015). However, the implementation of the FTLRP resulted in the disruption of the farm workers' livelihoods and sources of income, especially for those who were formerly working for the targeted white commercial farmers (Sachikonye 2003). In fact, approximately half of the farm workers lost their employment by early 2002 and then nearly 65% by the month of February in 2003 (Chilunjika and Uwizeyimana 2015). Sachikonye (2003) claims that the FTLRP resulted in negative effects on farm workers, including displacements, food and nutrition insecurity and increased poverty. Social protection is being ushered as a better strategy to the growing problem of food and nutrition insecurity among the poor and vulnerable like farm workers (Devereux 2016). If properly designed, social protection systems address several interrelated issues, including breaking the vicious cycle of poverty, inequality and food and nutrition insecurity, thus facilitating the attainment of several SDGs, including SDG 2 on Zero Hunger (Devereux 2016; Fanzo 2019).

There is paucity of scientific information with regard to the role of social protection in addressing hunger and malnutrition among farm workers in Masvingo. Understanding this information will assist the government and other key actors in designing and implementing effective social protection interventions in order to address the food and nutrition insecurity situation of the farm workers and other poor and vulnerable groups of the society. In this study, the author uses the case study of Masvingo in Zimbabwe to investigate ways of designing and implementing effective social protection for food and nutrition security among farm workers. The aim is to come up with effective social protection programmes that may contribute to improve food and nutrition security, break the vicious cycle of poverty, as well as achieve several related SDGs, especially SDG 1, SDG 2 and SDG 5 that focus on eliminating poverty, ending hunger and gender.

2 Literature Review

2.1 Conceptualising the Term Social Protection

While social protection has been a commonly used term in development literature, there are several scholarly definitions of it. For instance, Sabates-Wheeler and Roelen (2011) define social protection as all state and non-state interventions that give income as well as consumption allocations to the underprivileged, cushion the vulnerable from livelihood threats and improve the social well-being and rights of the poor with the aim of addressing their vulnerability. For the World Bank, social protection refers to the set of public policies and programmes that strive to assist individuals and communities to be able to prevent risk and vulnerability and cushion them against poverty through policy instruments that build resilience, fairness and justice (World Bank 2014; Alderman and Yemtsov 2012). Furthermore, UNICEF (2012) states that social protection refers to all private and government policies and programmes designed to reduce, prevent or eradicate all forms of vulnerabilities to poverty and marginalisation. While the definitions above seem to be neutral on the meaning of social protection, they are indirectly created on normative assumptions that means all programmes are designed to address all levels and forms of vulnerability, marginalisation, risk and poverty. Definitions of social protection clearly have a normative content, and one has to know what the organisation or writer offering the definition deems correct. All measures designed to support households in crisis situations, including helping them to deal with risks and invest in more profitable livelihoods, are shared characteristics in various definitions of social protection.

Hidrobo et al. (2018) claims that social protection interventions can be in any one of the following main categories: (i) social safety nets or social assistance such as direct non-contributory schemes like financial or in-kind transfers, public work programmes or humanitarian assistance; (ii) social insurance such as contributory interventions that cushion against shocks like employment retirement or ill-health; and (iii) labour market interventions like job training. Therefore, this research focuses on social safety nets as forms of social protection for food and nutrition security among farm workers in Masvingo.

2.2 Social Protection, Food and Nutrition Security and the Fulfilment of SDGs

There exists a strong link between social protection, food and nutrition security and attainment of SDGs (Rohregger 2017; Devereux 2016). Social protection combats hunger and malnutrition (SDG 2) and poverty (SDG 1), reduces inequality (SDG 10), cushions vulnerable groups against risks and promotes socio-economic transformation as stated in SDG 8 (Plagerson and Ulriksen 2016). Social protection systems are therefore a key strategy in addressing poverty and deprivation by providing people in crisis situations with immediate access to food and nutrition security, including other critical consumer goods (Brugh et al. 2018). Marginalised and poor groups of people like farm workers suffer hunger and malnutrition if there are variations in food supplies or prices. Thus, various instruments can be put in place to cushion them (Devereux 2012). Devereux (2016) argues that social protection can enhance food and nutrition security among poor and vulnerable groups like farm workers by (a) improving incomes (managing seasonal stress and cushioning against risks and shocks), (b) increasing incomes (stimulating agricultural production and improving livelihoods) and (c) improving social justice (capacitating poor farmers, marginalised, poor and landless people). For instance, social safety nets like public work programmes (PWPs) have been used to address food and nutrition insecurity among the poor and vulnerable in most developing countries (Devereux 2016). McCord (2013) claims that if PWP wages are paid by means of cash, food or in kind that will promote access to food. On the other hand, the assets created by PWPs in the form of roads and markets, schools and clinics or dams and irrigation systems are likely to promote food availability (Devereux and Sabates-Wheeler 2004). However, caution should be taken when designing PWPs as paying the poor households below-market wages has been deemed unethical and counter-productive because energy is lost in doing manual labour (Devereux 2016). PWPs are also affected by competing household labour requirements especially during farming seasons where demand for labour is high (Devereux 2016). In addition, some food and nutrition insecure groups are left out, for example, child-headed households, disabled persons, ageing people and the chronically ill (Devereux 2010).

Giving food aid or cash to buy food has the potential to end hunger and malnutrition among the poor and vulnerable households like farm workers. For instance, food aid can cushion the poor and vulnerable families during food price volatility and food shortages (Porter and Goyal 2016). However, some scholars argue that food aid does not end hunger and malnutrition but rather worsen them as it creates a dependency syndrome (Devereux 2016). Therefore, when designing and implementing such social protection systems, it is imperative for the policymakers and development partners to build the capacity of the deprived and vulnerable groups like farm workers to take care of themselves (Devereux 2016). Moreover, social assistance mechanisms like cash transfers have the potential to increase household disposable income, create employment opportunities and ultimately boost economic growth as stipulated in SDG 8 (Brugh et al. 2018). However, the main challenge with most cash transfer programmes is that they do not put into consideration inflation and price instablity (Hidrobo et al. 2018). In situations where food markets are not performing and food access is low, giving money to households may result in rising cost of food, for instance, the 2007–2008 worldwide food price crisis demonstrated the shortcomings of delivering cash as a solution to food price volatility (Devereux 2016).

2.3 Integrating Gender in Social Protection for Food and Nutrition Security

Mainstreaming gender (SDG 5) into social protection delivery systems has a substantial impact on addressing food and nutrition insecurity and other related sustainable development variables (De Schutter 2013). Gender is often an important, yet frequently ignored, aspect when designing and implementing social protection systems (WFP 2017). Gender-targeted social protection systems cater for multiple deprived groups including women and girls, who are traditionally disadvantaged and encounter unique challenges and needs (De Schutter 2013). Gender inequalities are also mirrored in poverty levels, with women having a higher risk of poverty than men (World Bank 2014). Women frequently lag behind their male counterparts when it comes to access to food and nutrition security (Bizzari 2017). Women and girls are historically counted among the most vulnerable groups with over 60% of the hungry and malnourished people of the world being either women or girls (Tinarwo et al. 2018).

The 2017 World Food Programme (WFP) report revealed that gender is linked to hunger and malnutrition with women-headed households experiencing a high degree of food and nutrition insecurity and lower shock resilience (WFP 2017). According to O'Campos (2015), targeting females when delivering cash transfers promotes additional spending on household goods and services and ultimately addresses nutritional requirements because females are usually more in charge of the household than male counterparts. Accordingly, social protection systems that are designed to empower women, on top of their reproductive and caretaking roles, considerably improve food and nutritional outcomes for the whole household (O'Campos 2015). Therefore, policymakers, stakeholders and those who design programme interventions need to close the gender gap in order to enhance food security and nutrition, boost economic productivity and ultimately achieve SDGs. Single women and households that are led by women are among the groups that are worst affected by poverty, hunger and malnutrition; hence, it is important to cushion them most when designing and implementing social protection systems (World Bank 2014).

2.4 Institutions Promoting Social Protection for Food and Nutrition Security in Zimbabwe

2.4.1 Ministry of Public Service, Labour and Social Welfare (MPSLSW)

The MPSLSW is a government institution responsible for coordinating and implementing social policy instruments with the objective of assisting the marginalised and vulnerable groups in Zimbabwe. This institutional framework is guided by the vision to ensure that social policies, including social safety nets, promote an accepted standard of living among Zimbabweans. The main function of this ministry is to ensure that social policies are effectively implemented by investigating and assessing cases of public assistance, promoting the welfare of the underprivileged and paying fees and grants to institutions that support this work. Other key functions of this institutional framework involve rehabilitation, private voluntary organisation (PVO) registration and processing of applications for the registration of welfare organisations. This MPSLSW also works with the support of other key stakeholders that include donors, development partners, research institutions, academia and the private sector in implementing social protection interventions.

However, the MPSLSW is confronted by a horde of challenges in delivering social protection systems in Zimbabwe in general and Masvingo in particular. In fact, this ministry lacks the adequate budgetary support to implement its social protection programmes. The lack of sufficient financial resources to this ministry and also other state institutions is attributed to the protracted socio-economic and political challenges that Zimbabwe is facing, including the deteriorating macro-economic environment (AfDB et al. 2019). Moreover, the institutional frameworks for the delivery of social protection systems in Zimbabwe are generally weak and highly fragmented and lack harmonisation (Republic of Zimbabwe and World Bank 2016).

2.4.2 The National Social Security Authority (NSSA)

NSSA is a statutory institution established under the Ministry of Public Service, Labour and Social Welfare in terms of Chapter 17:04 of the NSSA Act of 1989. NSSA is mandated by the Zimbabwean government to oversee the implementation of social security schemes and is currently managing two mandatory schemes, namely, the national pension scheme and the accident prevention and workers' insurance contribution. The establishment of NSSA was motivated by the fact that since the country attained its independence in 1980, the majority of workers in the country did not have satisfactory and all-inclusive social insurance (Chirisa 2013). Previously, employees who were accessing these social protection systems were not obligated to do so, and it was very limited in terms of its coverage. As a result, very few workers had social security cover and lacked transferability and portability. The main challenge of occupational schemes was that they were fragmented and could not adequately pool risks in times of crisis (Chirisa 2013). The other shortcoming of occupational pension schemes was that they were limited in terms of lacking risk sharing across generations because state financial support only provided cover for risks associated with retirement.

2.4.3 Food and Nutrition Council (FNC)

Food and Nutrition Council is a government institution housed in the Zimbabwean Office of the President and Cabinet (OPC) with the mandate to coordinate multi-sectoral response to food and nutrition insecurity in the country (Tinarwo et al. 2018). Furthermore, FNC play a watchdog function mainly responsible for monitoring results as well as mobilising resources to address hunger and malnutrition in the country. The other roles of the FNC include providing policy advice, research, advocacy, ensuring food standards compliance, capacity building and monitoring and evaluation. Notwithstanding its important role in coordinating multi-sectoral interventions for hunger and malnutrition in Zimbabwe, FNC faces a number of challenges in executing its functions. These include weak institutional capacity and shortage of adequate human, material and financial resources (Sadza et al. 2015).

2.4.4 Development Partners

There are a number of development partners working on social protection for food and nutrition security issues in Zimbabwe (Tinarwo et al. 2018). Every year, the World Food Programme (WFP) delivers food and nutrition assistance, and it works with the government and other stakeholders to reduce the number of hungry people throughout the country. WFP supports social protection programmes in the country including Productive Asset Creation (Cash/Food for Asset Programme) which focuses on supporting longer-term recovery and resilience building to address the underlying causes of hunger and undernutrition instead of the usual food aid (Sadza et al. 2015). Food and Agriculture Organization (FAO) supports the Government of Zimbabwe and other non-governmental organisations (NGOs) with the financial and technical capacity to design and implement food and nutrition security interventions to improve agricultural productivity and livelihoods for the poor and disadvantaged people. Ensuring food and nutrition security has been the mandate of FAO in Zimbabwe for the past decade. In support of the Zimbabwean government, the United Nations Development Programme (UNDP) has been key in efforts to eradicate poverty and inequality with a strong focus on building the resilience of vulnerable communities (AfDB et al. 2019). United Nations International Children's Emergency Fund

(UNICEF)'s social protection work responds to the organisation's goal to support the rights of children, that is, the right of every child to benefit from social protection. UNICEF is currently implementing community-based nutrition programmes in all the provinces of the country. Several donor agencies work and support the UN, the Government of Zimbabwe and other stakeholders in implementing social protection systems to enhance food security and nutrition, and these include the United Kingdom's Department for International Development (DFID), the United States Agency for International Development (USAID) and the World Bank, among others (Chirisa 2013). However, the weak and polarised political relationship between the Zimbabwean government and the main foreign donor governments following the contested FTLRP has witnessed a sharp decrease in donor support for social protection programmes (Muchadenyika 2016). In fact, in recent years, most donors only contribute towards humanitarian support than development assistance to Zimbabwe (Muchadenyika 2016).

2.5 Legal Frameworks Underpinning Social Protection for Food and Nutrition

2.5.1 Constitution

The Zimbabwean constitution enshrines, on top of other issues, address the delivery of social protection systems to all groups within the country that include children, the elderly and disabled people. Chapter 2 (15) of the Zimbabwean constitution obliges the state to address hunger and malnutrition among all Zimbabweans. In its national objectives, the government of Zimbabwe is compelled by the constitution to implement policy measures to ensure that basic social amenities, including good nutrition, health, food and water and sanitation, are accessed by everyone including disadvantaged groups. Chapter 2(21) of the constitution compels the Zimbabwean government and all institutions and agencies to take reasonable actions, including policy instruments, to protect the vulnerable people like elderly persons. Moreover, the constitution mandates the Zimbabwean government at both national and subnational levels, operating within the confines of the available resources, to deliver social protection systems, including those designed to alleviate hunger and malnutrition among all marginalised groups. Additionally, the state is also required by the constitution's Chapter 4 (77) to ensure safe, clean and portable water and food for all; food itself is regarded as a basic human right.

2.5.2 National Social Protection Policy Framework (NSPPF)

The National Social Protection Policy Framework was crafted by the government of Zimbabwe in 2016 with funding from UNICEF and technical support from other development partners. The overall goal of NSPPF is to eradicate all levels of poverty and vulnerability by enhancing access to social services. Moreover, the food and nutrition security objective of the NSPPF is to eradicate hunger and malnutrition from all vulnerable people in Zimbabwe with a particular focus on children, pregnant women and lactating mothers and other vulnerable groups (e.g. the elderly). The NSPPF is also designed to support the implementation of the Zimbabwean National Food and Nutrition Security Policy with a particular focus on the following programme activities: school feeding programmes, household food security, provision of public education on nutrition and promotion of dietary diversity.

2.5.3 Food and Nutrition Security Policy (FNSP)

The government of Zimbabwe launched the Food and Nutrition Security Policy in 2013 with the main objective of stimulating and guaranteeing sufficient food and nutrition for everyone and, at all times, mainly among the most disadvantaged (FNC 2012). Fundamentally, the FNSP provides the government of Zimbabwe with sustainable mechanisms for the protection of its population against the shocks related to hunger and malnutrition posed by a horde of risks such as climate change, food price volatility, unemployment and a fragile global food security situation. The policy commits the government of Zimbabwe to a programme of spearheading broad and unified multi-sectoral agricultural interventions to promote food and nutrition security for everyone (FNC 2012). Underpinning the FNSP are four core commitments, namely, food security, social safety nets, nutrition security and food safety and standards. The development of the FNSP in Zimbabwe was led by the Food and Nutrition Council under a government-mandated task force with support from various stakeholders.

2.5.4 National Action Plan for Orphans and Vulnerable Children (NAP for OVC)

This policy framework was crafted in 2003 by Zimbabwe's MPSLSW in collaboration with development partners (National Aids Council, UN agencies, donors, children and other stakeholders). The NAP for OVC had a vision of providing social protection services to all orphans and vulnerable children in Zimbabwe. It recognises unique humanitarian challenges such as those brought about by HIV and AIDS as well as opportunities to build on nationally, regionally and internationally. The goal of this policy framework is to give guidance to the various stakeholders towards a harmonised implementation of social protection interventions and increase access by orphans and vulnerable children in Zimbabwe (Chirisa 2013). The implementation of the specific objectives of NAP for OVC is being done according to the guidelines that were approved by the Zimbabwean government in July 2006. The guidelines provide for a two-pronged implementation modality with clear roles and responsibilities for state actors as spelt by the Private Voluntary Organisations Act, Chapter 17.05.

3 Materials and Methods

This study essentially follows a qualitative methodology based on purposively sampled farms in order to appreciate the efficacy of social protection systems in addressing food and nutrition insecurity among farm workers in Masvingo. The chapter attempts to trace food and nutrition security as a global challenge that makes it imperative for policymakers to deliver social protection systems in order to achieve the SDGs (Tinarwo et al. 2018). The concept of social protection for food and nutrition security has gained momentum among development institutions. Consequently, this concept is now globally accepted in such a way that it is possible to study it using systematic examination and inquiry (Hammarberg et al. 2016). This systematic examination and analysis primarily take the form of literature and document assessment. In addition, FGDs and KIIs were used to gather data from the purposively sampled key stakeholders that deal with social protection for food and nutrition security among farm workers, and these include government, development partners, academia and the farm workers themselves. Two FGDs were conducted (one with academia and the other one with farm workers), while 16 key informant interviews (nine males and seven female) were conducted with the research participants who held managerial or supervisory roles in the respective organisations during the period of August and September 2019. Data collected from KIIs and FGDs in vernacular language were transcribed to English and were analysed thematically.

4 Results

Findings from both KIIs and FGDs revealed that while Zimbabwe has adopted the social protection system, it has less impact on food and nutrition security, particularly on the poorest and vulnerable groups such as the farm workers, as a result of low coverage and inadequacy of social protection systems mainly in the form of social safety nets. Social protection for food and nutrition security programme interventions in Zimbabwe, including Masvingo province, is facing serious financial challenges due to the withdrawal of support by the major foreign government donors following a decades-long political stalemate with the government of Zimbabwe. The macro-economic challenges in Zimbabwe that include high unemployment, accelerating inflation, looming recession and shortage of foreign currency are constraining the country from fully supporting its social protection systems for food and nutrition security among farm workers without external support. This study also found that social protection systems in Zimbabwe are inadequate with none of them reaching more than 7% of the population individually and collectively, leaving 84 percent of Zimbabweans and 80 percent of the poorest quintile unprotected by any programme. These include the majority of the farm workers in Masvingo. The following quotes from participants' verbal statements illustrate the above:

> …The Harmonised Social Cash Transfer Program (HSCT,) which is the key social protection intervention implemented by the Zimbabwean government with support from development partners targeting food and nutrition insecure people only targets 65,441 households in 23 rural districts across the country. This coverage is insignificant considering that over 8.5 million individuals (half of Zimbabwe's population) are food insecure and in urgent need of food aid…. (M&E Advisor with a big donor agency in the country)

The research also found that the statutory and policy as well as the institutional framework for social protection for food and nutrition security in Zimbabwe are characterised by fragmentation, duplication and poor coordination and harmonisation. For example, the mandate for social protection lies with MPSLSW, while the Ministry of Lands, Agriculture, Water, Climate, and Rural Resettlement is responsible for food security. More so, the responsibility for nutrition security falls under the Zimbabwean National Nutrition Unit in the Ministry of Health and Child Care. Although individual programme interventions by the above institutions are intended to meet the requirements of social protection systems for food and nutrition security among poor and vulnerable groups like farm workers, poor governance and weak institutional arrangements tend to hinder their effectiveness. The verbal statement quoted below illustrates the observation above:

> Policy and institutional frameworks guiding social protection programmes in Zimbabwe are highly fragmented and lack harmonisation. This is worsened by the fact that governmental institutions with the mandate to implement social protection systems have several challenges that include limited budgetary support, weak institutional support and shortage of human and material resources. (Principal Director with a Zimbabwean government ministry)

KIIs and FGDs with the academia indicated that the implementation of social protection systems for food and nutrition security among farm workers in Masvingo is confronted by poor targeting. In fact, the research found that there has been no retargeting exercise in Zimbabwe's major social protection programmes like the HSCT since 2012, and as a result, there is a possibility of inclusion and exclusion errors in social protection for food and nutrition security implementation processes. The verbal statement quoted below illustrates the observation above:

> Targeting in the delivery of social protection systems, including food and nutrition assistance among farm workers, is very complicated, especially with the ongoing socio-economic crisis in the country. A retargeting process is required. It is imperative to adopt community-based verification mechanisms to allow for quicker and easier reassessments of eligibility. Involvement of the farm workers themselves in the targeting process is likely to enhance targeting understanding and transparency to different stakeholders in the designing and delivery of social protection programmes for food and nutrition security. (Associate Professor specialising in poverty and inequality with one of the state universities in Zimbabwe)

It was noted through KIIs that social protection for food and nutrition security among farm workers in Masvingo has also overlooked nutrition goals in its delivery. Mainstreaming nutrition in the delivery of social protection is central in order to reduce malnutrition especially among children. The verbal statement quoted below illustrates the observation above:

> It is worrying to note that while the social protection for food and nutrition security interventions among farm worker's children such as the School Feeding Programme have clear nutrition objectives within the broad Zimbabwean Social Protection Policy Framework, they lack dietary diversity in

> their meals. In fact, the 2019 Zimbabwe Vulnerability Assessment Committee Report revealed that 78 percent of the people in Masvingo, the bulk of them being poor and vulnerable groups like farm workers, are consuming poor diets and have a general low dietary diversity score. It is thus imperative for the Zimbabwean government and its stakeholders to mainstream nutrition security in the designing and carrying out of social protection interventions. In addition, social protection for food and nutrition security programmes among farm workers must be designed to ensure that cross-cutting groups like children, pregnant mothers and the ageing have extra support in order to meet their nutrition requirements. (Nutritional Advisor with one NGO operating in Masvingo)

A focus group discussion with the farm workers in Masvingo revealed that hunger and malnutrition are some of the greatest problems they are encountering. Most farm workers, especially those who worked in the former white commercial farmers before the FTLRP, indicated that they get their food through engaging in casual labour for the newly resettled farmers after FTLRP, selling firewood, selling agricultural inputs donated by government and development partners, artisanal mining or, in rare situations, through begging. The verbal statement quoted below illustrates the observation above:

> …Putting food on the table is one of the greatest hurdles of our days. We actually survive on one meal for the whole day or in extreme scenarios spedt the whole day without eating. These days we survive mainly through engaging in casual labor, artisanal mining or selling firewood to those in town as electricity is now a challenge and most people in urban areas prefer firewood as a substitute. In some instances, we sell maize seeds or fertilizers when government or NGOs such as Care International donate to us…. (Farm worker in one of the districts in Masvingo)

The same focus group discussion with the farm workers in Masvingo revealed that the distribution of agricultural inputs and food is characterised by a lot of segregation especially on the basis of political affiliation and patronage. Farm workers believed to be aligned to opposition political parties like Movement for the Democratic Change are usually excluded from government-led or government-coordinated social protection for food and nutrition security programmes. It was again noted by farm workers in this focus group discussion that social protection systems led by the Zimbabwean government are affected by a lot of corruption and lack of transparency in their delivery. The verbal statement quoted below illustrates the observation above:

> …The challenge with some of the government initiated social protection systems is that they encounter a lot of political interference from local political authorities like ruling party (ZANU PF) leaders, particularly ward councilors and members of parliament, and if you are not aligned to the ruling party you will be automatically left out from the list of beneficiaries. There is also a lot of corruption and lack of transparency in programmes like food aid by local political and traditional leaders in most districts in Masvingo…. (Farm worker in one of the districts in Masvingo)

The next section discusses the emerging findings.

5 Discussion

The research found that designing and implementing social protection is key in enhancing food and nutrition security for poor and vulnerable groups like farm workers and is key in the attainment of the global SDGs (Devereux 2016; Fanzo 2019). In fact, findings from the literature, KIIs and FGDs show that farm workers in Masvingo are perpetually poor and, as a result, are vulnerable to food and nutrition insecurity. It is thus imperative for policymakers to urgently address the hunger and malnutrition challenges that farm workers in Zimbabwe in general and Masvingo in particular face in order to achieve the SDGs.

Strong legal, institutional and administrative frameworks are crucial in delivering social protection for food and nutrition security among poor and marginalised groups like farm workers (AfDB et al. 2019). This is particularly true at the local administrative level, where implementation of social protection interventions takes place, including the targeting and registration of beneficiaries. The findings from literature were the same as those from the interviews in that the legal and institutional frameworks of social protection are weak and lack the necessary adequate human and financial resources (Republic of Zimbabwe and World Bank 2016). It is imperative for the government to ensure that there is coordination and harmonisation of institutions that deal with social protection systems to ensure food and nutrition security in Zimbabwe. This is important to avoid duplication and to minimise conflicts. It is again important to build functional administrative units and implement capacity development programmes in order to create strong and resilient structures capable of delivering social protection systems that address hunger and malnutrition, particularly at the subnational level. In addition, despite the commitment of Zimbabwe to extending social protection coverage, the citizens' rights to social protection designed to address food and nutrition security remains unfulfilled among the most vulnerable groups like farm workers. There is need to include the people's rights in the delivery of social protection interventions to incorporate food and nutrition security into the national social protection policy frameworks and its realisation as a fundamental human right.

The study found that lack of good governance is one of the major obstacles in the delivery of social protection for the improvement of food and nutrition security (Devereux 2016). The findings from the literature were the same as responses from the KIIs and FGDs as there were alleged reports of discrimination of farm workers in their bid to access food aid on the basis of political affiliation and patronage in Masvingo. Therefore, ensuring transparency, accountability and participation of primary beneficiaries is central in the delivery of social protection for food and nutrition security programmes. Good governance ensures that the occurrence of corruption is minimised with the voices of minorities, poor and the most vulnerable in society being heard in decision-making (FAO 2015). Moreover, it is important to strengthen reporting, monitoring and evaluation to avoid abuse and manipulation during the implementation of social protection systems that seek to enhance food and nutrition security (Devereux 2016). Therefore, procedures to improve transparency and accountability in social protection for food and nutrition security interventions among the poor and vulnerable groups like farm workers in Masvingo are needed, including clarifying targeting criteria, clear communication to communities, providing receipts to beneficiaries of delivered rations and improved monitoring and reporting.

Lack of adequate budgetary support is one of the factors that stifle the effective implementation of social protection for food and nutrition security among farm workers in Zimbabwe (AfDB 2019; Republic of Zimbabwe and World Bank 2016). Government needs to strengthen domestic resource mobilisation to complement donor support in implementing interventions meant for social protection for food and nutrition security. Currently, most social protection for food and nutrition security programmes is largely funded by development partners, and their project-based approach to implementation is not in line with sustainable development (Muchadenyika 2016). Fostering public-private partnerships and promoting multi-sectoral resource mobilisation is in the best interest of ensuring effective delivery of social protection for food and nutrition security interventions. Since food and nutrition insecurity is multidimensional and has multiple causes that cannot be addressed by a single stakeholder, multi-sectoral approaches become imperative, and they should go beyond agricultural production and involve a variety of policy areas including social protection (Burchi and Strupat 2016). Countries that have managed to sustain their social protection programmes invest over 22% of their gross domestic product towards interventionist programmes (OECD 2014).

Finally, the research revealed that social protection programmes to farm workers in Masvingo in particular are negatively affected by poor targeting. Targeting the underprivileged and vulnerable groups of the society, particularly women, in the delivery of social protection for food and nutrition interventions is important since they are more in charge of the household than their male counterparts (WFP 2017; O'Campos 2015). It is thus imperative for the government to ensure that a retargeting exercise is done from time to time to avoid inclusion and exclusion errors.

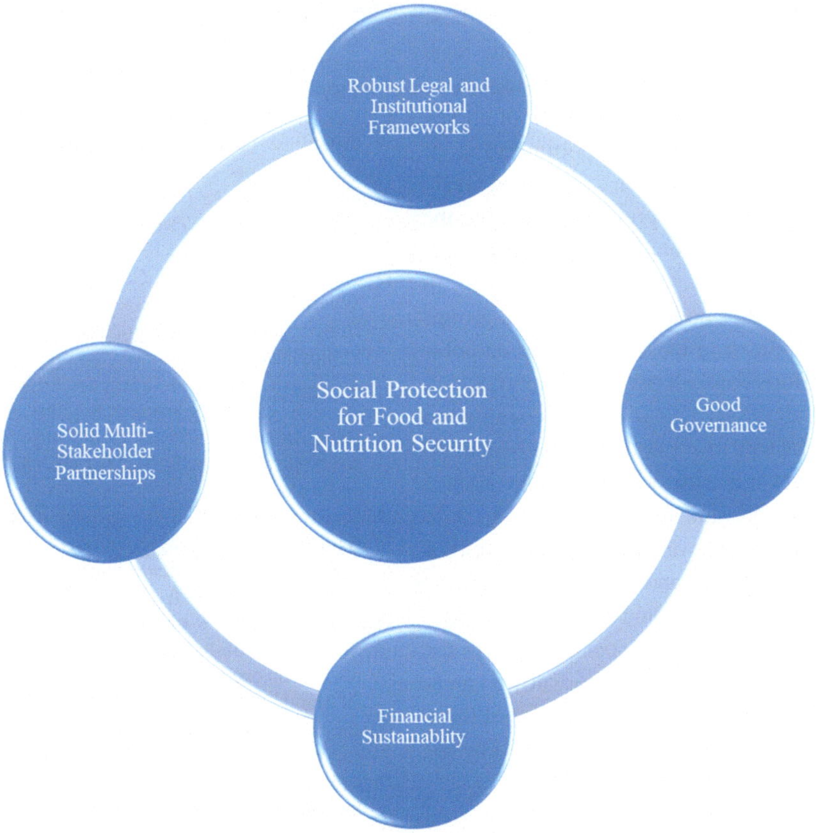

Fig. 10.1 Proposed conceptual framework for designing and implementing social protection for food and nutrition security among farm workers. (Source: Author's own compilation)

From the findings obtained from both the KIIs and FGDs as well as the discussion of these results, the author proposes a conceptual framework for designing and implementing social protection for food and nutrition among farm workers (Fig. 10.1). In the design and implementation of social protection for food and nutrition security among farm workers, strong legal frameworks and capacitated and resilient institutional structures are essential, particularly at the subnational level where interventions take place (FAO 2015). This must be supported by adequate financial, material and skilled human resources to ensure effective response. In addition, multi-stakeholder partnerships must be fostered between key stakeholders including the beneficiaries themselves to ensure social protection for food and nutrition security and ultimately achieve the SDGs (Tinarwo et al. 2018). Lastly, good governance must be practised by ensuring transparency and accountability as well as elimination of corruption during the designing and implementation of social protection for food and nutrition security among farm workers.

6 Conclusions

Social protection systems could address the hunger and malnutrition challenges of the poor and vulnerable groups like farm workers and contribute to Zero Hunger and other related SDGs. Nevertheless, the design and delivery of social protection for food and nutrition interventions are not a panacea. Several factors are essential to ensure the effectiveness of social protection systems that aim to reduce the hunger and malnutrition plights of farm workers and other deprived and vulnerable groups of society, and these include adequate budgetary support, strong legal and institutional architecture, practising good governance as well as the participation of all key stakeholders. In addition, the politicisation of social assistance, corruption and inadequate resources are some of the problems affecting the designing and implementation of social protection programmes among farm workers in Masvingo and other poor and vulnerable groups elsewhere. Addressing these challenges calls for concerted efforts from all the key stakeholders in pooling the necessary resources, practising good governance as well as capacitating the beneficiaries so that they move away from the dependence syndrome and become self-reliant which is in line with sustainable development. Future studies are needed to understand the impact of the COVID-19 pandemic on food and nutrition insecurity among farm workers especially due to the implementation of protracted lockdown measures, travel restrictions and enforcement of these measures by the government as prescribed by the World Health Organization.

References

African Development Bank (AfDB). (2019). Zimbabwe Infrastructure Report. Available at: https://www.afdb.org/fileadmin/uploads/afdb. (Accessed on 29 March 2021).

Alderman, H., & Yemtsov, R. (2012). Productive role of safety nets: background paper for the World Bank 2012-2022 social protection and labor strategy (No. 67609). https://ideas.repec.org/p/wbk/hdnspu/67609.html. Accessed 9 August 2019.

Bizzari, M (2017). Gender–Sensitive Social Protection for Zero Hunger, World Food Programme/WFP Regional Bureau for Latin America and the Caribbean. https://documents.wfp.org/stellent/groups/public/documents/liaison . Accessed 12 August 2019.

Brugh, K., Angeles, G., Mvula, P., Tsoka, M., & Handa, S. (2018). Impacts of the Malawi social cash transfer program on household food and nutrition security. *Food Policy*, 76, 19-32.

Burchi, F. & Strupat, C. (2016). The impact of cash transfers on food security in sub-Saharan Africa: evidence, design and implementation. Bonn: German Development Institute (GDIDIE). https://www.die-gdi.de/en/briefing- paper/ article/. Accessed 12 August 2019.

Chilunjika, A. & Uwizeyimana, DE. (2015). Shifts in the Zimbabwean Land Reform Discourse from 1980 to the Present. *African Journal of Public Affairs*, 8(3), 130:144.

Chirisa, I. (2013). Social Protection amid Increasing Instability in Zimbabwe: Scope, Institutions and Policy Options. In: Devereux, S & Getu, M. Informal *and Formal Social Protection Systems in Sub-Saharan* Africa. *Fountain Publishers*. https://books.google.com/books? Accessed 12 August 2019.

De Schutter, O (2013). Gender Equality and Food Security: Women's Empowerment as a Tool against Hunger, the Food and Agriculture Organisation/FAO, Philippines. http://www.fao.org/wairdocs/ar259e/ar259e.pdf . Accessed 12 August 2019.

Devereux, S., & Sabates-Wheeler, R. (2004). Transformative social protection. Available at: https://opendocs.ids.ac.uk/opendocs/handle/20.500.12413/4071. (Assessed on 29 March 2021).

Devereux, S. (2010). Building social protection systems in southern Africa. *Background paper for the European Report on Development*. http://citeseerx.ist.psu.edu/viewdoc/download . Accessed 09 August 2019.

Devereux, S. (2012). Social Protection for Enhanced Food Security in Sub-Saharan Africa. Available at: https://www.undp.org/content/dam/rba/docs/. (Accessed on 29 March 2021).

Devereux, S. (2016). Social protection for enhanced food security in sub-Saharan Africa. *Food Policy*, *60*, 52-62.

Devereux, S., & Tavener-Smith, L. (2019). Seasonal Food Insecurity among Farm Workers in the Northern Cape, South Africa. *Nutrients*, *11*(7), 1535.

Fanzo, J. (2019). Healthy and Sustainable Diets and Food Systems: The Key to Achieving Sustainable Development Goal 2? *Food Ethics*, 4(2), 159-174.

FAO. (2015). Regional overview of food insecurity: African food security prospects brighter than ever. http://www.fao.org/3/a-i4635e.pdf&ved=2. Accessed 4 August 2019.

Food and Nutrition Council (FNC). (2012). Policy – Food and Nutrition Security for Zimbabwe in the Context of Economic Growth and Development, pp. 1–47. Harare. https://extranet.who.int/nutrition/gina/en/node/14829 . Accessed 20 June 2019.

Hammarberg, K., Kirkman, M., & de Lacey, S. (2016). Qualitative research methods: when to use them and how to judge them. *Human Reproduction*, 31(3), 498-501.

Hidrobo, M., Hoddinott, J., Kumar, N., & Olivier, M. (2018). Social protection, food security, and asset formation. *World Development*, 101, 88-103

Hurst, P., Termine, P., & Karl, M. (2005). Agricultural workers and their contribution to sustainable agriculture and rural development. http://agris.fao.org/agris-search/search.do?recordID=GB2013203664. Accessed 09 December 2019.

McCord, A. (2013). Public works and resilient food systems. https://www.odi.org/sites/odi.org. Accessed 18 October 2019.

Muchadenyika, D. (2016). Multi-donor Trust Funds and Fragile States: Assessing the Aid Effectiveness of the Zimbabwe Multi-Donor Trust Fund. *Journal of International Development,* 28(8), 1337-1357.

O'Campos, A. (2015). Empowering rural women through social protection, the Food and Agriculture Organisation/FAO. http://www.fao.org/3/a-i4696e.pdf . Accessed 7 August 2019.

Organisation for Economic Cooperation and Development (OECD). (2014). Social Expenditure Update. Directorate for Employment, Labour and Social Affairs. http://www.oecd.org/social/soc/. Accessed 5 August 2019.

Plagerson, S., & Ulriksen, M. S. (2016). Can social protection address both poverty and inequality in principle and practice? *Global social policy*, 16(2), 182-200.

Porter, C., & Goyal, R. (2016). Social protection for all ages? Impacts of Ethiopia's Productive Safety Net Program on child nutrition. *Social Science & Medicine*, 159, 92-99.

Rohregger, B. (2017). The role of social protection for food and nutrition security: Examples from practice and international discussion. https://www.snrd-africa.net/wpcontent/uploads/2018/02/T. Accessed 2 August 2019.

Sabates-Wheeler, R., & Roelen, K. (2011). Transformative social protection programming for children and their carers: a gender perspective. *Gender & Development*, 19(2), 179-194.

Sachikonye, L. M. (2003). The situation of commercial farm workers after land reform in Zimbabwe. *Harare: Farm Community Trust of Zimbabwe*. Available at: http://archive.niza.nl/d. Accessed 09 September 2019.

Sadza, H.C., Nherera, C.M., Nhenga-Chakarisa, T., Tagwireyi, M.J. and Munyuki-Hungwe, M., (2015). ZIMBABWE Zero Hunger Strategic Review. https://documents.wfp.org/stellent. Accessed on September 2019.

Tinarwo, J, Babu, S.C & Karunya, I. (2018). Improving Food System Governance: Lessons from Multi-Stakeholder Partnerships in Zimbabwe. IFPRI Discussion Paper. Washington, DC: International Food Policy Research Institute (IFPRI). http://www.ifpri.org/publication/improving-food-system-resilience-through-better-governance-lessons-multistakehold. Accessed 09 August 2019.

United Nations Children's Fund (UNICEF). (2012). Integrated social protection systems: Enhancing equity for children. http://socialprotection.org/discover/publications/integrated-social-protection-systems-enhancing-equity-children&ved. Accessed 08 August 2019.

World Bank (2014). Analysis of the Safety Nets in Tajikistan: Considerations for the Transition Towards Greater Consolidation and Stronger Poverty-Focus. http://documents.worldbank.org/cu. Accessed 09 August 2019.

World Food Programme (WFP). (2017). Tajikistan Food Security Monitoring Bulletin. https://www.wfp.org/publications/tajikistan-food-security-monitoring-system-2017&ved. Accessed 09 August 2019.

Zimbabwe & World Bank. (2016). Zimbabwe Public Expenditure Review focusing on Social Protection. Available at: http://documents.worldbank.org/curated/en/7030614976330626. (Accessed on 29 March 2021).

Part III

Climate Action for SDGs

11 Mitigating Climate Change Through Carbon Sequestration for Sustainable Development: Empirical Evidence from Cameroon's Forest Economy

Ernest L. Molua

Abstract

Climate change is a wicked problem that requires urgent integrated approach for progress across multiple goals. This chapter invokes the connectivity of three United Nations Sustainable Development Goals (SDGs), including SDGs 12, 13 and 15 to highlight the need for concerted efforts to protect, restore and promote sustainable use of forest ecosystems while mitigating climate change in Cameroon. The fulcrum is on the potentials of forest serving as the nexus for climate action. Few economic assessments on carbon supply and sequestration have been done on Africa's forests. Beyond the direct provision of wood, the country's forests within the Congo Basin play different roles in the carbon cycle, from net emitters to net sinks of carbon, and possibly stand to benefit from the emerging global carbon market. The case study examines a carbon supply model and reveals that the short-run sequestration potential increases with rise in expected carbon revenues, forest density and government expenditures for better management of the forest sector. Increases in wood prices, fossil fuel price, timber harvest and consumption potentials have negative and statistically significant effects on carbon supply. In the long run, wood price and forest expenditure have a positive effect on carbon capture and supply. These results have interesting implications for carbon policy for both Cameroon and other developing countries in the sub-continent. Policy will have to address broad-ranged socioeconomic and political impediments for the promotion of carbon supply and sequestration. Affordable, scalable solutions must therefore be sought to enable countries to leapfrog to cleaner, more resilient economies.

Keywords

SDGs · Sustainable development · Climate change · Forest · Carbon sequestration

1 Introduction and Background

The United Nations' ambition to transform the world through the 2030 Agenda for Sustainable Development adopted by all member states in 2015 for the period 2016–2030 "focuses on time-bound targets for prosperity, people, planet, peace, and partnership" (UN 2015). The agenda of the UN Sustainable Development Goals

E. L. Molua (✉)
Department of Agricultural Economics and Agribusiness, University of Buea, Buea, Cameroon
e-mail: emolua@cidrcam.org

G. Nhamo et al. (eds.), *Sustainable Development Goals for Society Vol. 2*, Sustainable Development Goals Series, https://doi.org/10.1007/978-3-030-70952-5_11

(SDGs) or Global Goals provide a holistic and multidimensional view on development (Schröder et al. 2019; Pradhan et al. 2017). Conserving forests and other ecosystems is one of the 17 Global Goals that make up the 2030 Agenda (Sachs et al. 2019). Achieving economic growth and sustainable development requires societies to urgently reduce their ecological footprint by changing the way production and consumption of goods and resources are undertaken. This paper interweaves the nexus of three pertinent SDGs, including SDG-12 on responsible consumption and production, SGD-13 on climate action as well as SDG-15 on life on land which aptly captures the interactive role of the forest economy in climate change and sustainable production and consumption.

Forest ecosystems play important roles in human existence, with a significant number of the SDGs having indicators related to forests for the actualization of the UN goals of human development. Forest as a renewable natural resource is crucial for tackling many of the issues identified in the "Future We Want", such as poverty, food security, climate change, biodiversity, sustainable production and consumption and social inclusion, particularly meeting the basic needs of vulnerable people and ensuring their wellbeing. For instance, more than 20% of the household income for local families (SDG-1) come from forests. While SDG-15 which deals with life on land extensively addresses the place of forests in sustainable development, however, better forest management is required for SDG-2 and SDG-6, respectively. In the SDG-2 to end hunger, achieve food security and improve nutrition and promote sustainable agriculture, target 2.3 expects that by 2030 countries should achieve an important indicator 2.3.1 relating to an increase in the volume of production per labour unit for forestry as well as for classes of farming and pastoral enterprises. Such an indicator sufficiently accounts for the target to double agricultural productivity and incomes of small-scale food producers, particularly for women, indigenous peoples, family farmers, pastoralists and fishers, including through secure and equal access to land, other productive resources and inputs, knowledge, financial services, markets and opportunities for value addition and non-farm employment. In same token, SDG-6 to ensure availability and sustainable management of water and sanitation for all has target 6.6 to protect and restore ecosystems and to assist the recovery of those already degraded including mountains, forests, wetlands, rivers, aquifers and lakes. Tropical forests which shelter most of the world's terrestrial biological diversity (SDG-15) are seen as safe, natural means for carbon capture and storage, as well as recognized as an essential element of any strategy to stabilize global climate (SDG-13). Some other SDG targets related to forests include target 1.4 on access to natural resources for SDG-1; target 2.5 on maintaining genetic diversity of seeds, plants and animals for SDG-2; target 11.7 on accessing green spaces for SDG-11; and target 12.2 on sustainable management of natural resources for SDG-12. These reveal that forests' contributions are not limited to local livelihoods and global environmental objectives, but that stopping deforestation contributes to many other development goals at scales in between.

Integrated approaches are thus needed for enhancing the multiple contributions of forests to SDGs as they will harness synergies and balance cross-sectoral trade-offs between forests and other closely interlinked development issues. Several international instruments and processes offer a range of goals, objectives, targets and indicators on forests, based on which forest-related targets and indicators for the SDGs can be developed, for example, the UN Convention on Biological Diversity and the Strategic Plan for Biodiversity 2011–2020, including its 5 Goals and 20 Aichi Targets; the UN Framework Convention on Climate Change, the Kyoto Protocol and decisions on reducing emissions from deforestation and forest degradation in developing countries (REDD+); the UN Convention to Combat Desertification, including the concept on land degradation neutrality embraced at Rio + 20; the UN Zero Hunger Challenge; and the Global Partnership on Forest Landscape Restoration (IPBES 2019; IPCC 2013; UN 2002, 2012, 2015).

On the heels of these developmental necessities and quest for human progress, climate change is a real and undeniable emerging threat to human civilization, affecting every country on every continent, disrupting national economies and affecting lives (IPCC 2019a, b). Weather patterns are changing, sea levels are rising and weather events are becoming more extreme. The effects are already visible and will be catastrophic unless we act now. This is important since human life depends on the Earth as much as the forests and ocean for sustenance and livelihood (IPBES 2018, 2019). In fact, forests account for 30% of the Earth's surface, providing vital habitats for millions of species and important sources for clean air and water, as well as being crucial for combating climate change. Promoting the sustainable management of forests and halting deforestations are also vital to mitigating the impact of climate change (Bellassen and Gitz 2008; Duinker 1990). There are calls for urgent action to be taken to reduce the loss of natural habitats and biodiversity which are part of mankind's common heritage (Hess 2016; Zapfack et al. 2014).

In the midst of many interwoven goals, SDG-15 therefore stands out not only as the fulcrum on which many SDGs interconnect (Morton et al. 2017; Dzebo et al. 2018) but a livewire for many developing countries especially south of the Sahara in which daily livelihood is anchored to the environment particularly the forest ecosystems. The UN General Assembly resolution identified Sustainable Forest Management (SFM) as a gauge for Goal 15 and target 15.1 as well as for target 15.2 and formally defines it as "a dynamic and evolving concept [that] aims to maintain and enhance the economic, social and environmental values of all types of forests, for the benefit of present and future generations" (UN 2020: 1). The FAO (2020: 1) notes that "this definition implies SFM is a concept which varies over time and between countries, whose circumstances – ecological, social and economic – vary widely, and always addresses a wide range of forest values, including economic, social and environmental values, and take intergenerational equity into account". The key results emerging from existing data[1] is that the world continues to make progress in all dimensions of SFM; although forests continue to be lost, the rate of loss has been cut by 25% since the period 2000–2005 (UN 2020). According to the FAO (2020:1), "the change in forest area within legally protected areas is a proxy for trends in conservation of forest biodiversity as well as cultural and spiritual values of forests and thus a clear indication of the political will to protect and conserve forests. This indicator is related to the CBD Aichi Target 11 which calls for each country to conserve at least 17 per cent of terrestrial and inland water areas". The proportion of protected forest area and forests under long-term management plans are increasing with steady progress for Cameroon within the Central African subregion. Both SDG indicators 15.1.1 and 15.2.1 ensure forests are efficiently managed, and a better balance is struck between conservation and sustainable use of natural resources.[2]

Despite this positive outlook, nonetheless, deforestation and forest degradation are still concerns in some regions, particularly in the tropics, indicating the need for more action to reduce deforestation and implement SFM practices. The challenge, however, goes beyond the tropics. For sub-Saharan Africa, there's a small change for aboveground biomass stock in forest, as well as

[1]The UN data series which contributes to the measurement of SDG indicator 15.2.1, classified as Tier I, is officially defined as follows: Proportion of important sites for terrestrial and freshwater biodiversity that are covered by protected areas shows temporal trends in the mean percentage of each important site for terrestrial and freshwater biodiversity (i.e. those that contribute significantly to the global persistence of biodiversity) that is covered by designated protected areas (UN 2020).

[2]SDG indicator 15.2.1 is composed of five subindicators that measure progress towards all dimensions of sustainable forest management. The environmental values of forests are covered by three subindicators focused on the extension of forest area, biomass within the forest area and protection and maintenance of biological diversity and of natural and associated cultural resources. Social and economic values of forests are reconciled with environmental values through sustainable management plans. The subindicator provides further qualification to the management of forest areas, by assessing areas which are independently verified for compliance with a set of national or international standards (UN 2020).

significant positive change for proportion of forest area located in legally established protected area and proportion of forest under long-term forest management plans. However, negative change is reported for forest area under independently verified forest management certification schemes. Overall, nonetheless, efforts are required across all subregions to promote the implementation of sustainable management of all types of forests, halt deforestation, restore degraded forests and substantially increase afforestation and reforestation globally.

Sensible sustainable production and consumption do not take place in a vacuum. These happen in a world in which action to combat climate change and its impacts are imperative. An integrated approach is crucial for progress across the multiple goals. The Johannesburg Plan of Implementation of the World Summit on Sustainable Development (UN 2002) and the "Future We Want" outcome document of the Rio + 20 Conference (UN 2012) both recognized that "poverty eradication, changing unsustainable and promoting sustainable patterns of consumption and production and protecting and managing the natural resource base of economic and social development are the overarching objectives for sustainable development". Ensuring Sustainable Consumption and Production (SCP) patterns is therefore an essential requirement for sustainable development (Schröder et al. 2019). SCP not only promotes conservation through resource-use efficiency but has a cross-cutting role in sustainable development and its targets as basis for future development. The High-level Panel of Eminent Persons on the Post-2015 Development Agenda, commissioned by the Secretary-General of the United Nations in 2013, found that the world's consumption and production patterns need to be managed in a more sustainable and equitable way and that only by mobilizing economic, social and environmental action together can we irreversibly reduce poverty (UN 2013). By its cross-cutting nature, SCP addresses interlinkages and adopts a holistic approach, taking into account the economic, social and environmental aspects of sustainable development in a balanced and integrated manner (Schröder et al. 2019).

Human life on land depends on the forest ecosystems (IPBES 2018, 2019). Forest covers provide vital habitats for millions of species of animals, plants and insects and are important sources for clean air and water, as well as being crucial for combating climate change (Schröder et al. 2019; Oyono et al. 2005). Every year, millions of hectares of forests are lost globally, while the persistent degradation of drylands leads to desertification. The severe damage to land through deforestation, loss of natural habitats and land degradation disproportionately affects poor communities. Land use changes, including deforestation, result in a loss of valuable habitats, a decrease in clean water, land degradation, soil erosion and the release of carbon into the atmosphere (Bellassen and Gitz 2008). In the advent of global warming and climate change, carbon storage to offset carbon emissions in the form of carbon dioxide (CO_2) is gaining currency in national and international policy measures (Richards 2004). While acknowledging warming across the continent between 0.2° and 0.5 °C per decade up to the year 2100, the Fifth Assessment Report of the Intergovernmental Panel on Climate Change (IPCC 2013) confirmed that this change will also come with more frequent events such as storms, floods, sea level rise or droughts. Climate change thus represents a fundamental challenge to the sustainability of Africa's growth momentum. Hence, limiting the effects of climate change is necessary to achieve sustainable development and equity. At the same time, some mitigation efforts could undermine action on the right to promote sustainable development (Guariguata et al. 2008; Alig et al. 1997).

Under the United Nations Framework Convention on Climate Change (UNFCCC) established to cooperatively work to prevent dangerous anthropogenic interference with the climate system while coping with inevitable impacts of climate change; mitigation efforts are implemented through various types of policies, strategies and initiatives with the aim of mitigating greenhouse gas (GHG) emissions. Examples include the Kyoto Protocol's market mechanisms such as the Clean Development Mechanism (CDM), the mechanism for Reducing Emissions from Deforestation and Forest Degradation

(REDD+), the Nationally Appropriate Mitigation Actions (NAMAs) and Intended Nationally Determined Contributions (INDCs) (Bellassen and Gitz 2008; Jung 2005; Roslan 1995). The REDD+ and more broadly the Land Use, Land Use Change and Forestry (LULUCF) provisions are perceived to hold key roles in achieving the UNFCCC's ultimate goal – a rise in average global temperature of no more than 2° C by 2100 (Olesen et al. 2017; Campbell 2009; Schlamadinger et al. 2007).

The central feature of the Kyoto Protocol is its requirement that countries limit or reduce their greenhouse gas emissions. To help countries meet their emission targets, and to encourage the private sector and developing countries to contribute to emission reduction efforts, negotiators of the Protocol included three market-based mechanisms – emissions trading, the Clean Development Mechanism (CDM) and Joint Implementation (JI) – to help countries with binding greenhouse gas emissions targets (the Annex I countries) meet their treaty obligations (Jung 2005). The CDM allows emission-reduction projects in developing countries to earn certified emission reduction (CER) credits, each equivalent to one tonne of CO_2 (Michaelowa 2003). These CERs can be traded and sold and used by industrialized countries to meet a part of their emission reduction targets under the Kyoto Protocol. The mechanism stimulates sustainable development and emission reductions while giving industrialized countries some flexibility in how they meet their emission reduction limitation targets. Under the JI, countries with commitments under the Kyoto Protocol are eligible to transfer and/or acquire emission reduction units (ERUs) and use them to meet part of their emission reduction targets. Under Article 6, any Annex I country can invest in a project to reduce greenhouse gas emissions in any other Annex I country [referred to as a "Joint Implementation Project", particularly economies in transition (the EIT Parties) noted in Annex B of the Kyoto Protocol], as an alternative to reducing emissions domestically. In this way countries can lower the costs of complying with their Kyoto targets by investing in projects where reducing emissions may be cheaper and applying the resulting Emission Reduction Units (ERUs) each equivalent to one tonne of CO_2, which can be counted towards meeting their commitment of the Kyoto Protocol.

Whether the CER, CDM or JI, they all recognize the important role of forest ecosystems in the global carbon cycle, absorbing large amounts of atmospheric CO_2 through photosynthesis and emission of CO_2 to the atmosphere through respiration, decomposition and disturbances such as timber harvesting, fire, pest infestations and land use change (Depro et al. 2008). The forest sector is therefore strategic to play a major role in climate change mitigation (Guariguata et al. 2008). The important role of forests was flagged in the agreements of the 16th Session of the Conference of the Parties (COP) to the UNFCCC in Cancún. The agreement was emphatic on protecting the world's forests, which are a major repository of carbon. On agreeing to launch concrete action on forests in developing nations, COP-16 imported the essential elements of the Copenhagen Accord of COP-15 which represented key steps forwards to reduce greenhouse gas emissions and help developing nations protect themselves from climate impacts and developed countries build their own sustainable futures through mitigation pledges, a new Green Climate Fund for developing countries and a system to help verify countries' actions. The Cancún pledge hinged on the fact that, depending on their characteristics and local circumstances, forests can play different roles in the carbon cycle, from net emitters to net sinks of carbon (Newell and Stavins 2000; Kotto et al. 1997). Forests can sequester carbon by taking in carbon dioxide (CO_2), a major contributor to greenhouse effect, from the atmosphere, and transforming it into biomass through photosynthesis (Oyono et al. 2005). Sequestered carbon is then accumulated in the form of woody biomass, deadwood and litter in forest soils. In sustainably managed forests, the amount of carbon that can be released as a result of harvesting is equal to or smaller than the amount taken from the atmosphere, making forests "carbon-neutral" or "carbon sinks" (Newell and Stavins 2000). Promoting the expansion of sustainably managed forests, increasing sound mobilization of wood as well as replacing carbon-intensive commodities through wood products and bioenergy would enlarge carbon sink potential and significantly contribute to offsetting greenhouse gas (GHG) emissions (Brack 2018; Youssoufa et al. 2011; Kotto et al. 1997). The release of carbon from forest ecosystems results from natural processes (respiration and oxidation) as well as deliberate or unintended results of human activities (i.e. harvesting, fires, deforestation) (Fry 2008; Oyono et al. 2005).

Stern (2008) notes that cost-effective carbon sequestration from agricultural land use change practices could sequester about 1Gt of CO_2. When soils are exposed to microbial activity, CO_2 emissions are released. These emissions can be reduced by disturbing the soil less, for example, by using conservation tillage techniques and turning land into permanent set-aside. Mitigation and adaptation can positively or negatively influence the achievement of other societal goals, such as those related to human health, food security and biodiversity (Oberthür 2016; Guariguata et al. 2008; Albrecht and Kandji 2003). Supporting this argument are several studies that point out that land use change, including deforestation and forest degradation, accounts for 17% to 29% of global GHG emissions (Sohngen and Mendelsohn 2003). Reducing emissions from deforestation and forest degradation and enhancing carbon sinks (REDD+) are taunted as a panacea by the Parties to the UNFCCC with several initiatives and programmes such as the Forest Carbon Partnership Facility and the Forest Investment Program that build on REDD+ as a climate change mitigation solution (Phelps et al. 2010; Campbell 2009). These programmes have implications of different GHG emission levels for the rate of CO_2 emission reductions from 2030 to 2050. According to the IPCC (2013), delaying mitigation efforts beyond those in place today through 2030 is estimated to substantially increase the difficulty of the transition to low longer-term emission levels and narrow the range of options consistent with maintaining temperature change below 2 °C relative to pre-industrial levels. The Cost-effective mitigation scenarios that make it at least about as likely as not that temperature change will remain below 2 °C relative to pre-industrial levels (2100 concentrations of about 450 to about 500 ppm CO_2eq) are typically characterized by annual GHG emissions in 2030 of roughly between 30 GtCO_2eq and 50 GtCO_2eq.

At the 21st Conference of the Parties (COP-21) in Paris, countries adopted a legally binding global climate deal, requiring all Parties to put forward their best efforts through "Nationally Determined Contributions" (NDCs) and to strengthen these efforts in the years ahead (Nhamo and Nhamo 2016a). The Paris Agreement is a bridge between today's policies and climate neutrality before the end of the century (Kinley 2017; Nhamo and Nhamo 2016b). Governments agreed a long-term goal of keeping the increase in global average temperature to well below 2 °C above pre-industrial levels, and to aim to limit the increase to 1.5 °C, since these would significantly reduce risks and the impacts of climate change. The expectation was for global emissions to peak as soon as possible, recognizing that this will take longer for developing countries, and to undertake rapid reductions thereafter in accordance with the best available science (Nhamo and Nhamo 2016b; Oberthür 2016). In COP-22 in Marrakech, governments welcomed the Paris Agreement with its ambitious goals, its inclusive nature and its reflection of equity and common but differentiated responsibilities and respective capabilities, in the light of different national circumstances.

The aim of this chapter is therefore to demonstrate the possibility of mitigating climate change through forest sector carbon sequestration, using information from a tropical African country which is part of the Congo Basin Forest. In achieving this, the chapter specifically interweaves the possibility of ensuring responsible exploitation and consumption of land-based forest resources in a manner that addresses the climate change imperative. The remainder of the chapter is divided into three distinct sections as follows. Section 2 examines theoretical developments connecting sustainable forest management for climate action to ensure sustainable natural capital of land. Section 3 presents a case study on Cameroon by assessing the politico-economic determinants of the country's forest exploitation. Some policy recommendations based on the empirical findings and analytical review are provided in the concluding Section 4.

2 Literature Review: Nexus of Sustainable Forests for Climate Action and Better Life on Land

The three SDGs 12, 13 and 15 are very useful connectors between economic progress and environmental perspectives. The 2030 Agenda embraces the three dimensions of sustainability – economic, social and environmental – in an integrated and interconnected manner. Comprehensive sustainability will require not only ecological but also economic and social sustainability. By adopting the 2030 Agenda and the Paris Climate Agreement, the UN effectively created a framework for national action and global cooperation on sustainable development, while the Paris Agreement committed signatory countries to achieving net-zero greenhouse gas emissions by the middle of the century (Kinley 2017; Nhamo and Nhamo 2016a). This means the SDG 13 on climate change specifically links the Paris Agreement (Bruce et al. 2018; Fawcett et al. 2015), noting that the UNFCCC "is the primary international, intergovernmental forum for negotiating the global response to climate change" (UN 2015). The toolbox of the UNFCCC has as spanner sustainable forest management.

Forests are not only important because they are valuable economic asset providing livelihood opportunities, but promoting a sustainable use of forest ecosystems and preserving biodiversity are key to human survival (Schröder et al. 2019; Zapfack et al. 2014). According to Brack (2019), forests play a critical role in the Earth's climate system, in a number of different ways, including capturing carbon dioxide from the atmosphere and converting it, through photosynthesis, into living biomass. Forests also store carbon in forest soils, absorbed through leaf litter, woody debris and roots. The complex interactions involving soil minerals, plants, soil organisms and organic components are influenced by local climatic conditions and forest management (IPBES 2018, 2019; Duinker 1990). The acknowledgement of such important forest services pushed the UN (2017) Strategic Plan for Forests 2017–2030 adopted by the Economic and Social Council on 20 April 2017 to specify Global Forest Goals to involve (a) reverse the loss of forest cover worldwide through sustainable forest management, including protection, restoration, afforestation and reforestation, and increase efforts to prevent forest degradation and contribute to the global effort of addressing climate change; (b) enhance forest-based economic, social and environmental benefits, by improving the livelihoods of forest-dependent people; as well as (c) increase significantly the area of protected forests worldwide and other areas of sustainably managed forests, as well as the proportion of forest products from sustainably managed forests.

Granted that natural assets are for man's exploitation and welfare, however sensible sustainable use calls for stewardship in consumption (Hess 2016). SDG-12 on sustainable consumption and production is of particular relevance to the supply of forest products, and the significant links between SCP and forests is beginning to receive attention among the expert and policy communities. SDG-12 is considered a major contributor to the protection and enhancement of natural resources, including forests (FAO 2020), and is seen to be particularly relevant to the supply of forest products (Brack 2018). For instance, Hess (2016) discusses the importance of natural resources for economic growth and sustainable development and asserts that while increases in the quantity or quality of natural resources available to an economy enhance the productive capacity of the nation, there is increasing evidence of environmental stress-threatening future livelihoods.

With respect to SCP on forests and their conservation, sustainable management and use, as well as forest livelihoods, Schröder et al. (2019, p. 386) remind us that "SCP has been part of the international policy discourse for more than four decades, but the uptake of SCP has not been smooth and has tended to be biased towards relatively weak measures". Although SDG-12 targets or indicators make no direct reference to forests or forest communities, SDG-12 targets can contribute positively to forest protection and conservation efforts. SDG-12 can contribute to creating enabling conditions for advancing more responsible and sustainable supply of timber and other forest commodities, also linked to more responsible demand.

Glass and Newig (2019) note that the achievement of the SDGs depends on effective governance arrangements. Different aspects of governance, namely, participation, policy coherence, reflexivity, adaptation and democratic institutions on SDG achievement at the national level, may serve to explain SDG achievement. Governance is important since as Morton et al. (2017:81) demonstrate that all the goals are intimately interconnected, and "a failure to appreciate this will perpetuate an approach which will be non-aligned at best and highly ineffective at worst". There is the need to identify and assess synergies between climate and sustainable development policies and avoid or manage trade-offs. It emerges that the interaction of SDGs 12, 13 and 15 enhance both human wellbeing and the ecological health of the planet (Morton et al. 2017). According to Dzebo et al. (2018), there is a great potential for greater policy coherence in the implementation of the 2030 Agenda for Sustainable Development and the Paris Agreement. To take advantage of it, there is a need to identify and assess synergies between climate and sustainable development policies and avoid or manage trade-offs. At the national and sub-national level, where the Paris Agreement is implemented through national climate action plans (or Nationally Determined Contributions – NDCs), there is a need for sufficient understanding of the potential coherence between the implementation of the NDCs and the SDGs.

The attainment of the SDG agenda will greatly depend on whether the identified synergies among the goals can be leveraged. IIASA (2019) considered some key interventions that would be necessary to achieve the SDG outcomes with their implementation being organized into a set of six transformations, namely, education, gender and inequality; health, wellbeing and demography; energy decarbonization and sustainable industry; sustainable food, land, water and oceans; sustainable cities and communities; and digital revolution for sustainable development. In their view, each SDG transformation describes a major change in societal structure (economic, political, technological and social) to achieve long-term sustainable development, with each contributing to multiple SDGs. Excluding any of them would make it virtually impossible to achieve the SDGs.

An important subsector through synergy that is required to transform economies is the forest, with its onerous role in the carbon cycle, yet affected by changing climatic conditions. Evolutions in rainfall and temperature can have either damaging or beneficial impacts on forest health and productivity, which are very complex to predict. Depending on circumstances, climate change will either reduce or increase carbon sequestration into forests, which causes uncertainty about the extent to which the world's forests will be able to contribute to climate change mitigation in the long term (Guariguata et al. 2008; Albrecht and Kandji 2003). Forest management activities have the potential to influence carbon sequestration by stimulating certain processes and mitigating impacts of negative factors (Duinker 1990). The development of a market for carbon emissions is a significant component of the UNFCCC's Kyoto Protocol. Parties with emission reduction targets, i.e. Annex B of the Protocol, are allocated "assigned amount units" (AAUs) that represent the total emissions permitted to meet these targets.[3] Perez et al. (2007) note that in theory, carbon markets present win-win opportunities for buyers and sellers of carbon stocks. This may promote better forest management.

Some other studies have indicated the plausibility of sequestering significant amount of carbon from properly managed forests (Stoffberg et al. 2010; Hennigar et al. 2008; Thomson et al. 2008; Benítez et al. 2007; Karjalainen 1996). For instance, Benítez et al. (2007) note that within 20 years and considering a carbon price of US$50/tC, tree planting activities could offset 1 year of global carbon emissions in the energy sector. Similarly, on assessing the contribution of terrestrial carbon sequestration to climate change

[3] Domestic reduction policies help bring actual emissions in line with the allocated AAUs. Parties then submit national greenhouse gas inventories annually to the UNFCCC that account for all emissions that occurred within that year.

mitigation, Thomson et al. (2008) show that terrestrial sequestration reaches a peak rate of 0.5–0.7 Giga tonnes of carbon per year (GtC $yr.^{-1}$) in mid-century with contributions from agricultural soils (0.21 GtC $yr.^{-1}$), reforestation (0.31 GtC $yr.^{-1}$) and pasture (0.15 GtC $yr.^{-1}$). According to Thomson et al. (2008), sequestration rates vary over time and with different technology and policy scenarios. The combined contribution of terrestrial sequestration over the next century ranges from 23 to 41 GtC. This makes it clear that the contribution of forests to carbon cycles has to be evaluated taking also into account the use of harvested wood, e.g. wood products storing carbon for a certain period of time, or energy generation releasing carbon in the atmosphere. In cases where the net balance of carbon emissions by forests is negative, i.e. carbon sequestration prevails, forests contribute to mitigating carbon emissions by acting as both a carbon reservoir and a tool to sequester additional carbon (Albrecht and Kandji 2003; Sedjo et al. 1995). In cases when the net balance of carbon emissions is positive, forests contribute to enhancing greenhouse effect and climate change.

3 Case Study: Carbon Sequestration in Cameroon's Forests

With a significant landmass covered by the humid tropical rainforest and being part of the Congo Basin, Cameroon is a major source of tropical wood and a reservoir for carbon (Somorin et al. 2012; Ndoye and Kaimowitz 2000). Aside from timber, Cameroon's forest ecosystems are the source of many direct and indirect benefits. They provide habitats for some rare terrestrial species, and they offer watershed protection, control of soil erosion and hence siltation. They also provide a wide range of non-timber products and recreational, cultural, spiritual and amenity benefits (Zapfack et al. 2014). As part of the larger ecosystem of the Congo Basin, the country's forest thus plays an important role in mitigating the emissions of (CO_2), the most important greenhouse gas (Brown et al. 2010; IPCC 2007). Different management regimes in the country affect the ability of Cameroon's forests to sequester carbon. With recent developments in the carbon market, forestry authorities are now recognizing the potentials from financial markets for the ecosystem services that national and community forests provide, such as biomass for renewable energy, clean water, clean air, habitat for wildlife (especially threatened and endangered species) and more importantly carbon sequestration (Zapfack et al. 2014; Brown 2006; Brown et al. 1993; Kotto et al. 1997).[4] Cameroon must therefore better manage its tropical forest resources to sequester carbon. Protecting forests as biodiversity habitat is important as well to mitigate climate change, since deforestation and forest degradation represent a major source of greenhouse gas emissions (Zapfack et al. 2014; Fry 2008). There is, however, inadequate technical information to assist policy-making processes to guide new shifts in the country's efforts towards the UNFCCC's Clean Development Mechanism (Jung 2005; Richards 2004; Brown 1997).

Balancing many different national priorities can be challenging, and so identifying areas of synergy, where more than one goal can be met at the same time, can help developing countries achieve their climate goals and other development goals at the same time. Cameroon like most countries in the Congo Basin has articulated its climate priorities in two major dossiers: the Nationally Determined Contribution to the Paris Agreement and the National Adaptation Programmes of Action (Somorin et al. 2012; Youssoufa et al. 2011). These dossiers define key policies to promote adaptation actions, mitiga-

[4]Carbon sequestration is the general term used for the capture and long-term storage of carbon dioxide. Capture can occur at the point of emission (e.g. from power plants) or through natural processes (such as photosynthesis), which remove carbon dioxide from the Earth's atmosphere and which can be enhanced by appropriate management practices. Carbon sequestration methods include (a) enhancing the storage of carbon in soil (soil sequestration), (b) enhancing the storage of carbon in forests and other vegetation (plant sequestration), (c) storing carbon in underground geological formations (geosequestration), (d) storing carbon in the ocean (ocean sequestration) and (e) subjecting carbon to chemical reactions to form inorganic carbonates (mineral carbonation).

tion actions and cross-cutting actions, particularly on its capacity to engage on mitigation with the REDD+ mechanism (Youssoufa et al. 2011; Campbell 2009; Brown 2006).

Numerous studies have analysed the carbon sequestration potential of forests and forest management, focusing on national and supra-national scales or on the project level, some in the context of the flexible mechanisms of the Kyoto Protocol (Maamoun 2019; Backéus et al. 2005; de Jong et al. 2000; Newell and Stavins 2000; Sohngen et al. 1999; Sedjo et al. 1995; Duinker 1990). A significant number of these studies have demonstrated the plausibility of forests to sequester carbon for both ecological benefits and financial gains (e.g. Gough et al. 2019; Sedjo and Sohngen 2012; Imai et al. 2009; Tonna and Marland 2007; Krcmar et al. 2001). On examining carbon sequestration estimates of indigenous trees, Stoffberg et al. (2010) observe that amelioration of global warming presents opportunities even for urban forests to act as carbon sinks and thereby could possibly be included in the potential future carbon trade industry. In their study on indigenous urban trees (e.g. *Combretum erythrophyllum*, *Searsia lancea* and *Searsia pendulina*), Stoffberg et al. (2010) estimate that the tree planting will result in 200,492 tonnes CO_2 equivalent reduction and that 54,630 tonnes of carbon will be sequestrated over a 30-year period (2002–2032). The carbon dioxide reductions could be valued at more than US$ 3,000,000. This illustrates that when future carbon trade becomes operational for urban forests these forests could become a valuable source of revenue for the urban forestry industry, especially in developing countries (Diaz-Balteiro and Romero 2008).

More research is needed, especially for countries in the Congo Basin, to more accurately capture the impact of either country or region-specific interactions between climate and management of forest resources for carbon sequestration, which are lost in global-level assessments. Being party to global conventions on climate change and signatory to regional environmental initiatives, the redefinition of Cameroon's forest policy now accounts for climate change mitigation and adaptation. Brown et al. (2010) note that climate change presents additional challenges to a diverse country like Cameroon that shares the Congo Basin rainforest. Not only is the population vulnerable to the direct effects of climate change, but forest-dependent communities are also vulnerable to changing environmental policy that may affect their access to forest resources (Brown et al. 2010). In sum, therefore, given the importance of Cameroon's forest to its immediate and future economy, and possessing features of an important laboratory for a green economy, it is pertinent to evaluate and quantify the factors that may motivate the sector for its sequestration potential. This paper thus sets out to evaluate the maximum carbon supply and carbon sequestration and assess the implications of forest management as a significant carbon sink. Optimizing carbon supply has been a matter of concern in the forestry literature (Olschewskia and Benítez 2010; Matheya et al. 2009; Sedjo et al. 1995). There are, however, few published studies addressing the issue in Cameroon under the dispensation of mitigating climate change. Most studies on the forest sector have either assessed forest management choices (e.g. Oyono et al. 2005) or evaluated the costs and benefits of reducing deforestation and forest degradation (e.g. Bellassen and Gitz 2008) and reviewed carbon dynamics in slash-and-burn agriculture (e.g. Kotto et al. 1997; Albrecht and Kandji 2003). In other African countries, analysis has hinged on the potential benefits of carbon sequestration markets and land tenure challenges which impede forest sector response to carbon potentials (Zapfack et al. 2002; Woomer and Palm 1998).

4 Methodological Orientation for the Case Study

4.1 Economic-Ecological Modelling of Carbon Sequestration

A myriad of approaches have been explored to evaluate the effect of forest management activities on the dynamics of the ecological resource stock (e.g. Seong-Hoon et al. 2019; Murphy et al. 2018; Favero et al. 2017; Plantinga 2015; Kim and McCarl 2015; Hernandez et al. 2014;

Yousefpour and Hanewinkel 2009; Lubowski et al. 2006; Benítez et al. 2007; Krcmar et al. 2001; Sedjo et al. 1995). Caparrós and Jacquemont (2003), for instance, used an optimal control model to analyse the choice between two types of forests: (i) one with high timber and carbon sequestration values but lower, or negative, biodiversity values and (ii) one with lower timber and carbon sequestration benefits but with high biodiversity values. To assess four alternative objective functions that maximized (a) volume harvested, (b) wood product C storage, (c) forest C storage and (d) C storage in the forest and products for a hypothetical forest, Hennigar et al. (2008) employed an optimizing forest management model (Remsoft Spatial Planning System). Partial equilibrium econometric methods have also been widely applied (Adam et al. 2020; Ayoade et al. 2018; Murphy et al. 2018; Chakir et al. 2017; Favero et al. 2017; Plantinga et al. 1999), though without attention to the time series properties of the variables to be tested. The current study is set to examine country-level information. The ensuing modelling framework inspired by Sohngen et al. (1999), Bateman and Lovett (2000) and Sohngen and Mendelsohn (2003) assume that carbon sequestration benefits are based on tonnes of carbon stored in the biosphere and the wood market. The short-run wood supply (Q_{wt}), i.e. annual roundwood harvest, forest stock and forest area, in the country is

$$Q_{wt} = f(H_t) \tag{11.1}$$

The harvest (H_{it}) reflects the volume or amount of natural forest stock (S_{it}) available for harvest, where

$$S_t = f(S_{t-1}, G_t, H_t) \tag{11.2}$$

with G_t being the annual change in forest stock per hectare, H_t is annual harvest and S_{t-1} previous stock levels. The country's harvest reflects the amount of forest stock available for harvest, i.e.

$$Q_{wt} \leq S_t \tag{11.3}$$

The carbon supply function depends on the wood supply (Eq. 11.1), the forest area and forest stock growth (Eq. 11.2). The forest area and stock are related as

$$S_t = Ae^{b_i t} \tag{11.4}$$

where b_i is the rate of growth in supply. Following developments in climate change mitigation efforts, the maintained forest stock is a function of the annual rent for carbon, R_t, assumed to have the following form:

$$S_t = \varphi R_t^{k1} \tag{11.5}$$

where k_1 is the rate of growth in carbon rents. With no climate change, $k(t)$ is 0. Between Eqs. (11.4) and (11.5), $\varphi R_t^{k1} = Ae^{b_i t}$, where $R_t^{k1} = \frac{A}{\varphi} e^{b_i t}$, which could be summarized as

$$R_t^{k1} = we^{b_i t} \tag{11.6}$$

The linear form of Eq. (11.6) accounting for the growth or decay of carbon supply takes the form:

$$\ln R_t = \frac{1}{k_1}\left(\ln w + b_{it}\right) \tag{11.7}$$

Since greenhouse gases are assumed to increase radiative forcing, emissions cause k(t) to be positive. Studies have shown that the annual relative change in forest area (A_t) is a function of income per capita, tree growth, forest density and other exogenous variables (Diaz-Balteiro and Romero 2008; Backéus et al. 2005; de Jong et al. 2000; Roslan Ismail 1995). The short-run carbon supply equation may thus be expressed as a function of the forest area, A_t; previous stock levels proxied by forest density, FOD_t; price of wood, P_{wt}; carbon rent R_t; level of infrastructural development, TR_t; and consumption proxied by income per capita, $GDPc_t$. The empirical equation is thus modelled as a log linear form:

$$\begin{aligned} \ln S_t = {} & \alpha_0 + {}^2{}_0 \ln R_t + {}^2{}_1 \ln P_{wt} + {}^2{}_2 \ln P_{fst} \\ & + \alpha_1 \ln FOD_t + \alpha_2 \ln G_t + \alpha_3 \ln H_t + \\ & + \alpha_4 \ln Z_{it} + \alpha_5 \ln GDPc_{it} \\ & + \alpha_6 \ln FOR_t + \alpha_7 \ln TR_{it} + \mu_t \end{aligned} \tag{11.8}$$

where P_{fst} is the price of fossil fuel energy, FOR_t is forest-related policy (e.g. subsidy and tax)[5] and the other variables are as previously defined. R_t is carbon price as proxy for supply of carbon services,

[5]Subsidy may increase the annual net return to forested land and reduces the annual net return to deforested land.

G_t is change in forest stock, H_t is harvest and Z_{it} is other determinants of supply, e.g. production, transport and marketing infrastructure and structural economic reforms. Substituting for ln R_t as in eqs. (11.7) and Eq. (11.8) then yields

$$\ln S_t = \eta_k + \beta_1 \ln P_{wt} + \beta_2 \ln P_{\text{fst}} + \alpha_1 \ln \text{FOD}_t + \alpha_2 \ln G_t + \alpha_3 \ln H_t + \alpha_4 \ln Z_{\text{it}} + \alpha_5 \ln \text{GDPc}_{\text{it}} + \alpha_6 \ln \text{FOR}_t + \alpha_7 \ln \text{TR}_{\text{it}} + \varepsilon_t \quad (11.9)$$

where, η_k is a constant that embodies $\ln R_t = \frac{1}{k_1}(\ln w + b_{it})$. The estimation of Eq. (11.9) therefore allows for possible estimation of the elasticity of carbon supply, the rate of growth in carbon rents and the intertemporal effects of climate change on carbon supply. We employ time series econometric techniques to establish the causation in Eq. (11.9). Parajuli et al. (2016) compared the estimation results obtained from the multivariate vector error correction (VECM) method with the traditional simultaneous equations' estimation approach and found that the traditional simultaneous equations' estimation approach produces similar demand and supply coefficients as the VECM method.

4.2 Nature and Source of Data

Secondary data is employed covering the period 1980–2018. Information for Cameroon on land use area (e.g. forest area, A_t), forest stock (S_t), annual harvest (H_t) as well as annual change in forest stock per hectare (G_t) is obtained from the Food and Agriculture Organization (FAO) database (FAOSTAT) and statistical accounts of the Ministry of Forestry and Wildlife. Information on forest density (FOD_t) and price of wood (P_{wt}) are obtained from the International Tropical Timber Organization (ITTO). Information on income per capita (GDP_{ct}), exchange rate and rate of discount are obtained from the Penn World Tables. Information for Cameroon on road infrastructure (TR_t), forest-related expenditure (FOR_t) and fossil fuel (annual fuel pump) prices are obtained from the World Bank. Carbon prices are obtained from pointcarbon.com, an online carbon price data repository.

Table 11.1 Summary statistics for variables used in estimating wood and carbon capture

Variable	Mean
Forest stock (S_{it}) (million m^3)	21.3
Forest area (A_t) (million ha)	18.8
Forest density (FOD_t) (million per ha)	17
Annual harvest (H_{it}) (million per m^3)	2.5
Annual change in forest stock per hectare (G_{it}) (%)	0.98
Forest expenditure (FOR_t) (million US$)	300
Forest carbon (ton of c per ha)	300
Annual rent for carbon (R_t) (US$ per ha)	6000
Price of wood (P_{wt}) (US$ per m^3)	175
Road infrastructure (TR_t) (km/1000 km^2 of land area)	70
Income per capita (GDPc_t) (US$)	1100
Price of fossil fuel energy (P_{fst})	610

Notes: Annual harvest = industrial roundwood and wood fuel. Road infrastructure is proxied with road network density which relates to total network which includes the primary, secondary and tertiary networks. Current carbon prices in the EU emission trading scheme are in the order of $20 per tonne. (*Source*: Author's computation, 2020)

Table 11.1 provides the descriptive statistics of the variables. The average forest stock including secondary forest, primary forest and agroforested areas amounts to about 21.3 million m^3, within a forest area of 18.8 million ha. Cameron's forest density averages 17 million per ha providing opportunity for 2.5 million per m^3 of annual wood harvest. The annual change in forest stock per hectare is 0.98%. Government's effort to manage the forest resource is gauged by annual expenditure of US$300 million which represents 1% of GDP in current values. The country's forest carbon potential is estimated at about 300 tonne of carbon per ha generating a possible annual carbon rent of 6000 US$ per ha or a national forest average of annual rent of carbon averaging US$ 112.8 billion. A road network density of 70 km/1000 km^2 of land area supports the exploitation of forest assets and market linkage.

5 Presentation and Discussion of Findings

5.1 Long-Run Determinants of Carbon Capture

The estimated parameters of the carbon supply equation are shown in Table 11.2. The coefficients report expected signs, although the parameters differed substantially in magnitude. This estimation suggests that wood price and forest expenditure have a positive effect on carbon supply. In the long run, the effects of forest area, forest density and GDP per capita on carbon supply were all positive and significant. For instance, it is deduced that a 1% increase in consumption is linked with 0.8668% decrease in carbon supply. Similarly, rise in price of fossil fuel leads to a 10.2956% decline in carbon supply. It can also be deduced that the model fits the data set well, $R^{-2} = 0.6319$. This means that 63.19% of the variations in carbon supply are explained by changes in forest stock, area, density, price, forest policy and road infrastructure. These results seem plausible and corroborate previous studies. For example, Guariguata et al. (2008) observe that better management of natural forest offers additional opportunities for implementing adaptation to climate change measures, at both industrial and smallholder levels. Härtl and Knoke (2014) show the price of oil to have significant changes in wood supply, tending towards an increase in wood graded for energy use with rising oil and timber prices. Daigneault et al. (2008) indicate that competitiveness in the forestry sector is sensitive both to strong monetary policies and to the weak currency policies pursued by competitive governments, as well as a weak dollar policy that is intended to improve competitiveness in the global timber market and reduce the large trade gap and account deficit.

Table 11.2 Parameter estimates for long-run carbon capture

Variable	Coefficients	t-Statistics	Prob-value
Harvest (ln H_{it-1})	0.6701	2.5422**	0.0741
Change in forest stock (ln G_{it})	0.5256	1.1754	0.0025
Rent for carbon (ln R_t)	0.2619	1.6451*	0.0735
Forest area (ln A_t)	0.6578	3.8901***	0.0541
Forest density (ln FOD_t)	0.3825	2.7614***	0.0421
Price of wood (P_{wt})	0.4917	2.1882**	0.0167
Road infrastructure (ln TR_t)	0.1535	1.9361**	0.0231
Income per capita (ln $GDPc_t$)	−0.8668	−2.4734**	0.0001
Price of fossil fuel energy (ln P_{fst})	−0.2956	−1.6870*	0.0049
Forest expenditure (ln FOR_t)	0.4566	3.5108***	0.0055
Intercept	0.8904	2.2990**	0.0637

Diagnostic tests:
Adj. R^2 = 0.6319 F-stats = 139.999; DW = 1.8314; χ^2_{auto} (B-G) = 0.7521 (0.6817); χ^2_{norm} (JB) =1.5389 (0.9854); χ^2_{RESET} = 1.3721 (0.3178); χ^2_{white} white = (2.8341) 0.046
Notes:
χ^2_{auto} is the Breusch-Godfrey LM test for autocorrelation; χ^2_{norm} is the Jarque-Bera normality test; χ^2_{RESET} is the Ramsey test for omitted variables; χ^2_{white} is the white test for heteroskedasticity; * and ** indicate statistical significance at the 5% and 1% levels, respectively. (*Source*: Author's computation, 2020)

The economic and ecological implications of the results are instructive for future forest management. The observation on forest expenditure implies that public resources and increased attention to the political economy underlie carbon policies (Hansen and Lund 2011; Marfo and Mckeown 2013). The economic and legal implications of the interrelationship between carbon sequestration programs and biodiversity (Caparrós and Jacquemont 2003), as well as forest management choices, have been shown to influence the levels of carbon sequestration (Duinker 1990). Seidl et al. (2007) investigated effects of alternative management strategies for secondary spruce forests (*Picea abies* (L.) Karst.) and showed that in situ carbon sequestration is sensitive to forest management with the highest amount of carbon stored in the unmanaged strategy, followed by the continuous cover regime. Stern (2008) reiterates that mitigation – taking strong action to reduce emissions – must be viewed as an investment, a cost incurred now and in the coming few decades to avoid the risks of very severe consequences in the future. If these investments are made wisely, the costs will be manageable, and there will be a wide range of opportunities for growth and development along the way. Opportunity cost estimates for carbon sequestration reveal that carbon sequestration through forest management can be a cost-efficient way to reduce atmospheric CO_2, but the achievable quantities are limited due to biological limitations and societal constraints. Seidl et al. (2007) and Diaz-Balteiro and Romero (2008) emphasize the importance of developing sustainable forest management strategies that serve the multiple demands on forests in the future.

5.2 Short-Run Determinants of Carbon Capture

The equation of unrestricted error correction model is specified as follows:

$$\begin{aligned}\Delta \ln S_t = \eta_o &+ \sum_{t=1}^{l} \omega_1 \Delta \ln S_{t-1} + \sum_{t=0}^{m} k_1 \Delta \ln R_{t-m} + \sum_{t=0}^{n} \beta_1 \Delta \ln P_{wt-n} + \sum_{t=0}^{p} \beta_2 \Delta \ln P_{fst-p} + \sum_{t=0}^{q} \alpha_1 \Delta \ln FOD_{t-q} \\ &+ \sum_{t=0}^{r} \alpha_2 \Delta \ln G_{t-r} + \sum_{t=0}^{s} \alpha_3 \Delta \ln H_{t-s} + \sum_{t=0}^{u} \alpha_4 \Delta \ln Z_{t-u} + \sum_{t=0}^{v} \alpha_5 \Delta \ln GDPc_{t-v} + \\ &+ \sum_{t=0}^{w} \alpha_6 \Delta \ln FOR_{t-w} + \sum_{t=0}^{y} \alpha_7 \Delta \ln TR_{t-y} + \mu_{t-1}\end{aligned} \quad (11.8)$$

Table 11.3 provides the details of short-run results. Regarding all other regressors, they exert a statistically significant effect in carbon supply and have the expected signs. The results show that a 0.3971% increase in carbon supply in current period is significantly linked with a 1% rise in carbon in previous periods. Similarly, sequestration potential increases with rise in expected carbon revenues, forest density and government expenditures in the forest sector. The positive coefficients of 0.45927 and 0.3652 for forest density and policy measures indicate a 1% increase in carbon supply by about 0.45927% and 0.3652%, respectively. On the other hand, the results show that a 0.3652% increase in carbon supply in current period is significantly linked with a 1% improvement in road infrastructure. Similarly, increases in wood prices, fossil fuel price, timber harvest and consumption potentials have negative and statistically significant effects on carbon supply in the short run. These findings are robust, corroborating similar experiences. van't Veld and Plantinga (2005) show analytically that if price increases over time, it becomes optimal to delay certain sequestration projects,

Table 11.3 Parameter estimates for short-run carbon capture (dependent variable =$\Delta \ln S_t$)

Variables	Coefficients	t-Statistics	Prob-value
Constant	0.9981	2.2990**	0.0517
$\Delta \ln S_{t-5}$	0.3971	2.3517**	0.0015
$\Delta \ln R_{t-2}$	0.3532	1.7541 *	0.0572
$\Delta \ln P_{wt-2}$	0.5341	3.1647 ***	0.0007
$\Delta \ln P_{fst-1}$	−0.3376	−2.2873 **	0.0013
$\Delta \ln FOD_{t-3}$	0.4592	3.706 ***	0.0351
$\Delta \ln G_{t-2}$	0.6883	2.3604**	0.0027
$\Delta \ln H_{t-3}$	0.7092	3.2581 ***	0.0413
$\Delta \ln Z_{t-4}$	−0.2714	−4.1473***	0.0002
$\Delta \ln GDPc_{t-3}$	−0.9354	−3.0947***	0.0001
$\Delta \ln FOR_{t-2}$	0.5767	2.1083**	0.0273
$\Delta \ln TR_{t-2}$	0.3652	1.8736**	0.0523
ECM (μ_{t-1})	−0.5906	−3.582***	0.0011

Diagnostic tests:
Adj. R^2 = 0.7516; F-stats = 98.263; DW = 1.8931; χ^2_{auto} (B-G) = 1.868 (0.1873); χ^2_{norm} (J-B) = 0.3145 (0.9568); χ^2_{RESET} = 0.0012 (0.8241); χ^2_{white} white = 0.3551 (0.8061)

Notes:
χ^2_{auto} is the Breusch-Godfrey LM test for autocorrelation; χ^2_{norm} is the Jarque-Bera normality test; χ^2_{RESET} is the Ramsey test for omitted variables; χ^2_{white} is the white test for heteroskedasticity; * and ** indicate statistical significance at the 5% and 1% levels, respectively. (*Source*: Author's computation, 2020)

whereas the optimal timing of energy-based abatement projects remains unchanged. As a result, the optimal share of sequestration significantly falls. Calibrating their analytical model, van't Veld and Plantinga (2005) find that a modest 3% rate of price increase results in about a 60% reduction in the optimal sequestration share relative to constant price projections.

The ECM parameter (μ_{t-1}) for the lagged error terms is negative and significant, indicating the existence of a long-run relationship between the variables. The ECM measures the speed at which equilibrium is restored to the model. The results indicate that some 59.06% of the change in Cameroon's carbon supply is attributed to disequilibrium. Tests for normality of residuals, serial correlation, heteroskedasticity and misspecification of functional form were applied to the ECM. Since none of these tests disclosed any significant evidence of departure from standard assumptions, the empirical validity of the model was confirmed by the various diagnostic tests.

These results have interesting implications for carbon policy for both Cameroon and other developing countries in the sub-continent. Policy must promote sound market signals, overcome market failures and have equity and risk mitigation at its core. Stern (2008) identifies three essential elements of policy for mitigation to include carbon price, technology policy and the removal of barriers to behavioural change. Achieving this could mean, for example, halting deforestation and substantial intensification of sequestration activities. It is therefore imperative for countries to invest in mechanisms that would mainstream climate change into their development strategies to stave off its possible negative impacts (Jung 2005; Michaelowa 2003). Similarly, commensurate efforts must also be made to identify and exploit the opportunities that climate policy presents. Policy promoting REDD+ is important since the AFOLU accounts for about a quarter (~10–12 $GtCO_2$ eq/yr) of net anthropogenic GHG emissions mainly from deforestation, agricultural emissions from soil and nutrient management and livestock. Most recent estimates indicate a decline in AFOLU CO_2 fluxes, largely due to decreasing deforestation rates and increased afforestation (Olschewskia and Benítez 2010; Albrecht and Kandji 2003). However, the uncertainty in historical net AFOLU emissions is larger than for other sectors, and additional uncertainties in projected baseline net AFOLU emissions exist. Nonetheless, in the future, net annual baseline CO_2 emissions from AFOLU are projected to decline, with net emissions potentially less than half the 2010 level by 2050 and the possibility of the AFOLU sectors becoming a net CO_2 sink before the end of century (IPCC 2013).

Policy will have to address broad-ranged socioeconomic and political impediments (Guariguata et al. 2008; Unruh 2008; Fry 2008; Michaelowa 2003). According to Guariguata et al. (2008), the relationship between tropical forests and global climate change must also focus on adaptation not only mitigation, with emphasis placed on how management activities may help forest ecosystems adapt to this change. Youssoufa et al. (2011) draw attention on the lack of awareness and poor flow of information on the potentials of forests for climate change adaptation and highlight the need for integrating forest for adaptation into national development programmes and strategies. They recommend a review of the existing environmental legislations and their implications on poverty reduction strategy and adaptation to climate change (ibid.). According to Unruh (2008), the prospect of using tropical forest to sequester significant amounts of atmospheric carbon as one mitigation approach to climate change under the Kyoto Protocol, the Clean Development Mechanism (CDM) and African land tenure is important and that instead it exists as a prohibitive obstacle to the implementation of afforestation and reforestation sequestration approaches. Five primary tenure problems could be examined and corrected: (1) the disconnect between customary and statutory land rights, (2) legal pluralism, (3) tree planting as land claim, (4) expansion of treed areas in smallholder land use systems and (5) the difficulty of using the "abandoned land" category. The pervasiveness of these tenure-related issues means that the prospects for successfully implementing afforestation and reforestation projects in Africa are in reality quite weak. This will mean that UNFCCC processes be significantly realigned with African reality in order for sequestration expectations to be practical. At COP-21 governments agreed to strengthen societies' ability to deal with the impacts of climate change and also provide continued and enhanced international support for adaptation to developing countries (Nhamo and Nhamo 2016b). The agreement also recognizes the importance of averting, minimizing and addressing loss and damage associated with the adverse effects of climate change. Furthermore, it acknowledges the need to cooperate and enhance the understanding, action and support in different areas such as early warning systems, emergency preparedness and risk insurance. There are positive signs after COP-22, where government's proclamation signals a shift towards a new era of implementation and action on climate and sustainable development, as well as the highest political commitment to combat climate change, as a matter of urgent priority via enhancing adaptive capacity, strengthening resilience and reducing vulnerability.

6 Conclusions

The time-bound targets of the SDGs and the Paris Climate Agreement combine to provide a framework for national action and global cooperation towards sustainable development while achieving net-zero greenhouse gas emissions by the middle of the century. Sachs et al. (2019) note that the SDGs and the Paris Agreement on Climate Change call for deep transformations in every country that will require complementary actions by governments, civil society, science and business. This will require stakeholders to have shared understanding of how all SDGs can be operationalized. This will require SDG transformations as modular building blocks of SDG achievement. For example, six clustered transformations may include (1) education, gender and inequality; (2) health, wellbeing and demography; (3) energy decarbonization and sustainable industry; (4) sustainable food, land, water and oceans; (5) sustainable cities and communities; and (6) digital revolution for sustainable development. Such transformations which come with priority investments and regulatory challenges will need actions by well-defined parts of government working with business and civil society. This implies that governments have an important role to play, with these transformations possibly operationalized within the structures of government while respecting the strong interdependencies across all the SDGs.

While Pradhan et al. (2017) reiterate that SDG-1 (no poverty) has synergetic relationship with most of the other goals, this will mean the other SDGs being studied in this paper, particularly SDG-13 and SDG-15 on climate action and life on land, have a coronary effect with possibility to reduce poverty and improve livelihoods. Though in developing their climate policies, countries like Cameroon have not explicitly mentioned the SDGs, there are however many areas of synergy that can be found between climate policies and the SDGs. This interconnectivity and clear aims of the global goals call for understanding by all stakeholders on how to promote prosperity while protecting the planet. By investing in approaches to sustainable forest management that better preserve and restore the natural resource base and increase the resilience of the ecosystem to a changing climate, developing country governments like that of Cameroon contribute to SDGs 12, 13 and 15.

Achieving healthy ecosystems to protect the planet and sustain livelihoods should be primordial in development policy-making, because of the synergistic effects emanating from the forest ecosystem through its myriad of environmental goods and services. In addition, forests and rangelands sustain a range of industries, generate jobs and income and act as a source of food, medicine and fuel for more than a billion people. A strong case can be made for investments in climate action since forest-related climate policy intersects with other societal goals creating the possibility of co-benefits or adverse side effects. These intersections, if well-managed, can strengthen the basis for undertaking climate action.

Overall, this case study employs partial equilibrium analysis to examine potential carbon supply from ancillary forest services. The rationale hinges on the premise that saving tropical forests as a global warming countermeasure is important not only for ecological benefits and proper forest management but also because of the opportunity for obtaining monetary gain that flows from the global carbon markets. The study finds that timber harvest in the previous period contributes to a decline in the per unit area of carbon supply. However, carbon supply comes from the management of new forest areas. Economic performance has a negative effect on long-run carbon capture and supply. This research thus contributes to policy decision-making within the climate change debate. Given the limitations of the current examination, areas for future research should include evaluating the distributional outcomes (how regions and population subgroups are affected) and understanding market adaptations for regions and individuals. Limiting the effects of climate change is necessary to achieve sustainable development and equity, including poverty eradication. At the same time, some mitigation efforts could undermine action on the right to promote sustainable development and on the achievement of poverty eradication and equity.

References

Adam, J., Daigneault, B.L. Sohngen & Sedjo, R. (2020). Carbon and market effects of U.S. forest taxation policy. *Ecological Economics*, 178, 106803.

Albrecht, A. & Kandji, S.T. (2003). Carbon sequestration in tropical agroforestry systems. *Agriculture, Ecosystems & Environment*, 99(1), 15–27.

Alig, Ralph J., Adams, D.M., McCarl, B.A., Callaway, J.M., & Winnett S. (1997). Assessing Effects of Mitigation Strategies for Global Climate Change with an Intertemporal Model of the U.S. Forest and Agriculture Sectors. *Environmental Resource Economics* 9, 259–274.

Backéus, S., Wikström P. & Lämås T. (2005). A model for regional analysis of carbon sequestration and timber production. *Forest Ecology and Management,* 216(1), 28-40.

Bateman I.J. & Lovett, A.A. (2000). Estimating and valuing the carbon sequestered in softwood and hardwood trees, timber products and forest soils in Wales. *Journal of Environmental Management*, 60(4), 301–323.

Benítez-Ponce, P.C., I. McCallum, M. Obersteiner & Yamagata, Y. (2007). Global potential for carbon sequestration: geographical distribution, country risk and policy implications. *Ecological Economics*, 60, 572–583.

Bruce M.C., James H., Janie R., Clare M S., Stephen T. & Eva, L.W. (2018). Urgent action to combat climate change and its impacts (SDG 13): transforming agriculture and food systems. *Current Opinion in Environmental Sustainability*, 34(1), 13–20.

Brack D. (2019) *Forests and Climate Change. Background study prepared for the fourteenth session of the United Nations Forum on Forests.* New York: United Nations Secretariat.

Brack, D. (2018). Sustainable consumption and production of forest products. Background study prepared for the thirteenth session of the United Nations Forum on Forests. www.un.org/esa/forests/wp-content/uploads/2018/04/UNFF13_BkgdStudy_ForestsSCP.pdf, Accessed on 19 June 2020.

Bellassen V., & Gitz V. (2008). Reducing Emissions from Deforestation and Degradation in Cameroon: Assessing costs and benefits. *Ecological Economics*, 68(2), 336–344.

Brown H.C.P, Nkem, J.N., Sonwa D.J., & Youssoufa B.(2010). Institutional adaptive capacity and climate change response in the Congo Basin forests of Cameroon. *Mitigation and Adaptation Strategy for Global Change,* 15, 263–282.

Brown D.R. (2006). Personal preferences and intensification of land use: their impact on southern Cameroonian slash-and-burn agroforestry systems. *Agroforestry Systems* 68, 53–67.

Brown S. (1997). Estimating biomass and biomass change of tropical forests: a primer. FAO *Forestry Paper*, Rome. p55.

Brown S., Hall C. A. S., Knabe W., Raich J., Trexler M. C., Woomer P.L. (1993). Tropical forest: their past, present and potential future roles in the world's carbon budget. *Water Air Soil Pollution*, 70, 71–94.

Campbell, B.M. (2009). Beyond Copenhagen: REDD+, agriculture, adaptation strategies and poverty. *Global Environmental Change*, 19, 397–399.

Caparrós A. & Jacquemont F. (2003). Conflicts between biodiversity and carbon sequestration programs: economic and legal implications. *Ecological Economics*, 46 (1), 143–157.

Chakir R., Stéphane De Cara & Vermont, B. (2017). Price-Induced Changes in Greenhouse Gas Emissions from Agriculture, Forestry, and Other Land Use: A Spatial Panel Econometric Analysis. Revue économique, 68 (3), 471.

Daigneault A.J., B. Sohngen & Sedjo, R. (2008). Exchange rates and the competitiveness of the United States timber sector in a global economy. *Forest Policy and Economics*, 10 (3), 108–116.

de Jong, B.H.J., Tipper R., Montoya-Gómez, G. (2000). An Economic Analysis of the Potential for Carbon Sequestration by Forests: Evidence from Southern Mexico. *Ecological Economics,* 33, 313–327.

Depro B.M, Murray B.C, Alig R.J. et al. (2008). Public land, timber harvests, and climate mitigation: Quantifying carbon sequestration potential on U.S. public timberlands. *Forest Ecology and Management* 255, 1122–1134

Duinker, P.N. (1990). Climate change and forest management, policy and land use. *Land Use Policy*, 7(2), 124-137.

Diaz-Balteiro, L. & Romero C. (2008). Making forestry decisions with multiple criteria: A review and an assessment. *Forest Ecology and Management*, 255(9), 3222–3241.

Dzebo, A., Janetschek, H., Brandi, C. & Iacobuta, G. (2018). The Sustainable Development Goals Viewed through a Climate Lens. *SEI Policy Brief.* Stockholm Environment Institute, Stockholm. www.sei.org/publications/the-sustainable-development-goals-viewed-through-a-climate-lens/ Accessed on 19 June 2020

Favero A., R. Mendelsohn & Sohngen, B. (2017). Using forests for climate mitigation: sequester carbon or produce woody biomass? *Climatic Change*, 144(2), 195–206.

Fawcett, A.A., Iyer, G.C., Clarke, L.E., Edmonds, J.A., Hultman, N.E., McJeon, H.C., Rogelj, J., Schuler, R., Alsalam, J., Asrar, G.R., Creason, J., Jeong, M., McFarland, J., Mundra, A., & Shi, W. (2015). Can Paris pledges avert severe climate change?. *Science* 350 (6265):1168–1169.

FAO (2020). *Sustainable Development Goals. Indicator 15.2.1 – Progress towards sustainable forest management.* Rome: Food and Agricultural Organisation of the United Nations. http://www.fao.org/sustainable-development-goals/indicators/1521/en/. Accessed on 24 August 2020.

Fry, I. (2008). Reducing emissions from deforestation and forest degradation: opportunities and pitfalls in developing a new legal regime. *Review of European Community & International Environmental Law*, 17(2), 166–182.

Glass, L. M. & eNewig, J. (2019). Governance for achieving the Sustainable Development Goals: How important are participation, policy coherence, reflexivity, adaptation and democratic institutions? *Earth System Governance 2*, 100031. https://doi.org/10.1016/j.esg.2019.100031

Gough C.M., Jeff W. Atkins, Robert T. Fahey, & Hardiman, B.S. (2019). High rates of primary production in structurally complex forests. *Ecology*, https://doi.org/10.1002/ecy.2864.

Guariguata M.R., Cornelius, J.P. Locatelli, B., Forner, C., & Sánchez-Azofeifa G.A. (2008). Mitigation needs adaptation: Tropical forestry and climate change. *Mitigation and Adaptation Strategy for Global Change,* 13, 793–808

Hansen C.P. & Lund, J.F. (2011). The political economy of timber taxation: The case of Ghana. *Forest Policy and Economics*, 13(8). 630–641.

Härtl F., & Knoke, T. (2014). The influence of the oil price on timber supply. *Forest Policy and Economics*, 39(1), 32–42.

Hernandez M., T. Gómez, J. Molina, M.A. León, & Caballero, R. (2014). Efficiency in forest management: A multiobjective harvest scheduling model. *Journal of Forest Economics*, 20(3): 236–251.

Hennigar, C.R., MacLeana D.A., & Amos-Binksa L.J. (2008). A novel approach to optimize management strategies for carbon stored in both forests and wood products. *Forest Ecology and Management*, 256(4), 786–797.

Hess, P.N. (2016). *Natural resources and climate change.* Routledge. 52 pp. ISBN: 9781315722467. https://doi.org/10.4324/9781315722467-12

Imai, N., Samejima, H., Langner, A., Ong, R. C., Kita, S., Titin, J., Chung, A. Y., Lagan, P., Lee, Y. F., & Kitayama, K. (2009). Co-benefits of sustainable forest management in biodiversity conservation and carbon sequestration. *PloS one*, 4(12), e8267.

IIASA (2019) Deep transformations needed to achieve Sustainable Development Goals. *Science Daily.* International Institute for Applied Systems Analysis (IIASA). Retrieved June 18, 2020 from www.sciencedaily.com/releases/2019/08/190826112705.htm

IPBES (2019): *Global assessment report on biodiversity and ecosystem services of the Intergovernmental Science-Policy Platform on Biodiversity and Ecosystem Services.* Brondizio, E.S. Settele, J., Díaz, S. and Ngo H.T. (editors). IPBES secretariat, Bonn, Germany. https://zenodo.org/record/3553579#.XsLOCkRKjtQ. Accessed on 19 June 2020

IPBES (2018) *The IPBES assessment report on land degradation and restoration.* Montanarella, L., Scholes, R., and Brainich, A. (eds.). Secretariat of the Intergovernmental Science-Policy Platform on Biodiversity and Ecosystem Services, Bonn, Germany. 744 pages. https://doi.org/10.5281/zenodo.3237392. Accessed on 19 June 2020.

IPCC (2019a). *Climate Change and Land: an IPCC special report on climate change, desertification, land degradation, sustainable land management, food security, and greenhouse gas fluxes in terrestrial ecosystems* [P.R. Shukla, J. Skea, E. Calvo Buendia, V. Masson-Delmotte, H.-O. Pörtner, D.C. Roberts, P. Zhai, R. Slade, S. Connors, R. van Diemen, M. Ferrat, E. Haughey, S. Luz, S. Neogi, M. Pathak, J. Petzold, J. Portugal Pereira, P. Vyas, E. Huntley, K. Kissick, M. Belkacemi, J. Malley, (eds.)]. Geneva: Intergovernmental Panel on Climate Change. https://www.ipcc.ch/srccl/. Accessed on 19 June 2020.

IPCC (2019b). *Global Warming of 1.5°C. An IPCC Special Report on the impacts of global warming of 1.5°C above pre-industrial levels and related global greenhousegas emission pathways, in the context of strengthening the global response to the threat of climate change, sustainable development, and efforts to eradicate poverty* [Masson-Delmotte, V., P. Zhai,

H.-O. Pörtner, D. Roberts, J. Skea, P.R. Shukla, A. Pirani, W. Moufouma-Okia, C. Péan, R. Pidcock, S. Connors, J.B.R. Matthews, Y. Chen, X. Zhou, M.I. Gomis, E. Lonnoy, T. Maycock, M. Tignor, and T. Waterfield (eds.)]. Geneva: Intergovernmental Panel on Climate Change. https://www.ipcc.ch/sr15/. Accessed on 19 June 2020.

IPCC. (2013). *Climate Change 2013: Agriculture, Forestry and Other Land Use (AFOLU)*. The Fifth Assessment Report (AR5) of the United Nations Intergovernmental Panel on Climate Change (IPCC), Cambridge University Press: Cambridge, UK.

IPCC. (2007). *Climate Change 2007.* The Fourth Assessment Report (AR4) of the United Nations Intergovernmental Panel on Climate Change (IPCC), Cambridge University Press: Cambridge, UK.

Jung, M. (2005). The role of forestry projects in the clean development mechanism.' *Environmental Science & Policy*, 8(2), 87–104.

Karjalainen, T. (1996). Dynamics and potentials of carbon sequestration in managed stands and wood products in Finland under changing climatic conditions. *Forest Ecology and Management*, 80(1), 113–132.

Kim Man-Keun, & McCarl, B.A. (2015). Uncertainty Discounting for Land-Based Carbon Sequestration. *Journal of Agricultural and Applied Economics*, 41 (1), 1–11.

Kinley, R. (2017). Climate change after Paris: from turning point to transformation, *Climate Policy* 17(1), 9–15. https://doi.org/10.1080/14693062.2016.1191009.

Kotto S.J., Woomer P.L., Moukam A., Zapfack L. (1997). Carbon dynamics in slash-and-burn agriculture and land use alternatives of the humid forest zone of Cameroon. *Agriculture, Ecosystem and Environment* 65, 245–256.

Krcmar, E., Stennes, B., Cornelis van Kooten, G., and Vertinsky I. (2001). Carbon sequestration and land management under uncertainty. *European Journal of Operational Research*, 135(3), 616–629.

Lubowski, R., Plantinga, A., and R. Stavins (2006). Land-use change and carbon sinks: Econometric estimation of the carbon sequestration supply function. *Journal of Environmental Economics and Management* 51,135–152.

Maamoun, N. (2019). The Kyoto Protocol: Empirical Evidence of A Hidden Success. *Journal of Environmental Economics and Management*, 95, 227–256.

Marfo E., & Mckeown, J.P. (2013). Negotiating the supply of legal timber to the domestic market in Ghana: Explaining policy change intent using the Advocacy Coalition Framework. *Forest Policy and Economics*, 32(1), 23–31.

Matheya, A.H., Nelson H. and Gaston C. (2009). The economics of timber supply: Does it pay to reduce harvest levels? *Forest Policy and Economics*, 11(7), 491–497.

Michaelowa, A. (2003). CDM host country institution building. *Mitigation and Adaptation Strategies for Global Change*, 8(3), 201–220.

Morton S., Pencheon D. & Squires N. (2017). Sustainable Development Goals (SDGs), and their implementation: A national global framework for health, development and equity needs a systems approach at every level. *British Medical Bulletin*, 124:81–90. https://doi.org/10.1093/bmb/ldx031

Murphy Rose, Dominique M. Gross & Jaccard, M. (2018). Use of revealed preference data to estimate the costs of forest carbon sequestration in Canada. *Forest Policy and Economics*, 97(1), 41–50.

Newell, R. & R. Stavins (2000). Climate Change and Forest Sinks: Factors Affecting the Costs of Carbon Sequestration. *Journal of Environmental Economics and Management*, 40(3): 211–235.

Ndoye, O., & Kaimowitz, D. (2000). Macro-economics, markets and the humid forests of Cameroon, 1967–1997. *The Journal of Modern African Studies*, 38(2), 225–253. https://doi.org/10.1017/S0022278X00003347.

Nhamo, G. & Nhamo, S. (2016a). One global deal from Paris 2015: Convergence and contestations on the future climate mitigation agenda. *South African Journal of International Affairs*, 23(3), 323–346. https://doi.org/10.1080/10220461.2016.1252281

Nhamo, G. & Nhamo, S. (2016b). Paris (COP21) Agreement: Loss and damage, adaptation and climate finance issues. *International Journal of African Renaissance Studies*, 11(2), 118–138. https://doi.org/10.1080/18186874.2016.1212479

Oberthür, S. (2016). Reflections on Global Climate Politics Post Paris: Power, Interests and Polycentricity. *The International Spectator: Italian Journal of International Affairs*, 51(4), 80–94. https://doi.org/10.1080/03932729.2016.1242256

Olesen Strange A., Lesschen J.P., Rayment M., Ebrahim N., Weiss P., Arets EJMM, Frelih-Larsen A., Sikirica N., Nabuurs G.J. & Schelhaas M. (2016). Agriculture and LULUCF in the 2030 framework. Luxembourg: European Union. http://edepot.wur.nl/405796.

Olschewskia, R. & Benítez P.C. (2010). Optimizing joint production of timber and carbon sequestration of afforestation project. *Journal of Forest Economics*, 16(1), 1–10.

Oyono, P.H., Kouna C. & Mala W. (2005). Benefits of forests in Cameroon. Global structure, issues involving access and decision-making hiccoughs. *Forest Policy and Economics*, 7(3), 357–368.

Phelps, J., Webb, E., & Agrawal A. (2010). Does REDD+ threaten to recentralize forest governance. *Science*, 328(5976), 312–313.

Plantinga, A.J., T.M. Mauldin & Miller D.J. (1999). An Econometric Analysis of the Costs of Sequestering Carbon in Forests. *American Journal of Agricultural Economics*, 81, 812–824.

Plantinga A.J., (2015). Integrating Economic Land-Use and Biophysical Models. Annual Review of Resource Economics, 7(1), 233–249.

Parajuli R., D. Zhang, & Sun, J.C. (2016). Modeling stumpage markets using vector error correction vs. simultaneous equation estimation approach: A case of the Louisiana sawtimber market, *Forest Policy and Economics*, 70 (1), 16–19.

Pradhan, P., Costa, L., Rybski, D., Lucht, W., & Kropp, J.P. (2017). A Systematic Study of Sustainable Development Goal (SDG) Interactions. *Earth's Future*, 5:1169–1179. https://doi.org/10.1002/2017EF000632

Perez, C., Roncoli, C., Neely, C., & Steiner, J. (2007). Can carbon sequestration markets benefit low-income producers in semi-arid Africa? Potentials and challenges. *Agricultural systems*, 94(2). https://doi.org/10.1016/j.agsy.2005.09.009.

Richards, K. (2004). A Brief Overview of Carbon Sequestration Economics and Policy. *Environmental Management*, 33(4), 545–558.

Roslan I. (1995). An economic evaluation of carbon emission and carbon sequestration for the forestry sector in Malaysia. *Biomass and Bioenergy*, 8(5), 281–292.

Sachs, J.D., Schmidt-Traub, G., Mazzucato, M. et al. (2019). Six Transformations to achieve the Sustainable Development Goals. *Nat Sustain* 2, 805–814. https://doi.org/10.1038/s41893-019-0352-9

Schlamadinger B., Bird N., Johns T., Brown S., Canadell J., Ciccarese L., Dutschke M. & Yamagata Y. (2007) A synopsis of land use, land-use change and forestry (LULUCF) under the Kyoto Protocol and Marrakech Accords. *Environmental Science and Policy*, 10(4), 271–282.

Sedjo R. & B. Sohngen (2012). Carbon Sequestration in Forests and Soils. *Annual Review of Resource Economics*, 4:127–144.

Seong-Hoon Cho, Moonwon Soh, Burton C. English, T., Yu E. & Boyer, C.N. (2019). Targeting payments for forest carbon sequestration given ecological and economic objectives. *Forest Policy and Economics*, 100, 214–226.

Schröder, P., Antonarakis, A., Brauer, J., Conteh, A., Kohsaka, R., Uchiyama, Y., & Pacheco, P. (2019). SDG 12: Responsible Consumption and Production – Potential Benefits and Impacts on Forests and Livelihoods. In P. Katila, C. Pierce Colfer, W. De Jong, G. Galloway, P. Pacheco, & G. Winkel (Eds.), *Sustainable Development Goals: Their Impacts on Forests and People* (pp. 386–418). Cambridge: Cambridge University Press. https://doi.org/10.1017/9781108765015.014

Sedjo, R., Wisniewski, J., Sample, A. & Kinsman J. (1995). The Economics of Managing Carbon via Forestry: Assessment of Existing Studies. *Environmental and Resource Economics* 6:139–165.

Seidl, R., Rammer, W., Jäger, D., Currie W.S. & Lexer M.J. (2007). Assessing trade-offs between carbon sequestration and timber production within a framework of multi-purpose forestry in Austria Forest. *Ecology and Management*, 248(1–2): 64–79.

Sohngen, B. & Mendelsohn, R. (2003). An Optimal Control Model of Forest Carbon Sequestration. *American Journal of Agricultural Economics* 85, 448–457.

Sohngen, B., Mendelsohn R. & Sedjo, R. (1999). Forest Management, Conservation, and Global Timber Markets. *American Journal of Agricultural Economics* 81, 1–13.

Somorin, O., Brown, H., Visseren-Hamakers, I., Sonwa, D., Arts, B., & Nkem, J. (2012). The Congo Basin forest in a changing climate: policy discourses on adaptation and mitigation (REDD+). *Global Environmental Change* 22, 288–298.

Stern, N. (2008). *Stern Review on the Effects of Climate Change*. http://webarchive.nationalarchives.gov.uk/+/http:/www.hm-treasury.gov.uk/sternreview_index.htm Accessed on 19 June 2020

Stoffberg G.H. , M.W. van Rooyen, M.J. van der Linde, Groeneveld, H.T. (2010). Carbon sequestration estimates of indigenous street trees in the City of Tshwane, South Africa. *Urban Forestry & Urban Greening*, 9(1): 9–14.

Thomson A.M., Izaurralde R.C., Smith S.J., & Clarke L. E. (2008). Integrated estimates of global terrestrial carbon sequestration. *Glob Environ Chang* 18(1):192–203.

Tonna, B. & Marland, G. (2007) Carbon sequestration in wood products: a method for attribution to multiple parties. *Environmental Science and Policy,* 10, 162–168.

Unruh, J.D. (2008). Carbon sequestration in Africa: The land tenure problem. *Global Environmental Change*, 18(4), 700–707.

UN (2020). *Progress Towards Sustainable Forest Management.* New York: United Nations Statistical Division. https://sdg.tracking-progress.org/indicator/15-2-1-progress-towards-sustainable-forest-management/. Accessed 24 August 2020

UN (2017). *Resolution adopted by the Economic and Social Council on 20 April 2017: United Nations strategic plan for forests 2017–2030 and quadrennial programme of work of the United Nations Forum on Forests for the period 2017–2020* (E/RES/2017/4, July 2017). New York: United Nations

UN (2015). *Transforming our World: The 2030 Agenda for Sustainable Development.* New York: United Nations.

UN (2013). *A New Global Partnership: Eradicate Poverty and Transform Economies Through Sustainable Development.* United Nations, New York. https://www.uneca.org/sites/default/files/uploaded-documents/Macroeconomy/post2015/post2015-hlp-report_en.pdf Accessed on 19 June 2020

UN (2012). *The Future We Want – Outcome document*, A/RES/66/288. New York: United Nations. http://rio20.net/wp-content/uploads/2012/06/N1238164.pdf Accessed on 19 June 2020

UN (2002). *Johannesburg Declaration on Sustainable Development*, World Summit on Sustainable Development, A/CONF.199/20, Chapter 1, Resolution 1, Johannesburg, September 2002

van't Veld, K. & A. Plantinga (2005). Carbon sequestration or abatement? The effect of rising carbon prices on the optimal portfolio of greenhouse-gas mitigation strategies. *Journal of Environmental Economics and Management*, 50(1), 59–81.

Yousefpour, R.. & Hanewinkel M. (2009). Eco-efficiency: From technical optimisation to reflective sustainability analysis. *Ecological Economics* 68(6),1711–1722.

Youssoufa M.B., O. Somorin, D.J. Sonwa, J.N. Nkem and B. Locatelli, (2011). 'Forests and climate change adaptation policies in Cameroon. *Mitigation and Adaptation Strategy for Global Change*, 16, 369–385.

Woomer, P.L., & Palm, C.A. (1998) An approach to estimating system carbon stocks in tropical forests and associated land uses. *Commonwealth Forestry Review* 77(3), 181–190.

Zapfack L., Kotto-Same J., Amougou A. & Achoundong, G. (2014). Biodiversity Conservation and Carbon Sequestration in Cocoa Agroforest in Southern Cameroon. In: Behnassi M. et al. (eds.), *Science, Policy and Politics of Modern Agricultural System*, https://doi.org/10.1007/978-94-007-7957-0_18, Dordrecht: Springer Science+Business Media.

Zapfack L., Englad S., Sonké B., Achoundong G., & Birang à Mandong (2002). The impact of land conversion on plant biodiversity in the forest zone of Cameroon. *Biodiversity and Conservation* 11, 2047–2061.

12 Private Sector Sustainable Development Goals' Localisation: Case of Kruger Mpumalanga International Airport, South Africa

Kaitano Dube and Godwell Nhamo

Abstract

There are increased calls for various industries to be more accountable to the communities they serve by giving back through the localisation of the 2030 Agenda for Sustainable Development, which embeds 17 intertwined Sustainable Development Goals (SDGs). The aviation industry is no exception and launched its sustainable future initiative by implementing a 2016 document titled 'On-Board a Sustainable Future'. The document provides a policy roadmap for the localisation of the SDGs within the aviation industry. The aviation industry in South Africa has not been left behind, particularly the private sector. Regardless of the efforts made by the aviation sector in South Africa in this regard, there is minimal documentation and knowledge available about what the airports and other econmic sectors are doing to localise SDGs. Therefore, this chapter documents what the Kruger Mpumalanga International Airport, a private entity in South Africa, is doing to localise the SDGs. Data was generated from fieldwork gathered through in-depth interviews, field observations and document analysis, that was conducted in 2019. The study shows that the airport is making strides in localising the SDGs, thereby responding to their customers and community's needs. Some of the significant projects include investments in education (SDG 4), zero hunger (SDG 2), clean water and sanitation (SDG 6) and building lasting partnerships (SDG 17). The study recommends the continuous monitoring and evaluation of progress made by the airport to address the challenges as it strives towards attaining the SDG targets by 2030.

Keywords

Airports · Aviation · SDGs · South Africa · Kruger Mpumalanga International Airport

K. Dube, PhD (✉)
Department of Hospitality, Tourism and PR, Vaal University of Technology,
Vanderbijlpark, South Africa
e-mail: kaitanod@vut.ac.za

G. Nhamo, PhD
Exxaro Chair in Business and Climate Change, Institute for Corporate Citizenship, University of South Africa, Pretoria, South Africa
e-mail: nhamog@unisa.ac.za

1 Introduction

In most areas, the growth in the aviation sector has resulted in the rapid growth of airports. These airports host a large number of passengers and other persons running into tens of millions annually at some major global airports. As a conse-

G. Nhamo et al. (eds.), *Sustainable Development Goals for Society Vol. 2*, Sustainable Development Goals Series, https://doi.org/10.1007/978-3-030-70952-5_12

quent of the rapid growth of the aviation sector (Lynes and Dredge 2006), airports have witnessed an upsurge in the scope and extent of environmental challenges that they must battle. As with any other subsector of the aviation industry, there has been a growing demand for the industry to be sustainable and more responsible and accountable. There are growing calls by the global community to reduce the environmental footprint (Dube and Nhamo 2019, 2020a; Upham et al. 2003). Such calls have grown louder as some stakeholders question the manner in which the sector handles the issue of greenhouse gas (GHG) emissions that result in global warming and ultimately climate change (Chapman 2007; Gössling and Scott 2018). The GHG emissions are linked to the sector's heavy reliance on carbon fuels. Climate change then leads to the manifestation of extreme hazards such as floods, droughts, wildfires, extreme frost, sea-level rise and heatwaves (World Meteorological Organization – WMO 2019). Such extreme hazards have increased and amplified global pressure for every sector to address the global challenge of climate change. According to Kim (2020), companies should relook their corporate social responsibility (CSR) to ensure that they address these emerging and prevalent weather extremes. Companies should also focus on their immediate communities utilising their CSR to raise awareness and educate consumers on the need to conserve the environment.

There have been various efforts by the airline industry and air manufacturers to address the sector's host of environmental problems. According to Lynes and Dredge (2006), the dynamic and volatile aviation industry is also associated with a high level of environmental problems, including high levels of resource consumption. The industry further faces challenges with noise and land pollution (Lu and Wang 2018). With regard to airports, Cosgrove (2018) notes that airports are particularly challenged as they have to deal with a whole host of environmental challenges from ground transport, waste from landing aircraft and waste that is generated by restaurants and other shops that are housed at airports. Given the foregoing and intense environmental scrutiny, there have been efforts by various aviation entities to align their operations to the global demands, which has resulted in them ensuring their operations follow a sustainable path.

In 2015, the world adopted 17 ambitious Sustainable Development Goals (SDGs) and 169 targets under the 2030 Agenda for Sustainable Development (United Nations (UN) 2015). The SDGs were available for adoption by every sector across the world. In response, the aviation sector, under the stewardship of the International Civil Aviation Organization (ICAO), reinforced the aviation industry's quest to go green. The ICAO produced the roadmap code-named 'On-Board a Sustainable Future' to make the sector sustainable and align with the SDG agenda (ICAO 2016). The roadmap lays out plans that each of the aviation's subsectors has to focus on by identifying problematic issues and setting up some targets. Part of the plans that were set out in 2016 included addressing aircraft noise; air quality issues; global emissions with a focus on climate change, among other issues; and airport planning, which introduced a raft of environmental measures (ibid.).

It is 4 years post the launch of the roadmap and 5 years post the adoption of the SDGs. With the SDGs set to lapse in 2030, there is very little documentation as to how the airport industry has embraced and localised the SDGs. This study comes as a response to that knowledge gap. The study, therefore, seeks to document how one of the biggest private airports in South Africa, the Kruger Mpumalanga International (KMI) Airport, has embraced the SDGs and sustainability agenda over the past couple of years.

2 Literature Survey

The demand for sustainability in the aviation sector has been a critical focus for the past couple of years. Such pressure has given rise to environmentally conscious travellers who demand businesses to be sustainable in the manner they conduct their business. Calls have also been made for businesses to be moral and ethical. There were already signs that customers tend to favour businesses that practise sustainable and responsible business. Hagmann et al. (2015) observe that customers prefer using airlines that have embraced green aviation practices. This is regardless of the same travellers being reluctant to pay more for green travel.

There are various reasons why aviation businesses engage in sustainability programmes. According to Lynes and Gibson (1998), organisations and companies adopt 'greening' programmes as part of efforts to reduce costs and increase efficiency by reducing waste and improving resource efficiencies. On the other hand, several companies adopt green initiatives because of regulatory demands (Khanna and Anton 2002) and to gain a competitive advantage. They also engage in green practices as part of the branding process and as a response to environmental pressure groups or to meet financier requirements (Céspedes-Lorente et al. 2003). Consumers of late have added pressure to force the aviation industry to reduce its environmental cost and especially the GHG emissions (interchanged in this chapter with carbon footprint). One of the programmes that are popular across the world, which is tailored to force the aviation industry to be more green, is the flight shame programme (Gössling et al. 2020).

In response to public concern, the aviation sector is making various efforts. Dube and Nhamo (2019) highlight a raft of measures being made by the sector to address, particularly, climate change. The study indicates that the airline industry is instituting measures, which include the use of, and investment in, biofuels, and changing flight manoeuvres that include a continuous and accelerated descent approach. The accelerated descent helps reduce fuel burn. Other measures include reducing the weight on aircraft and also retrofitting old aircraft to reduce the amount of carbon emissions, among other such measures (ibid.). According to Dube and Nhamo (2020a, b), the air manufacturing industry has been making frantic efforts to address climate change issues and other issues espoused in Agenda 2030. The two leading aircraft manufacturers, namely, Boeing and Airbus, have set their eyes on addressing SDGs. To this end, some progress has been made whereby the aircraft manufacturers tackle SDGs, including those dealing with climate change (SDG13), by investing in more fuel-efficient aircraft. Other SDGs that seem to have been harnessed include SDG6 (water and sanitation), SDGs 4 (focussing on quality education), SDG 5 (gender equality), SDG 9 (industry innovation and infrastructure), SDG 12 (responsible consumption and production) and SDG 17 (partnering for SDGs) (ibid.).

ICAO's Committee on Aviation Environmental Protection (CAEP), which comprises more than 600 international experts dealing with various environmental aspects such as noise, air quality, climate change mitigation and adaptation and aircraft end-of-life recycling, proposed several environmental initiatives to steer the sector on a sustainability path (ICAO 2016). The CAEP also recommended that several measures be taken by airports to address the various environmental issues, which included a new plan for airport design that addresses climate change issues to ensure climate change mitigation and adaptation are included in airport design. In addition, there was a further recommendation for airports to work closely with local communities to address environmental challenges caused by airports to ensure sustainability. It is the communities that are close to the airports that suffer the most from the airports' various environmental challenges such as noise and air pollution.

Boons et al. (2010) argue that achieving sustainability at airports requires a balance between economic, ecological and social demands. In order to respond to social demands, airports have to be responsive to the social demands of host communities and allow for economic benefits to accrue to local communities. On the other hand, ecological demands state that airports must have minimal ecosystem demands, minimal resource use and install buildings with minimal ecological impact (ibid.). Furthermore, airports had to contribute to the economic wellbeing of host communities. Bąk (2018) mentions that as part of aligning to the SDGs, airports have to focus on installing green walls and roofs and investing in renewable energy and devices that save water and energy.

Given the preceding, various airports seem to have been responding to the demands for sustainability, particularly in response to a matter relating to climate change. The Airport Carbon Accreditation, which celebrated its 10th year of existence in 2019, marked the expansion of the programme to 14 airports in Africa, 23 in Latin America, 39 in North America, 54 in Asia and 144 airports in Europe (Airport Carbon Accreditation 2019). These airports were at various accreditation namely; mapping, reduction, optimisation and neutrality. The programmes mainly address issues relating to waste and energy reduction, including green energy and carbon offsetting initiatives. In 2019 alone, at least 322,297 tonnes of carbon dioxide equivalent (CO_{2e}) emissions were reduced at the 274 accredited airports, while 710,673 CO_{2e} was offset at participating airports. Of the participating airports, 52 airports had achieved carbon neutrality (ibid.). The Airports Council International (ACI) also has a Green Airports Recognition, which looks at various environmental aspects that need to be addressed by an airport to be accredited (ACI 2020). These initiatives also speak to various other SDGs and include the following: environmental policy and management (including green procurement), noise, air quality, waste, water (including both municipal water usage and wastewater discharge), energy, carbon, biodiversity, ground transportation and land and water contamination.

According to Dube (2020), in South Africa, public airports are in the process of embracing SDGs by mainly investing in green technology with also a number of them being accredited under the global Airport Accreditation Programme. Such green technology included investing in solar farms at some of the Airport Company of South Africa-run airports. This directly responds to SDG13 that focusses on climate action, particularly the mitigation agenda. Some of the buildings, such as hotels at airports, were also following the green building route to address environmental challenges associated with travelling (Dube and Mearns 2019). In Colombia, one of the airports uses clean development mechanisms to create sustainable infrastructure as well as better environmental conditions for the airport (González-Ruiz et al. 2018). In as much as various airports might be striving to do different things to address sustainability matters, there is very little documentation of these efforts, especially in Africa. The next section presents the methodological orientation of the chapter.

3 Research Design

A case study approach was utilised, which focussed on one of the two largest private international airports in Southern Africa, namely, the Kruger Mpumalanga International (KMI) Airport. A case study research design was followed as it allowed for in-depth insight into the case that was under investigation (Ridder 2017). The data generation methods included in-depth interviews with key informants, detailed field observations and document analysis. The KMI Airport is located in the Mpumalanga Province of South Africa. Its construction began in 2001, and it opened its doors to business in 2002. The KMI Airport is a gateway to the iconic Kruger National Park shared between South Africa and Mozambique. The airport is located about 27 kilometres north-east of Nelspruit (Mbombela) and 42 kilometres south-east of the Kruger National Park. Other major tourist iconic sites that can be linked through the KMI Airport include the third largest canyon in the world, the Blyde River Canyon and God's window, and cultural attractions such as the Sudwala caves.

The airport services about 260,000 passengers annually on scheduled and also private charter flights with the capacity to handle 600,000 passengers annually. Private aircrafts can also fly directly in and outside of the airports with daily scheduled international flights to Victoria Falls and domestic flights to a number of airports in South Africa, including Cape Town International Airport and OR Tambo International Airport. The major users of the airport are SA Airlink and other private small aircrafts. The location of the KMI and other major airports in South Africa is presented in Fig. 12.1.

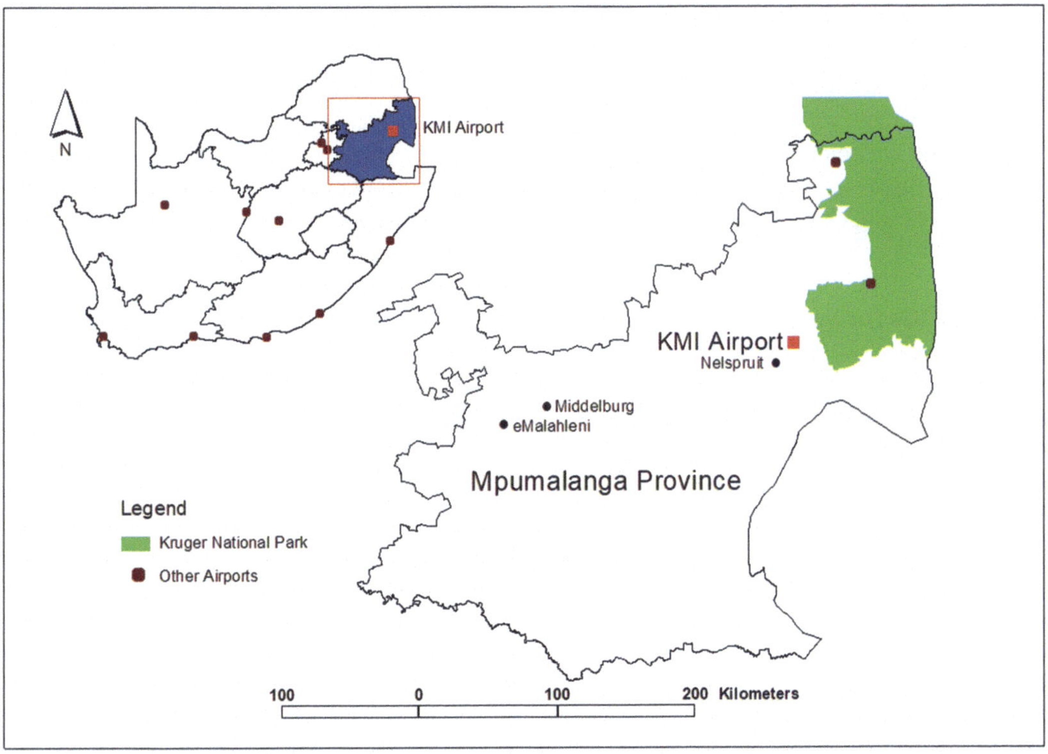

Fig. 12.1 Location of KMI and other major airports in South Africa. (Source: Authors (2020))

The study is a culmination of fieldwork that took place between June and July 2019. During the fieldwork, 25 key informants granted interviews from the KMI Airport and its two farms. The key informants were identified through the facilitation and coordination by the safety, security and health officers. Included in the interviews were junior and middle managers, as well as the top executives that understood various aspects that were affecting airports with a focus on the environment and climate, and how the company was responding to the SDG agenda. Among the 25 stakeholders granting interviews, were other tenants and airport support services from the meteorological organisation, car rental companies, airline representative air traffic and navigation controllers and tourism organisations. On average, interviews lasted 20 min. During the interviews, notes were taken, and audio recordings were made with the express permission of the respondents for onward analysis. Follow-up questions were made to further probe and investigate the issues under study.

Field observations were conducted, and they focussed on various aspects of the airport activities. The observations included, among others, a visit to the airport, the waste and water facilities, runway facilities and installations, control towers, airport offices, bathrooms and various airport community projects. The field observations facilitated learning and observations of various operations at the airport to see how these responded to sustainability issues. The KMI Airport relevant reports were also retrieved for analysis. Post-fieldwork data cleaning was conducted, which allowed for content and thematic analysis to be made in line with the study themes that were under investigation. Data analysis followed the dictates of content and thematic analysis as espoused by Vaismoradi et al. (2013) and Nowell et al. (2017).

4 Results and Discussion

The following section looks at how the KMI Airport is dealing with SDGs for society and SDG for the environment. The results are presented and discussed, highlighting some of the programmes being undertaken.

4.1 SDGs for Society

The study shows that KMI Airport is involved in a number of community and environmental projects aimed at driving its sustainability agenda. It emerged that from a community perspective, the KMI Airport naturally has an obligation to respond to the community as it has a stake in the airport business. Part of the airport's 364 hectares (ha) site came from the Dwaleni community. As a consequence, the community holds 10% in the KMI Airport through the Mbuyane Communal Property Association (MCPA). It is through the MCPA that the KMI Airport has been able to contribute to the community that hosts it. The community has two members that represent it on the board of directors with voting rights at a board level. Apart from community projects that address community needs, which range from community infrastructure development needs to addressing water needs, the community also gets a certain percentage of the landing fees that accrue at the airport.

Agenda 2030 addresses the issue of infrastructure a number of times with a call to improve rural, urban and transboundary infrastructure. Infrastructure development is the pinnacle of economic development in many areas. In particular, SDG 9 targets 9a calls for the facilitation of sustainable and resilient infrastructure development in developing countries (United Nations (UN) 2015). To this end, the airport through the MPCA had funding for the development, improvement and surfacing of roads and pathways in the Dwaleni community. The non-surfaced roads were particularly problematic during the rainy season as they become impassable, making it difficult for communities to access other services in and out of the community. The intense storms and rain activities that are becoming a feature in the community were making these non-gravelled roads problematic in a province that battles to deal with infrastructure problems (City of Mbombela 2018).

Dwaleni, like any other informal settlement in South Africa, faces inadequate water and sanitation supply challenges. The water supply and sanitation challenges are well acknowledged by the City of Mbombela's Spatial Development Plan of 2018. This is regardless of the South African constitution guaranteeing water as a critical human right imperative. As highlighted earlier, SDG 6 calls for the provision of sustainable water and sanitation. SDG 6 target 6.4 specifically calls for the sustainable supply of freshwater to address water scarcity and to substantially reduce the number of people suffering from water scarcity. To this end, the KMI Airport installed boreholes and a water tank to ensure that the community had access to fresh, safe water as the community was having challenges regarding adequate safe water supplies. Water is a critical commodity that helps the community to address its health, sanitary and socio-economic aspirations.

Given the levels of poverty in communities surrounding the airport, the KMI Airport also presents several employment opportunities to assist in efforts to reduce poverty, unemployment and inequality. The Mpumalanga province has high levels of poverty, unemployment and inequality. The unemployment rate for the province in 2018 stood at 32.4%, poverty levels at 42% and inequality levels stood at 0.61 Gini coefficient (Mpumalanga Provincial Treasury 2018). In a bid to address the above challenges addressed in SDG 1, calls for the need to institute poverty reduction initiatives and create decent work and sustainable employment (SDG 8) geared at women (SDG 5), immigrants and youths with a specific mention about the tourism industry, have been growing. The KMI Airport indicated that through the MCPA, it is embarking on community projects that are tailored to assist the community with skills transfer (SDG 4), thereby raising prospects of the participants to obtain employment. The revenue from departure fees is also aimed at reducing poverty (SDG 1). As an airport, the KMI, directly and indirectly, contributes to employment creation by offering employment to the locals. The KMI Airport also indicated that it had plans to create shadow trainees to allow for skills transfer and create a pool where it can have access to employees when a vacancy arises from the local populace.

The KMI Airport management indicated that although it had the desire to create and raise the number of women within the organisation, which stood at about 20% of its employees, it was faced with various challenges to achieve this. At the executive level, the KMI Airport management indicated that it had managed to create a 50/50 ratio. Of the four executives, two were women. The challenges were at lower levels. Given the nature of the aviation industry, the airport had failed to recruit women into its fire services department and baggage handling department, which affected the company's desire to achieve the desired goal to achieve SDG 5, which calls for gender equality and more female participation. The company indicated that women were mostly interested in clerical and other office jobs. This confirms the findings that achieving SDGs had numerous challenges that needed to be addressed if the aspirations set out in Agenda 2030 were to be achieved (Dube and Nhamo 2020b).

The poverty levels in the Mpumalanga province are notably higher in peri-urban and rural areas; these also translate to high levels of food insecurity. It emerged that the KMI Airport could be one of the few if not the only airport company to venture into agricultural production. The company executives indicated that given the availability of land within the vicinity of the airport, they had acquired 720 ha of land for farming. Out of that land, they use 160 ha for citrus trees, with 70 ha of the 720 haused for macadamia trees as commercial ventures. This venture links directly into SDG 2, focussing on ending hunger and malnutrition, thereby also directly addressing elements of SDG 3 (health and wellbeing).

The KMI Airport further indicated that it sought to create more jobs and wealth for local communities. To this end, the airport took a portion of the airport land amounting to 5 ha and use it for vegetable production. The high-value vegetable variety resulted in a commercially viable company that has contracts with some of the leading retailers in the country, such as Woolworths. The farms employ a number of general workers and agricultural specialists who work there on a full-time and part-time basis, thus providing jobs for skilled and semi-skilled local people. Figure 12.2 shows a portion of the garden that produces high-value vegetables. Given the desire and necessary inputs and drive, the workers working on vegetable production can transfer the acquired skills to establish their own gardens and produce for their families. This could assist in reducing malnutrition and hunger in the local community.

4.2 SDGs for the Environment

Besides addressing various SDGs for society, the airport has initiatives to address several environmental challenges that affect its business operations in line with the expectation from the community and the ICAO's roadmap for SDGs. Since the KMI Airport is located in a very hot province, the months between November and March have temperatures ranging above 25 °C (Fig. 12.3). Due to climate change, there has been an increased number of heatwaves in the area, which can make life unbearable during the day and also at night. The airport design had to take into consideration the local climate. As such, the KMI Airport was designed by using locally available resources to ensure a lower construction carbon cost that is associated with the transportation of building materials. It is one of the very few airports, to our knowledge, which is thatched using grass. In South Africa, besides the KMI Airport, the Skukuza terminal at the Skukuza camp in Kruger National Park is also thatched. SDG 13 on climate change action calls for the construction of climate-resilient and climate-adapted buildings.

The use of double thatching in the construction of the KMI Airport terminal building ensures

Fig. 12.2 Part of the farm that is used to produce high-value vegetables by Kruger Mpumalanga International Airport. (Source: Authors from fieldwork (2019))

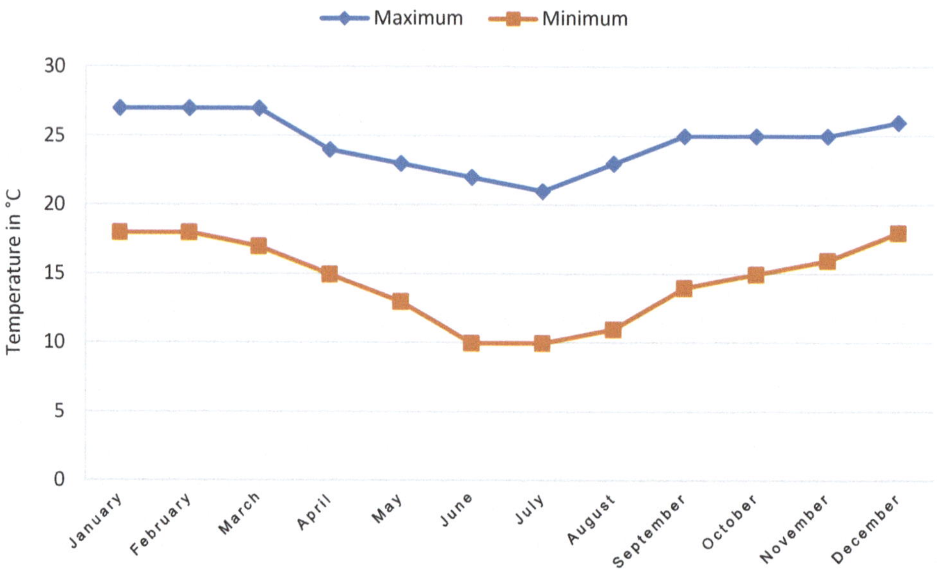

Fig. 12.3 Monthly average temperatures for Kruger Mpumalanga International Airport. (Source: Authors, data from the National Oceanic and Atmospheric Administration (2019))

that the building is warmer in winter and cooler in summer. This results in the limited use of air conditioners that are synonymous with airports. Airports use a lot of electricity in heating and cooling, which has an enormous carbon footprint. The KMI Airport's high roof further allows for air to flow freely and for natural light to be used during the day in most parts of the airport, which curtails the use of electric lighting during the day. The passengers and airport staff, therefore, enjoy the natural fresh air as trees are also scattered around the airport, which also assists in freshening and regulating air temperature during warmer days (Fig. 12.4). The use of LED lights at the KMI Airport at night keeps the energy usage at a minimal. The bathrooms have also been fitted with motion detection sensors that switch on and off depending on occupancy, which also assists in rationalising energy usage at the airport.

Airports also use a lot of ground vehicles, which add to their carbon footprint (Postorino and Mantecchini 2014). The KMI Airport's design and infrastructure work to its advantage. It does not make use of ground vehicles and buses like other bigger airports. The airport provides a linkage for airport shuttles and transfers, which is hosted on its website, and this facilitates transport sharing. Transport sharing is one of the advised measures that has been promoted by the tourism industry as part of the measures that can be employed by tourists to reduce the carbon footprint of the transport sector (Cohen and Hopkins 2019; Jones and Leibowicz 2019). Tourists walk from the departure lounge and arrivals to the terminal building or to the aircraft, which curtails a lot of carbon footprint. One of the restaurants at the airport uses gas in a bid to reduce their carbon emissions. As such, the

Fig. 12.4 Kruger Mpumalanga is one of the few thatched roof airports in the world. (Source: Based on fieldwork (2019))

airport has made some steps to respond to the Paris Agreement and aligns to the calls made under SDG 13 on climate change action.

One of the biggest concerns among global players is the issue of environmental pollution. Airports as transport hubs generate much pollution given the numbers that pass through the airports on a daily or annual basis. SDG 14 deals with the protection, conservation and sustainable use of the oceans, seas and marine resources for sustainable development. Of interest to this work is target 14.1 that calls for a reduction of pollution from land sources. To this end, the KMI Airport instituted a programme of point source separation of waste from aircraft and tenants at the airport. This however excludes international waste which is dealt with separately in line with industry best practice and national legislation. The restaurants also adopted the use of bamboo straws to deal with the challenge of single-use plastic pollution. Straws have been blamed for causing challenges in the oceans, which causes impacts on marine life (Rochman et al. 2016; Viera et al. 2020) and the growth of the blue economy. Plastic substitution is one of the measures being adopted by many restaurants to address the magnitude of marine pollution across the world. Dube and Nhamo (2020b) reported that Grootbos was getting rid of using plastic in all its lodges and restaurant bars so as to save the marine life and promote biodiversity, which are already being threatened by other stressors such as climate change and over-harvesting.

Airports and aviation, in general, have been blamed for threatening biodiversity. Aviation threatens birds with the KMI Airport reporting at least two bird strikes a month. Bird strikes are often experienced during take-off and landing and can be costly to both the airlines and the environment (Pennell et al. 2017; Roca-González et al. 2020). The airport was instituting biological control measures to deal with bird strikes through continuous trimming or cutting of vegetation and grass near the runway. SDG 15 calls for the protection of biodiversity particularly that of mountain ecosystems. In as much as there are efforts to clear vegetation to reduce bird strikes, there is a need to evaluate the effectiveness of measures put in place to protect birds from aircraft going forward in light of the number of reported strikes with a view to either complement or cement such measures.

The KMI Airport also has trouble with an alien specie invasion, which threatens the local vegetation species. Lantana threatens the biodiversity in the area, increases the risk of fires and threatens the grazing for the wildlife that is kept at the airport. The KMI Airport is currently using chemical and mechanical means to deal with the challenge of the species invasion in line with SDG target 15.8, which calls for the eradication of alien species to protect the natural forest and water resources. It was observed that more effort was needed in this regard given the extent of the observed invasion, as only two people are dedicated to dealing with this challenge.

The KMI Airport is located in a province that is rich in biodiversity given its proximity to one of the largest national parks in Africa, the Kruger National Park, and hundreds of game lodges and conservancies. This means that the issue of poaching is a recurrent and real challenge that the airport has to deal with. To this end, the KMI Airport instituted measures to protect wildlife from poachers. The KMI Airport works with the International Fund for Animal Welfare and an anti-poaching unit that makes use of sniffer dogs to detect rhino, ivory and arms, so as to prevent poaching of wildlife through the airport. This contribution is welcome given the gravity of poaching in the province, particularly of endangered species in the Kruger National Park, which have widely been documented for years (Geldenhuys 2016; Koen 2017).

5 Conclusions and Recommendations

The study investigated how KMI Airport is localising the SDGs. The study found that the airport has managed to make some progress in addressing selected SDG agendas and targets. The study found that with regard to SDGs for society, the airport leveraged the existing relationship with the community. This is through the 10% stake that is owned by the community. Hence, the KMI Airport instituted a number of initiatives that have assisted the community to improve their quality of life with direct feedback and shared value creation. The study also found that the airport's infrastructure development assistance resulted in the improvement of the host community's road system and water supply challenges, which is a problem for many rural communities in South Africa. Other projects that were conducted by the airport further assisted in reducing unemployment levels in areas where unemployment levels are very high.

The diversification of the KMI Airport into agriculture assisted in creating a platform for learning labs for neighbouring communities. Given the capacity for airports to generate resources, it might be worth investing further into value addition of the agricultural products that can assist in creating employment and add to the gross domestic product of the region. The study also noted that the airport's initiatives were more focussed on one community, although there are other communities that are affected by the airport's business. Hence, it is worth involving other communities close to the airport. Community projects are often associated with arguments around control and resource use and priorities. This study did not examine, in detail, how the community perceives resource use from the airport as a factor that may warrant further research to understand if all the people are benefiting equitably.

The KMI Airport was also engaged in a number of initiatives to take care of the environment that it operates in. It emerged that the KMI Airport had initially designed the airport to be constructed in an advantageous position so as to address SDG 13 on climate change action. The use of double-thatched roofing ensures that the airport cuts back on energy use, which reduces the environmental and financial cost of energy at the airport. Several other initiatives have been factored in to improve the environmental standards of the airport by employing renewable energy and energy-saving technology. Besides focussing on climate change, the KMI Airport also addresses other critical environmental challenges such as efforts made to curb poaching in a bid to protect the already endangered species.

The implementation of the SDGs by airports and the aviation sector, in general, has been a contentious global issue, which has not been adequately articulated. This study is critical in that it documented the effort that is made by airports to tackle some of the global environmental challenges, thus providing essential lessons that can be emulated by other airports across the globe. It also highlights some of the challenges that airports experience in achieving sustainability. The study recommends the replication of similar studies, particularly at bigger airports with a view of taking stock and ensuring that some progress is made before the lapse of the SDGs in 2030. Without such studies, it is challenging to ascertain if airports are making any progress or are in a position to address various challenges regarding the achievement of the set goals and targets.

References

ACI. (2020). *Green Airports Recognition*. ACI. Retrieved from http://www.aci-asiapac.aero/services/main/42/. (Accessed 30 May(Accessed 6 February 2020). 2020).

Airport Carbon Accreditation. (2019). *Annual Report 2018-2019*. Airport Carbon Accreditation. Retrieved from https://www.airportcarbonaccreditation.org/library/annual-reports.html. (Accessed 30 May 2020).

Bąk, J. (2018). The possibility of implementing solutions for selected sustainable development goals at passenger airports. E3S Web of Conferences (Vol. 45, p. 00007) EDP Sciences. https://doi.org/10.1051/e3sconf/20184500007

Boons, F., Van Buuren, A., & Teisman, G. (2010). Governance of sustainability at airports: Moving beyond the debate between growth and noise. *Natural Resources Forum, 34*(4), 303-313. https://doi.org/10.1111/j.1477-8947.2010.01314.x

Céspedes-Lorente, J., de Burgos-Jiménez, J., & Álvarez-Gil, M. J. (2003). Stakeholders' environmental influence. An empirical analysis in the Spanish hotel industry. *Scandinavian journal of management, 19*(3), 333-358.

Chapman, L. (2007). Transport and climate change: a review. *Journal of transport geography, 15*(5), 354-367.

City of Mbombela. (2018). *The Ciity of Mbombela local municipality Spatial Development Framework Review 2018*. City of Mbombela. Retrieved from http://www.

mbombela.gov.za/draft%20sdf%20document.pdf. (Accessed 31 May 2020).

Cohen, S. A., & Hopkins, D. (2019). Autonomous vehicles and the future of urban tourism. *Annals of tourism research, 74*, 33-42.

Cosgrove, C. (2018). Green initiatives for sustainable airport services. *Journal of Airport Management, 12*(4), 350-358.

Dube, K., & Mearns, K. (2019). Tourism and Recreational Potential of Green Building A case study of Hotel Verde Cape Town. In G. Nhamo, & V. Mjimba (Eds.), *The Green Building Evolution* (pp. 200-219). Pretoria: Africa Institute of South Africa.

Dube, K., & Nhamo, G. (2019). Climate change and the aviation sector: A focus on the Victoria Falls tourism route. *Environmental Development, 29*, 5-15.

Dube, K. (2020). TOURISM AND SUSTAINABLE DEVELOPMENT GOALS IN THE AFRICAN CONTEXT. *International Journal of Economics and Finance Studies, 12*(1), 88-102.

Dube, K., & Nhamo, G. (2020a). Major Global Aircraft Manufacturers and Emerging Responses to the SDGs Agenda. In G. Nhamo, O. Odularu, & V. Mjimba, *Scaling up SDGs Implementation Emerging Cases from State, Development and Private Sectors* (pp. 99-113). Springer Nature Switzerland AG: Springer International Publishing.

Dube, K., & Nhamo, G. (2020b). Sustainable Development Goals localisation in the tourism sector: lessons from Grootbos Private Nature Reserve, South Africa. *Geojournal*. https://doi.org/10.1007/s10708-020-10182-8

Geldenhuys, K. (2016). SAPS investigations to rhino poaching cases in the Kruger National Park. *Servamus Community-based Safety and Security Magazine, 109*(9), 44-46.

González-Ruiz, J. D., Duque, E. A., & Restrepo, J. C. (2018). Clean Development Mechanism in Airports: The Colombian Case. In C. K, C. I, L. W, & W. C. (Eds.), *Proceedings of the 21st International Symposium on Advancement of Construction Management and Real Estate* (pp. 481-490). Springer.

Gössling, S., & Scott, D. (2018). The decarbonisation impasse: global tourism leaders' views on climate change mitigation. *Journal of Sustainable Tourism, 26*(12), 2071-2086.

Gössling, S., Humpe, A., & Bausch, T. (2020). Does 'flight shame' affect social norms? Changing perspectives on the desirability of air travel in Germany. *Journal of Cleaner Production, 266*. https://doi.org/10.1016/j.jclepro.2020.122015.

Hagmann, C., Semeijn, J., & Vellenga, D. B. (2015). Exploring the green image of airlines: Passenger perceptions and airline choice. *Journal of Air Transport Management, 43*, 37-45.

ICAO. (2016). *On Board a Sustainable Future*. ICAO. Retrieved from https://www.icao.int/environmental-protection/Pages/env2016.aspx. (Accessed 29 May 2020).

Jones, E. C., & Leibowicz, B. D. (2019). Contributions of shared autonomous vehicles to climate change mitigation. *Transportation Research Part D: Transport and Environment, 72*, 279-298.

Khanna, M., & Anton, W. R. (2002). Corporate environmental management: regulatory and market-based incentives. *Land economics, 78*(4), 539-558.

Kim, Y. L. (2020). Taking Another Look at Airline CSR: How Required CSR and Desired CSR Affect Customer Loyalty in the Airline Industry. *Sustainability, 12*(10), 1-19.

Koen, H. S. (2017). *Predictive policing in an endangered species context: combating rhino poaching in the Kruger National Park*. Pretoria: Doctoral dissertation, University of Pretoria.

Lu, J. L., & Wang, C. Y. (2018). Investigating the impacts of air travellers' environmental knowledge on attitudes toward carbon offsetting and willingness to mitigate the environmental impacts of aviation. *Transportation Research Part D: Transport and En, 59*, 96-107.

Lynes, J., & Gibson, R. (1998). Voluntary corporate initiatives for environmental improvement. *Alternatives Journal, 24*(2), 18-20.

Lynes, J. K., & Dredge, D. (2006). Going green: Motivations for environmental commitment in the airline industry. A case study of Scandinavian Airlines. *Journal of sustainable tourism, 14*(2), 116-138.

Mpumalanga Provincial Treasury. (2018). *Presentation to the Select Committee on Finance Mpumalanga Economic and Development Perspectives*. Provincial Treasury Mpumalanga Province. Retrieved from http://pmg-assets.s3-website-eu-west-1.amazonaws.com/180612mpumalanga.pdf. (Accessed 31 May 2020).

Nowell, L. S., Norris, J. M., White, D. E., & Moules, N. J. (2017). Thematic analysis: Striving to meet the trustworthiness criteria. *International Journal of Qualitative Methods, 16*(1), 1-13.

Pennell, C. G., Rolston, M. P., Van Koten, C., Hume, D. E., & Card, S. D. (2017). Reducing bird numbers at New Zealand airports using a unique endophyte product. *New Zealand Plant Protection, 70*, 224-234.

Postorino, M. N., & Mantecchini, L. (2014). A transport carbon footprint methodology to assess airport carbon emissions. *Journal of Air Transport Management, 37*, 76-86.

Ridder, H. G. (2017). The theory contribution of case study research designs. *Business Research, 10*(2), 281-305.

Roca-González, J. L., Vera-Lopez, J. A., & Rodriguez-Bermudez, G. (2020). Organisational and costing aspects to prevent wildlife strikes on airports: a case study of Spanish airport security managers. *Safety science, 122*. https://doi.org/10.1016/j.ssci.2019.104520

Rochman, C. M., Browne, M. A., Underwood, A. J., Van Franeker, J. A., Thompson, R. C., & Amaral-Zettler, L. A. (2016). The ecological impacts of marine debris: unraveling the demonstrated evidence from what is perceived. *Ecology, 97*(2), 302-312.

United Nations- UN. (2015). *TRANSFORMING OUR WORLD:THE 2030 AGENDA FOR SUSTAINABLE DEVELOPMENT A/RES/70/1*. United Nations. Retrieved from https://sustainabledevelopment.un.org/content/documents/21252030%20Agenda%20for%20Sustainable%20Development%20web.pdf. (Accessed 24 August 2020).

Upham, P., Thomas, C., Gillingwater, D., & Raper, D. (2003). Environmental capacity and airport operations: current issues and future prospects. *Journal of Air Transport Management, 9*(3), 145-151.

Vaismoradi, M., Turunen, H., & Bondas, T. (2013). Content analysis and thematic analysis: Implications for conducting a qualitative descriptive study. *Nursing & Health Sciences, 15*(3), 398-405.

Viera, J. S., Marques, M. R., Nazareth, M. C., Jimenez, P. C., & Castro, Í. B. (2020). On replacing single-use plastic with so-called biodegradable ones: The case with straws. *Environmental Science & Policy, 106*, 177-181.

WMO. (2019). *2019 concludes a decade of exceptional global heat and high-impact weather*. Retrieved from https://public.wmo.int/en/media/press-release/2019-concludes-decade-of-exceptional-global-heat-and-high-impact-weather. (Accessed 6 February 2020).

13 Scaling up University Engagement with the Water SDG for General Environmental Stewardship and Climate Change Resilience

Godwell Nhamo and David Chikodzi

Abstract

Universities remain at the forefront of scaling up the implementation of the 2030 Agenda for Sustainable Development and its interlinked 17 Sustainable Development Goals (SDGs). This chapter investigates how the University of South Africa (UNISA) has localised the water and sanitation SDG to equip itself with eventualities regarding climate change (SDG 13), water shortages, general groundwater conservation as well as environmental stewardship. The mixed methods research design was adopted as the strategy of inquiry. Results show that UNISA developed a Water Master Plan which includes strategies for rainwater harvesting, water efficiency and wetland rehabilitation. The master plan strategies have potential for saving huge volumes of water and reduction of water-related bills. This is possible through the minimisation and efficient utilisation of water from both boreholes and the municipal systems. The UNISA's engagements remain a proactive move in a country that is fighting extreme and frequent droughts, with terms like 'Day Zero' having emerged in reference to a situation where the City of Cape Town was predicted to run out of water in 2018. Although some of the initiatives are not yet fully operational, the study recommends that other institutions of higher education in South Africa and elsewhere across the globe join this movement of global water stewards and building climate resilience.

Keywords

UNISA · SDG 6 · Water efficiency · Wetlands · Rainwater harvesting

1 Introduction

Since water is life, there is a need for humanity to make every drop of water count (UN Water 2018). On 25 September 2015, global leaders adopted the 2030 Agenda for Sustainable Development (AfSD), which is a shared blueprint containing a set of 17 interlinked Sustainable Development Goals (SDGs) (United Nations 2015). The SDGs guide the world in ending poverty, fighting for inequality and injustice, looking after the environment and, above all, tackling climate change by 2030. The SDGs are universal, meaning that they are accepted and apply to every country in the world. The SDGs also balance all

G. Nhamo (✉) · D. Chikodzi
Exxaro Chair in Business and Climate Change, Institute for Corporate Citizenship, University of South Africa, Pretoria, South Africa
e-mail: nhamog@unisa.ac.za; chikod@unisa.ac.za

G. Nhamo et al. (eds.), *Sustainable Development Goals for Society Vol. 2*, Sustainable Development Goals Series, https://doi.org/10.1007/978-3-030-70952-5_13

the three dimensions of sustainable development which are the social, environmental and economic pillars commonly referred to as people, planet and profits. The 17 SDGs and their 169 targets and 230 indicators were defined and developed through an extensive dialogue among United Nations Member States, local authorities, civil society, the private sector and many more stakeholders (United Nations 2015). These SDGs are supposed to be achieved at global, national and subnational levels and are driven by the motto 'let no one be left behind' (UN Water 2018).

The realisation of the SDGs has been observed to require action at a local level, hence the need for local application of high-level agendas (Tan et al. 2019). Localisation involves landing the SDGs into appropriate subnational contexts, from the setting of similar and/or aligned goals and targets and determining the means of implementation, including the selection of appropriate indicators to measure and monitor progress (United Nations 2016). It relates to how local and regional governments as well as individual organisations can support the achievement of the SDGs through action from the grassroots and to how the SDGs can provide a framework for local development policy. Tan et al. (2019) present the SDG localisation agenda as incorporating the following main ingredients: (1) local agenda setting, (2) decision-making and (3) monitoring and reporting using locally adapted indicators. This move generates the ownership necessary for successful SDG implementation at local scale.

Universities remain at the forefront of scaling up the implementation of the 2030 AfSD and the 17 SDGs. The universities' role in meeting the sustainable development challenges goes beyond advancing the training and skills required to achieve the SDGs. They also articulate tangible solutions that resonate with the communities in which they operate (Nhamo and Mjimba 2020). Universities around the world are increasingly facing a growing need to redefine their activities, strategies and organisation along with the issues of sustainability (Beynaghi et al. 2016). This is being done through initiatives such as the international sustainable campus network, the Green Campus Initiatives and SDG hubs created at universities around the world. Under these initiatives, universities are implementing the SDGs both internally and externally through partnerships with government and communities to promote and support their adoption in society as well (Findler et al. 2019). Sharp (2009) advocates for university campuses to be transformed into living laboratories for the demonstration and practice of environmental sustainability. Also, Kestin et al. (2017) argue that universities occupy a unique position within society and have long been powerful drivers of global, national and local innovation, as well as societal wellbeing. Universities, therefore, have a critical role to play in the attainment of the SDGs and will also significantly benefit from engaging with the global goals.

The University of South Africa (UNISA) is the largest open distance learning institution in Africa. It is also the longest standing university dedicated to distance education in the world. Close to a third of South African university students are enrolled at UNISA, which registers more than 400,000 students from across South Africa, Africa and other parts of the world in a single academic year (UNISA 2016). The qualifications offered by UNISA range from certificates to degrees from diverse fields of study. The university has six campuses around South Africa which are Muckleneuk, Sunnyside, Florida, Midrand, Cape Town and Durban (UNISA 2016). Through its sustainability policy, in particular the water efficiency and rainwater harvesting plan, UNISA has fully committed itself to be a sustainable entity, thereby opening the way to get involved in the localisation and institutionalisation of the SDGs. The university undertook to address all the SDGs through the SDG Declaration in November 2019 (UNISA 2019). The approach will focus on operations; research, development and innovation; teaching and learning; as well as community engagement and outreach perspectives. Although all the targets of SDG 6 can be directly and indirectly linked to UNISA initiatives, targets 6.3, 6.4 and 6.6 are those with direct link and relevancy to this chapter (Box 13.1).

Box 13.1 SDG 6 Targets with direct link to this chapter

- Target 6.3: By 2030, improve water quality by reducing pollution, eliminating dumping and minimising release of hazardous chemicals and materials, halving the proportion of untreated wastewater and substantially increasing recycling and safe reuse.
- Target 6.4: By 2030, to substantially increase water use efficiency across all sectors and ensure sustainable withdrawals and supply of freshwater to address water scarcity.
- Target 6.6: By 2020 to, protect and restore water-related ecosystems, including wetlands, rivers and aquifers.

Source: United Nations 2016:35

The chapter investigates how UNISA has localised mainly the water and sanitation SDG and in the process readying itself for eventualities regarding climate change (SDG 13), water shortage and general groundwater conservation. This comes against a background of the country facing recurrent drought episodes, increasing population, rapid urbanisation and widespread degradation of ecological infrastructure such as wetlands that deliver key ecosystem services (Ramsar Convention Secretariat 2016; SANBI 2014). South Africa is a water-scarce country that is vulnerable to climate change and variability (SANBI 2014), yet it has an above-average water consumption per capita. It has an average per capita water consumption of 235 litres per day compared to the global average of 185 litres per capita per day (South Africa Department of Water and Sanitation 2017).

2 A Literature Survey

This literature review comes in two main parts. The first part focuses on the overview on SDG localisation in general. The second part narrows down to addressing SDG localisation in institutions of higher education with an emphasis on SDG 6 and limited reference to other SDGs as appropriate.

2.1 Overview on Localisation of the SDGs

Several authors like Owens (2017) and Bhowmik et al. (2017) have identified the steps that universities can follow in the process of localising the SDGs. These steps include, but are not limited to: mapping existing knowledge on the SDGs; building internal capacity and the identification of champions for buy-ins; some form of SWOT (strength, weaknesses, opportunities and threats) analysis; integrating, embedding and implementing the SDGs within a university's policy strategies and plans; and MRV (monitoring, reporting and verification) actions undertaken on the SDGs. These identified steps are also closely aligned with those identified for localising the SDGs at local government level, large business campuses and other large organisations. The Global Taskforce of Local and Regional Governments, the United Nations Development Programme (UNDP) and United Nations Habitat developed a roadmap for localising the SDGs for cities and regional governments. They identified the following steps as being the best to take in the process of localising the SDGs through raising awareness, advocacy, implementation, and monitoring and evaluation (United Nations 2016). Additionally, Kestin et al. (2017) argue that in their 'business-as-usual' actions in education, research and community outreach, universities already make many important contributions to the attainment of the SDGs. However, for the SDGs to be successful at universal level, universities need to become champions of sustainable development and play a leading role in innovation and implementation of the SDGs. They also suggested five steps that a university can follow in localising the SDGs as follows:

1. Map what the university is already doing.
2. Build capacity and ownership of the SDGs.

3. Identify priorities, opportunities and gaps.
4. Integrate, implement and embed the SDGs.
5. Monitor, evaluate and communicate.

Sharp (2013) discusses her experiences of institutional change with reference to the Green Campus Initiative. In her view, successful approaches include buy-in from the top management, having effective and efficient coordination mechanisms, maximising the role of face-to-face communication and galvanising both formal and informal support. Other steps may focus at enhancing the path of least resistance. This involves the identification and running with ideas that attract the most support. There will also be the need to start with pilot and much smaller projects and build on their success. Lastly, there is the need for continuity that allows 2–3 years of implementation to establish relationships and build skills, especially internally. Those involved in the projects also need to develop an information system to capture and present such information in digestible formats for all levels of management.

As highlighted in UNISA (2016), the legal and policy framework through which UNISA can localise the SDGs especially SDG 6 that addresses issues of water and sanitation is provided for under several national, subnational and international laws, policies and conventions. These provide an enabling framework through which the localisation can be done and include but not limited to the Constitution of South Africa, 1996 (as amended), especially the Bill of Rights; the National Health Act, Act 61 of 2003; the National Environmental Management Act No. 107 of 1998 (as amended) and its subsidiary Acts; the 2012 National Development Plan: Vision 2030; the National Climate Change Response Policy White Paper of 2011; the National Water Act 36 of 1998; the Draft National Climate Change Adaptation Strategy of 2019; the First Nationally Determined Contributions of 2018; the Climate Change Bill of 2019; and the 2019 Voluntary National Review (VNR) under the United Nations High-Level Political Forum of the SDG implementation and the Ramsar Convention on wetlands. The provisions given in the above legislation and policies therefore show that there is legally no handicap that can limit UNISA from effectively localising the water and sanitation as well as other SDGs.

However, significant problems exist in deciphering high-level SDGs into the local contexts as shown by the gaps in literature between global indicators and their local relevancy and the manner in which stakeholders understand such. Global indicators enable comparison between and among countries and continents. However, these may be too coarse for local validity and the ability to motivate action by reflecting local values (Klopp and Petretta 2017). Moreover, global indicators may be unusable at local scales where data, or the resources and capacity to obtain such data, are limited (Patel et al. 2017). It has emerged that developing locally relevant SDG indicators is constrained by resources that include both individual and institutional capacities, as well as funding (Tan et al. 2019). The next sub-section deliberates specifics regarding localising SDG 6 and other SDGs related to water and sanitation.

2.2 Focus on SDG 6 and Higher Education Institutions

Universities the world over, particularly those from the United Kingdom, have embraced the water and sanitation SDG. To this end, some examples will be drawn that provide evidence and insights on how universities have been working towards water conservation and efficiency to address SDG 6. Biswas and Dash (2019) highlight that the management of water and wastewater on university campuses encompasses two critical dimensions: control through proper maintenance and installation of low-flow water systems, and proper water conservation methods. Water conservation measures include putting in place efficient systems in reporting, responding and repairing leaks to minimise wastage. It also includes measures to retrofit buildings with water-saving installations like water-efficient shower roses, waterless urinals, undertaking minimal irrigation, protecting underground water, harvesting rainwater as well as diverting rainwater for watering plants. With adequate pretreatment, harvested rainwater can also be used for flushing toilets and other appropriate uses. Rainwater has been captured at Christ University, Bengaluru, for recharging underground water. At Kalyani University, West Bengal, several ponds were constructed for collecting and using rainwater.

A study to assess the level of awareness of students on the importance of water resources and their water conservation practices at ten selected residential colleges at the University of Kebangsaan in Malaysia revealed good insights. The authors found that students had a high awareness level on water resource conservation, although this was not converted to action (Augustine and Hanafiah 2019). In a study on implementing climate change (SDG 13) research at universities that sampled 82 responses, Filho et al. (2018) found that matters dealing with water conservation and efficiencies ranked first.

Focusing on the global implementation of sustainability strategies in universities, Nunes et al. (2018) found that there were 32 universities that were seeking to reduce water use. Out of these, 12 were either investigating on implementing wastewater recycling. Universities of Ottawa and Nottingham were cited as good examples implementing a closed water use loop. Under its Emissions and Discharges theme, Bath Spa University (2018: 12) has a target of 'zero spillages of harmful materials reaching surface/groundwater or main sewer every year'. Overall, from the water theme, the university undertakes detailed five-yearly water audits on its main campus to identify any opportunities for water saving. This measure is done to supplement regular annual monitoring. Water efficiency has been embedded in the new designs of the university's Newton Park campus, and rainwater harvesting to supply water to all toilets was integrated into the Commons Academic Building. Lastly, an automated water metering network throughout the main campus has been installed with the view to provide site-wide water main and building-level monitoring. The focus is on early detection of leaks to enable quick response and the detection of other unusual consumption patterns (Nunes et al. 2018). Rainwater harvesting and other water efficiency and saving devices including flush reductions, installation of half-hourly meters, tap timers, greywater use and tap flow restrictions have been applied for long at Oxford University (2011).

The University of Surrey (2019) Water Use Policy clearly links its actions to the need to address the United Nations SDGs, specifically target 6.4. Both existing and new sub-metering systems are viewed as appropriate for attaining the targets. Equipment and facilities that consume substantial amounts of water are also targeted for monitoring. Metering and monitoring will assist in determining the trends and patterns as well as abnormal usage that helps in leak detections. A typical example includes washrooms with old installations that are not water efficient and would require retrofitting with modern water-efficient installations. Retrofitting and/or new procurements of water-efficient installations would also require bringing on board procurement teams. Furthermore, the University of Surrey plans to capture and reuse greywater. Similar measures are also embedded into the Nottingham Trent University (2019) Water Policy. Table 13.1 shows the universities' commitments to the water and sanitation SDG.

The University of Plymouth (2018), with an estimated 21,000 students as of 2018, takes water efficiency and conservation seriously. Plans were put in place to work towards significant water use reductions in line with its carbon and energy plan. As of 2017/2018 academic year, the university had reduced water consumption by 50% from the 2005/2006 consumption levels – the absolute use of water decreased from 116,433 m^3 in 2005/2006 to 70,191 m^3 in 2017/2018. The percentage of recycled water also increased from 0.57% in 2015/2016 to 1.05% in 2017/2018. The use of rainwater from tanks has been improving, with 423 m^3 harvested in 2015/2016 and 740 m^3 in 2017/2018.

At the University of Bristol (2017), there has been significant progress made in water conservation through the detection of leaks and infrastructure upgrading. From the baseline year of 2007/2008, water consumption has gone down

Table 13.1 University commitments to the water SDG

University	Policy	Water use target(s)
University of Surrey	University of Surrey Water Use Policy (2019–2021)	15% water reduction target on a 2017/2018 academic baseline year which should be delivered by the end of the 2020/2021 academic year
Manchester Metropolitan University	Manchester Metropolitan University Environmental Strategy 2016	A 25% total water reduction target set for 2020/2021 based on 2013/2014 base year
University of Plymouth	University of Plymouth Sustainability Strategy (as revised) (2016–2020)	Reduce annual water consumption to below 3.0 m^3/student by 2020 from 7.3 m^3 in 2005–2006
Bath Spa University	Sustainability Strategy 2018	Reduce water loss by 50% by 2020 and reduce water consumption per student and staff full-time equivalent by 5% by 2020 using 2017 baseline of 0.55 m^3 full-time equivalent
University of Bristol	University of Bristol Carbon & Water Strategy – November 2017	To cap water consumption at 2016/2017 levels to 2020/2021 in the face of increasing staff and student numbers
University of Oxford	University of Oxford Water Management Strategy January 2011	Water use to decrease by 11% by 2014/2015 against 2009/2010 levels (347,364 m^3)

Source: Authors

by 28% from 491,473 m^3 per year to 352,111 m^3 in 2016/2017. Over 50% of water is consumed in student residences and the remaining half by academic staff and administration buildings. The trends in water use over the said years are shown in Fig. 13.1. Other interesting facts from the university water audit and monitoring are that four buildings (Physics, Biomedical Sciences, Chemistry and the Richmond Building) accounted for half of all non-residential water consumption. Swimming pools and student hygiene and cooking are activities that consume huge amounts of water. Given the foregoing, the following are some of the water efficiency and saving measures proposed and being undertaken: managing urinal controls, identifying saving measures in commercial kitchens, fixing leaking taps and continuously raising awareness on water saving.

Edge Hill University (2016) documents findings from a water efficiency audit undertaken by the United Utilities Pty Ltd. in December 2015. Should water efficiency measures identified be implemented, these were likely to save up to 5% of water consumption annually. An automatic meter reading (AMR) logging device was attached to the main and single supply point water meter to the campus to check consumption activities on a 24-h basis. The AMR device is accessible remotely through a website portal. The Edge Hill University water consumption rose from 56,446 m^3 in 2013 to a peak of 96,668 m^3 in 2014, before declining to 77,251 m^3 in 2015 and going up again to 91,759 m^3 in 2016 (base year). Although typically university facilities consumed on average 180 litres of water per student per day, this was different from Edge Hill University where this came to 131 litres per person per day. From the survey, it emerged that there were already good practices that included the use of waterless urinals and self-closing non-concussive basin taps.

The Edge Hill University (2016) water efficiency audit observed shower flow rates, tap flow rates, urinals, WC cistern capacities and swimming pool facilities. The findings on showers were that these exceeded industry-acceptable standards of 7.6 litres per minute. As such, the recommendation was to retrofit water-saving showerheads or inline flow restrictions. Regarding the taps and washbasins, the flow rate averaged about 10 litres per minute. The recommendation

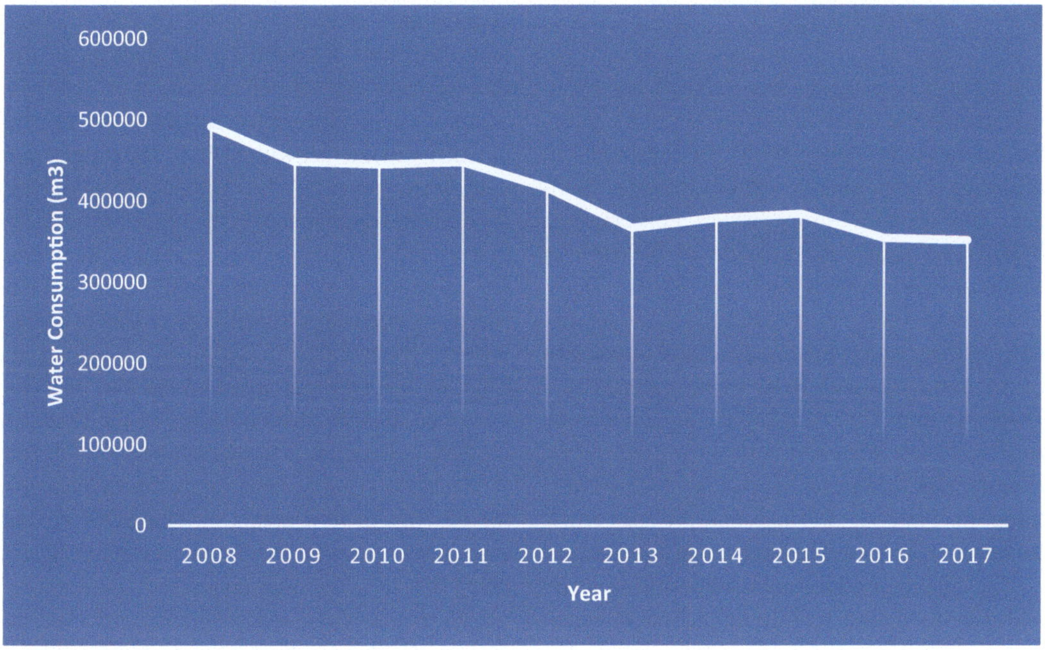

Fig. 13.1 Trends in University of Bristol water consumption (2007/2008–2016/2017). (Source: Authors, based on University of Bristol (2017: 3))

was to retrofit them to reduce flow rates to an average of 4 litres per minute. Similar observations were made for kitchen taps that averaged flow rates of 15 litres per minute. Retrofitting was suggested to reduce flow rates to about 10 litres per minute for both cold and hot taps. Some urinals were observed not to have any control systems on flows and as such operated 24 h. As for the WC flushing cisterns, most had the capacity of 8 litres per flush, and a recommendation was made to reduce this by one litre per flush. Lastly, site sub-metering was recommended as this would assist in identifying leakages, wastage and/or undue water consumption by users. Leakages are a serious matter. Manchester Metropolitan University (2016) observed a 21% increase in their water consumption in the 2015/2016 year from their 2010/2011 baseline year. This was caused by two factors:

- A substantial underground leak at All Saints Campus, ultimately repaired, saving approximately 70 m^3 water per day. This further triggered the installation of auto leak detection systems.
- The Business School borehole, normally providing greywater to flush the building toilets, was out of commission for 8 months of the year.

Urine separation at source is emerging as one of the water-efficient measures in university campuses. Ishii and Boyer (2016) researched on student support for such. The authors surveyed 8800 on-campus students at the University of Florida. Overall, 570 students responded with valid survey responses after watching a video on the expected benefits and risks of urine separation at source. The key findings were that an estimated 84% would vote in favour of urine source separation in residence halls. However, this support was less apparent when measured by their willingness to pay, as 33% of the respondents were unwilling to pay for the implementation of urine source separation. An estimated 40% were only willing to pay between $1 and $10 per semester.

3 Methodology

The chapter investigates how UNISA has localised mainly the water and sanitation SDG and in the process readying itself for eventualities regarding climate change (SDG 13), water shortage and general groundwater conservation. Figure 13.2 shows the location of the UNISA campuses and centres included in the study. These campuses are distributed across most provinces in the country; hence, successful interaction of the university with SDG 6 can be used as a good case study for scaling up its implementation as it will have been tested in most parts of the country. The campuses from Fig. 13.2 include those in Pretoria, Johannesburg, Cape Town, Rustenburg, Polokwane, Ekurhuleni and Durban.

The chapter used a combination of case study and systems approach in the collection and analysis of data. The UNISA case study was used to analyse how the university has been interacting with SDG 6. The case study was done through documentary analysis of UNISA's water master plan. This was followed by an analysis of the time-series water volume utilisation on different campuses and other interventions done by the university to interact with SDG 6. The time-series data were provided by UNISA Estates, the department which manages the university's water system and is the custodian of SDG 6. Realising the interlinkages and integrated nature of the SDGs has been emphasised to be crucial in ensuring that sustainable development is achieved. This implies that systems thinking and analysis are other suitable methodologies in the study of the SDGs (Bai et al. 2016; Kestin et al. 2017; Tan et al. 2019). The systems approach emphasises the analysis of how action on one SDG, in this case SDG 6, can in turn impact on other SDGs given that they are inter-connected and indivisible. Remote sensing and geographic information system (GIS) were then used to mark wetland boundaries and assess the state of wetlands (SDG 15) on university properties and how they were

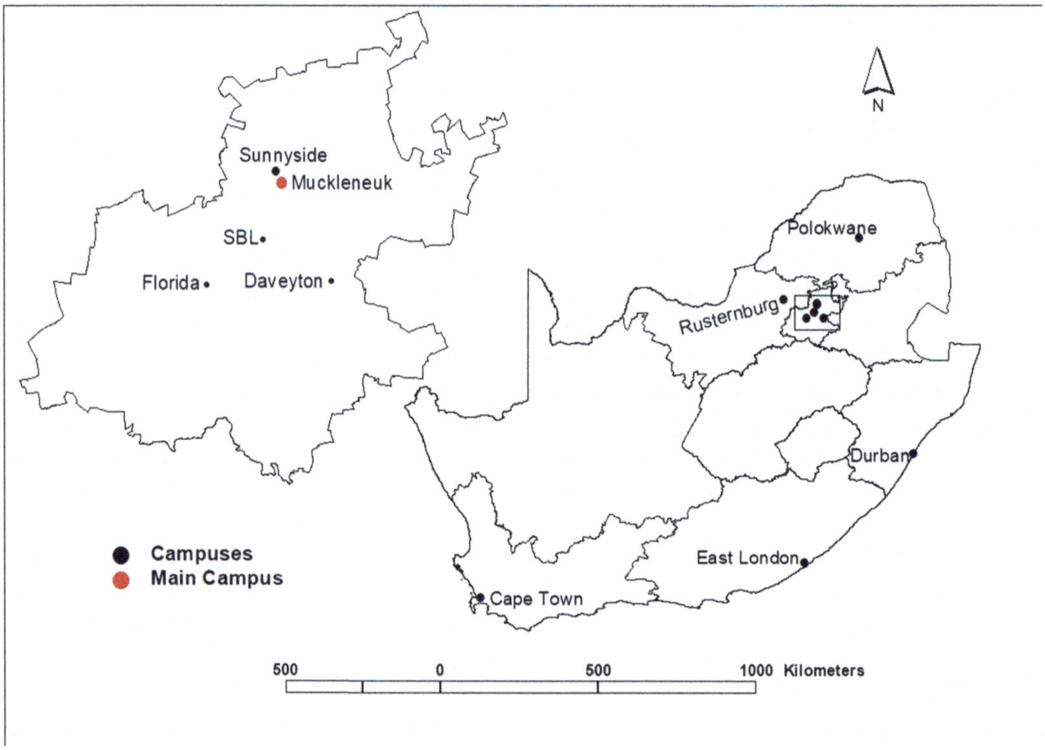

Fig. 13.2 Study area. (Source: Authors)

behaving over time in terms of moisture regime, vegetation condition as well as changes in surrounding land-use practices. Field excursions to the wetlands were done at three campuses, namely, the School of Business Leadership (SBL), Muckleneuk and Daveyton, to determine their physical state, perform biodiversity assessments (absence and presence of species) as well as validate remotely sensed maps. Document analysis and critical document analysis were other methods applied. The main document analysed was the integrated water efficiency and rainwater harvesting master plan. In-depth interviews with key informants from UNISA's Estates department were carried out to assist with in-depth analysis of the strategies being implemented by the university to localise SDG 6 as well as the challenges being faced in the process.

4 Presentation of Data and Discussion of Key Findings

4.1 Managing Water Consumption

In localising SDG 6 and as part of its sustainability journey, UNISA developed integrated water efficiency and rainwater harvesting master plans for each of its campuses. These plans were formulated using the templates recommended on the localisation of the SDGs at universities and also taking into account all applicable existing international, national, provincial and local legislation and policies. The master plans required a participatory evaluation of the current water efficiency and rainwater harvesting practices at each of the six campuses; participatory formulation of detailed objectives for the plan; detailed implementation plans for infrastructure changes required to optimise the use of grey and potable water; detailed reporting systems, including measuring templates; and the identification of opportunities for improvement (UNISA 2016).

The first step in formulating an integrated water efficiency and rainwater harvesting master plan was to perform a gap analysis and opportunity assessment. This basically involved the review of the current situation of water utilisation and needs, as well as the current wastewater generation and management (UNISA 2016). This also enabled the identification of baseline information, where available, and the identification of areas that required attention, including gaps in information and an assessment of potential opportunities. This was followed by the development of the draft master plan by assessing the gaps and opportunities identified within the current water and wastewater systems. A review of the best practice methods to develop an implementation guideline was done, which led to recommendations on suitable alternatives. These recommendations were based on national and international good practice guidelines. In the process, urgency and ownership of the process were instilled by assigning timeframes and responsibilities. These also acted as a monitoring framework in order to assess the implementation of the plan.

The highlighted primary objective of the plan is to apply the water efficiency measures, on UNISA campuses (UNISA 2016). Subsequently, these measures would lead to improved water efficiency, having alternative water source supply from rainwater harvesting, borehole water, stormwater, runoff reduction and associated stormwater pollution load reduction. All these are integral components of the 'Water Efficiency and Alternative Supply Plan'. Practically the water efficiency and alternative water supply plan include doing the following: improving collection and quality of water data, reducing demand on municipal supply system and the installation of rainwater water collection and harvesting infrastructure on campus. It also includes ensuring borehole systems are operational, evaluating the viability of stormwater harvesting from detention ponds, installing water-efficient toilets and urinals and reducing the annual water consumption through leak prevention. There would also be a need to suggest additional water-sensitive and alternative water supply initiatives to be considered in the future (i.e. permeable pavement installation), considering storm runoff as a resource and realising its value wherever possible, and encouraging, educating and influencing all campus users (workers, students, servicers and visitors) to follow the principles of responsible water-sensitive management (UNISA 2016). Similarly, Nunes et al. (2018) observed that a substantial number of universities around the world seeking to implement sustainability strategies were seeking to enhance efficiency in water utilisation. This means that water is a key commodity at university setups, hence the need to devise strategies to use it parsimoniously.

After the draft master plan was done for each campus, it had to have official status in the university. This meant that it had to be sold to all the university stakeholders and have a buy-in of the top management including the University Council and the Senate (UNISA 2020). There is also the Institutional Forum that advises the Council and the Management Committee that assists the Principal and Vice-Chancellor in his day-to-day management of the university. This committee consists of the Principal and Vice-Chancellor, the Vice Principals and the Registrar (UNISA 2020). Other stakeholders of the University are the academic, administrative and professional employees, students and alumni, donors and two recognised unions. The Students' Representative Council is properly constituted and is governed by its constitution, which is approved by the Council (UNISA 2016). The acceptance and education of these stakeholders were done in order to make the implementation much easier as everyone would be aware of the master plan. In terms of the actual implementation of the water master plan, for rainwater harvesting and water use efficiency, the operational responsibility largely lies within the University Estates, which contains the following directorates: Property Management, Project Management, Maintenance, Support Services and Science Campus. Drawing from an example of the SBL, Table 13.2 shows some of the identified water efficiency and rainwater harvesting opportunities.

As a limitation, the derived master plans for each campus, in the end, were based on information available or provided by each of the campuses, and in some cases, adequate information was not available. Hence, assumptions had to be made. Some of the most critical information that was not provided or available included the utility bills for almost all of the campuses. This information was critical in determining the baseline water consumption levels and also the total cost of the commodity.

It was observed that the main uses of water at UNISA's campuses were for irrigation, portable and sanitation purposes. All these uses are included in the highlighted university's water management master plan. The university put in place a smart system to monitor in near real time the inflow and consumption of water on its campuses as well as a system of monitoring hardware installed in the water distribution system. Figure 13.3 shows an example of the recorded daily consumption of water at the Muckleneuk Campus from the 12th to the 18th of February 2020. The almost constant rate of water use shows that the water system is intact and has no leakages or bursts, which may otherwise result in the excessive use of water on the date of such an incident. The system of water consumption measurement can be done on an hour-by-hour basis in order to provide insights into the nighttime water use which is a more robust measure of the possibility of leakages given that very few people will be on campus during this period. Such technologies have, to a large extent, increased water use efficiency on UNISA campuses as abnormal uses and leakages are quick to be identified.

The utilisation of water at UNISA campuses is also being metered on a building-by-building basis. This is an even more effective way of improving efficiency on water use compared to just measuring at the inlet valve. The technical ability to account for water received and how its use is distributed across buildings makes it much easier to trace specific areas of overuse and possible inefficiencies. Figure 13.4 shows water consumption at the OR Tambo, Kgorong and Simon Radipere buildings at the Muckleneuk Campus from the 10th to the 16th of February 2020. It can be noted that the OR Tambo building was by far the largest user of water and its peak use started on Tuesday the 11th and dropped on Saturday the 15th of February 2020.

Table 13.2 Identified water efficiency and rainwater harvesting opportunities at the SBL

Opportunity assessment	Status	Comments
Internal water metering and routine leak checking	Applicable and beneficial	–
Leak prevention through pressure control	Possible, but not beneficial	Nighttime pressures could be monitored to establish possible benefit
Education and awareness tags in all toilet blocks	Applicable and beneficial	–
General water awareness campaign	Applicable and beneficial	–
Procurement/future renovations to include water-efficient design specifications	Applicable and beneficial	–
Install rainwater harvesting system	Applicable and beneficial	The large flat roof provides good potential to capture rainwater for use. The possibility of capturing stormwater from the eastern outfall channel should also be investigated
Install greywater reuse system	Possible, but not beneficial	
Install borehole system	Exists – completed	Borehole system in place. All irrigation makes use of groundwater.
Install low-flow showerheads	Possible, but not beneficial	The number of showers on campus was not provided
Install water-efficient faucets in bathrooms	Applicable and beneficial	Existing faucets are standard open/close type; some without aeration
Install dual-flush cisterns	Not applicable	Flush master system in place
Install water-efficient urinals	Applicable and beneficial	
Install waterless urinals	Possible, but not beneficial	Pilot project recommended
Irrigate early or late (not between 1000 h and 1600 h)	Exists – completed	Automated irrigation system in place
Install water-efficient sprinkler system	Exists but more scope required	Efficiency of existing system to be investigated
Use compost regularly for water retention in gardens	Applicable and beneficial	This could be linked to waste management
Place layer of mulch around trees and plants	Applicable and beneficial	–
Use rainwater/greywater/borehole water for irrigation	Exists – completed	Borehole system in place. All irrigation makes use of groundwater
Include water efficiency into landscaping agreements and contracts	Applicable and beneficial	–

Source: Authors, based on UNISA (2020)

Within a single building, there are also installed smart meters to measure the consumption of water at various sections of the building. In huge buildings, measuring water inflow at a single point may not be enough to understand and manage water use efficiency and also to trace potential losses. Hence, various measurements at key sections of the building help to pin-point areas of excessive use or loss. The Simon Radipere building, for example, has got metres at the East End Roof Tank Bypass, East End Roof Tank Outlet, West End Roof Tank Main Inlet and West End Roof Tank Outlet. The CAS van Vuuren building has water meters at the East End Roof Tank 1 Outlet, East End Supply to Cooling Towers, Main Tank Outlet 1 and West End Roof Tank Outlet. This makes it easy for the University Estates to monitor water transactions and act at a much-localised level within a single building.

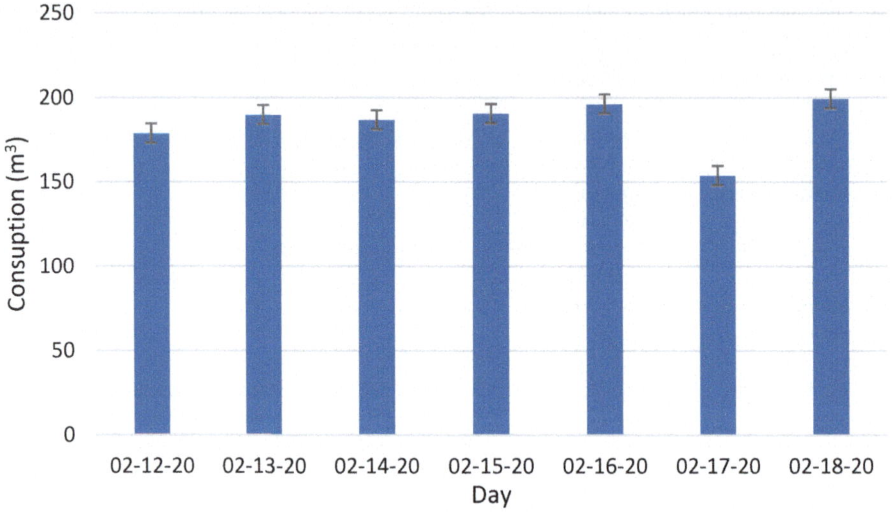

Fig. 13.3 Daily consumption at Muckleneuk. (Source: Authors, data from University Estates)

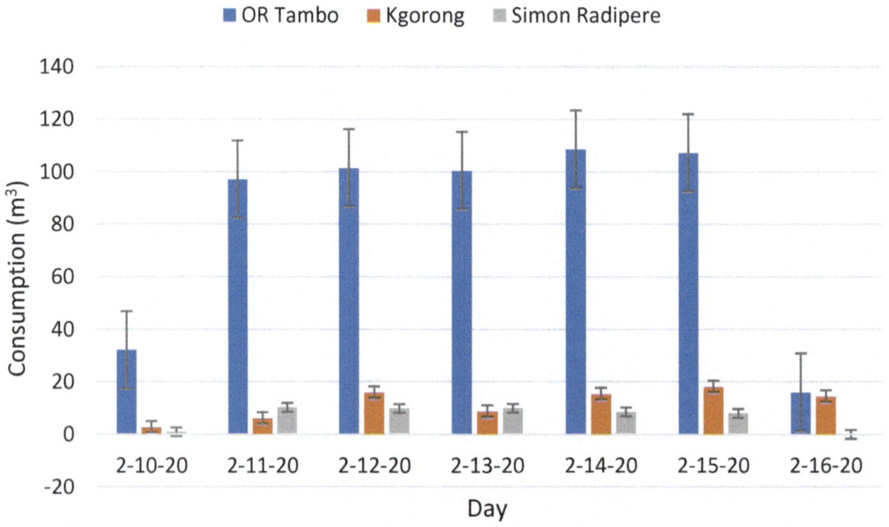

Fig. 13.4 Weekly water consumption in selected buildings. (Source: Authors, data from University Estates)

The continuous recording of water consumption by UNISA can also assist in the long-term management of water resources by the university. The analysis of long-term trends in water consumption can give an indication of the achievement of reduced water use and sustainability at a specific campus. This will inform the need to further improve on the designed system if there are no improvements or maintain it if it is achieving the desired outputs. Figure 13.5 shows the long-term trends in the consumption of water at the UNISA's SBL campus. There has been on average a reduced consumption of municipal water on the campus from 2016 to early 2020 as shown by the negative trendline equation $y = -0.0038x + 175.46$ on the time-series data obtained from accurate IP-based smart meters and available information from the Meter Data Management System (MDMS). The reduction can be attributed to the installation of efficient water use technologies, exclusive use of borehole water for irrigation and increased consciousness

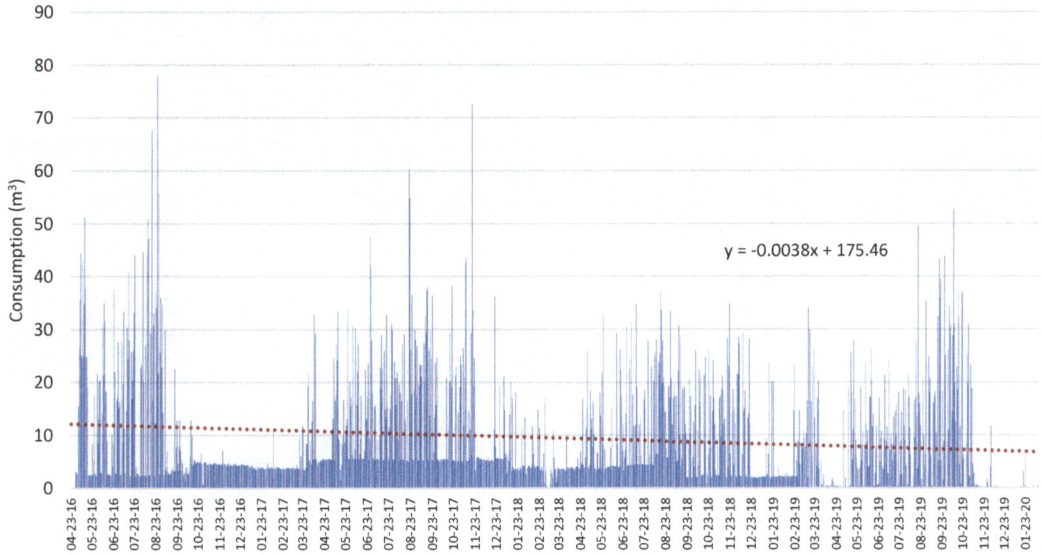

Fig. 13.5 Time-series water consumption at UNISA's SBL (2016–2020). (Source: Authors, data from University Estates)

of campus users on the need to conserve water. Such a declining trend would be very difficult to observe if there was no systematic measurement of consumption, and hence the impact of water efficiency technologies instituted at the campus is also difficult to note. Systematic recording of consumption is also in a way a cost-cutting measure because the university will be paying for exactly what it has consumed rather than the estimated consumption which municipalities frequently rely on. The means for doing that are available through the smart systems' vending platforms. In literature, the University of Surrey's (2019) Water Use Policy emphasises efficient metering systems as being key to achieving successful implementation of this policy and for attaining set water efficiency targets. Efficient metering facilitates the identification of facilities consuming abnormal amounts of water, helps to determine trends and patterns of use as well as assists in leak detections.

The smart water meter system in place at UNISA campuses can also produce alarms from the data collected. The alarms can range from high night flow, which assists in identifying areas of high leakage, to the status of a metering gauge. The alarms can be sent through an email or via text message. The alarms provide capacity for timely interventions to prevent excessive water loss in the case of leakages and also timely attention on faulty equipment to minimise loss of data collected during the downtime.

The need for monitoring and evaluation is very important in the attainment of the SDGs. The University Estates produces weekly and monthly reports on water consumption at UNISA campuses. The reports are produced in the form of tables and graphs that show both the flow profile and consumption pattern of water at different scales from an individual building to the entire campus. The process of plan-do-check is greatly enhanced in the process due to the availability of quantitative data on the resource that is being planned for. Areas of weakness and areas of strength can therefore be noted from the reported figures.

Rainwater harvesting and water-efficient equipment have been installed in several UNISA campuses (Figs. 13.6 and 13.7). Figure 13.6 shows part of the infrastructure at Sunnyside Campus in Pretoria. Each of the harvest tanks installed has a capacity to store 5000 litres of water. The only downfall at the time of fieldwork in February 2019 was that the infrastructure was

Fig. 13.6 Rainwater harvesting infrastructure at Sunnyside Campus. (Source: Authors)

Fig. 13.7 Water-efficient installations in RR Maluleke Building. (Source: Authors)

not being used. The tanks were harvesting rainwater and were full, but there is no connection to irrigation piping of other facilities like bathrooms. As such, the investment remains a white elephant for over a year since their installation. On further investigation, it emerged that the challenge may be emanating from the lack of communication and working together between line departments from University Estates from design, through procurement and operation.

As presented in the literature section, water-efficient technologies remain at the forefront of addressing the water SDG in many higher education campuses across the world. To this end, there has been great work done to retrofit old water technologies with water-efficient technologies at all campuses. Figure 13.7 shows automated bathroom taps. The same applies to urinals that are also fitted with automated equipment.

4.2 Focus on Wetlands

UNISA has embarked on wetland rehabilitation projects on its campuses. Its campuses with wetlands visited include the Muckleneuk, SBL and Daveyton. On the 4th of September 2019, through its sustainability office, UNISA launched the SBL wetland rehabilitation project. The project was launched by the Vice-Chancellor of the university. This showed commitment on the part of the university management to achieve target 6.6, which aims by 2020 to protect and restore water-related ecosystems, including wetlands, rivers and aquifers. During the launch, those in attendance including staff, students and government representatives were made cognisant of the importance of wetland ecosystems and the services which they provide. At the time of the rehabilitation launch, the SBL wetland was in a degraded state and is still threatened by landscaping done on it to drain its water and surrounding development projects that have replaced part of the wetland outside the UNISA property. The rehabilitation objectives of the wetland include maximising its moisture retention, removing alien species and the re-establishment of natural ecosystems. Figures 13.8, 13.9, and 13.10 show the Daveyton and SBL Campus wetlands.

Figure 13.10 shows that the Daveyton Campus wetland is still relatively healthy when compared to the one at SBL. A comparison of the SBL wetland in 2004 and 2019 for the same period of the year (September) shows the level of construction around the SBL may be affecting the health and water regime of the wetland. In 2004 there were only two major buildings surrounding the SBL compared to seven in the year 2019. Construction on and around wetlands together with prolonged droughts has been noted to be one of the prime causes of their shrinkage and eventual disappearance in many parts of the world. Some of the ecosystem services being offered by the SBL and Daveyton wetlands include the provision of flood attenuation during intense rainfall episodes, provision of water through boreholes drilled on them, reduction of pollution levels in storm runoff and being used as a habitat by several species including migratory and indigenous birds, snakes, frogs and rabbits. Given the commitment of UNISA to limit construction on its wetlands and allow natural processes to re-establish themselves, the university is not only addressing SDG 6 by indirectly

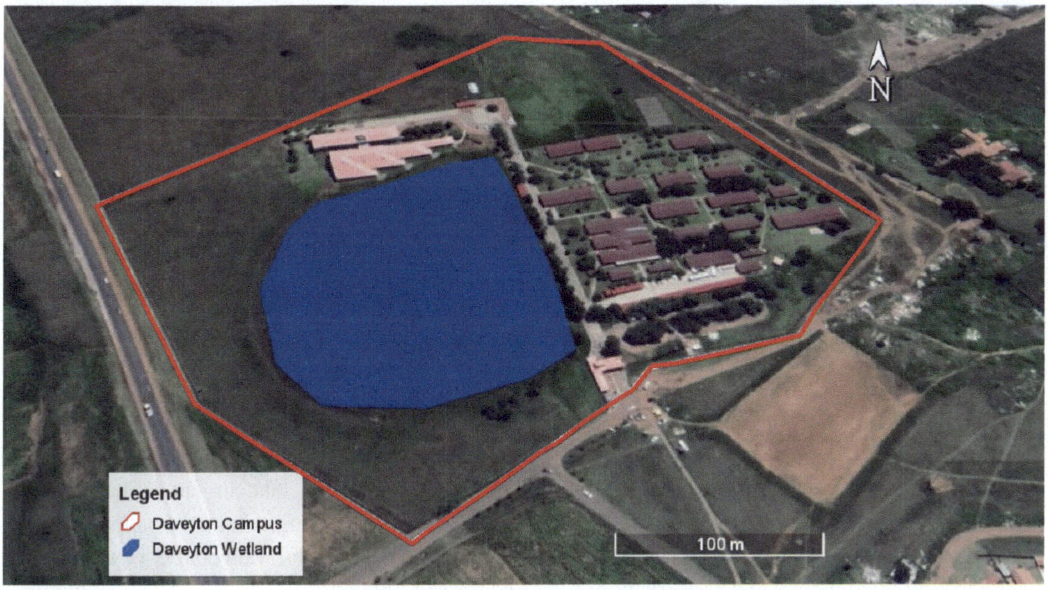

Fig. 13.8 Daveyton Campus wetland. (Source: Authors, image from Google)

Fig. 13.9 SBL Wetland 2004. (Source: Authors, image from Google)

Fig. 13.10 SBL wetland 2019. (Source: Authors, image from Google)

interacting with SDG 15, which aims at protecting, restoring and promoting sustainable use of terrestrial ecosystems and also reversing land degradation and biodiversity loss. Specifically, the university is addressing target 15.1, which aims to ensure that by 2020 there is the conservation, restoration and sustainable use of terrestrial and inland freshwater ecosystems and their services, in particular wetlands (United Nations 2015). Target 15.5, which aims to take urgent and significant action to reduce the degradation of natural habitats and halt the loss of biodiversity, is also being addressed. Figures 13.11 and 13.12 show pictures of the SBL and Daveyton wetlands.

Fig. 13.11 SBL wetland. (Source: Authors)

Fig. 13.12 Daveyton wetland. (Source: Authors)

Looking ahead, UNISA is implementing and using alternative water source supply such as rainwater harvesting, borehole water, stormwater, runoff reduction and associated stormwater pollution load reduction. This is meant to reduce the demand on the municipal supply system, hence contributing to an eventual reduction in water demand at the municipal level. UNISA installation of rainwater water collection and harvesting infrastructure at most of its campuses and the high total roof area at most of its campuses can be an important source of alternative water supply for the different water needs of the university community. Further irrigation on most campuses relies on either borehole water or captured stormwater. Irrigation is also time based, normally done early in the morning or late in the evening. The amount of water applied is also adjustable to meet seasonal needs with the irrigation equipment fitted drip systems and with water-efficient nozzles or sprinklers that have uniform application. However, the major handicap is the fact that the irrigation systems are not automated and that workers report to work between 7 am and 3 pm. Irrigation sometimes is done during the daytime. The desire for automated irrigation equipment is also low, given what the Estates department termed high maintenance costs and downtime compared to manual systems. Efficiency in water utilisation during irrigation is also sometimes compromised by very long procurement channels that have to be followed in the event of an identified leakage or maintenance issue.

Water efficiency in irrigation at UNISA campuses has also been enhanced by the selection of water-wise plants and flowers. As a golden rule of procuring landscaping plants, the university gives the highest priority to the indigenous plant species that do not require huge volumes of water for their sustenance. To this effect, the University Estates department in partnership with the College of Agriculture and Environmental Sciences has embarked on a project to start a thriving nursery of over ten different indigenous plant species that the university requires to landscape the different parts of its campuses.

The university also plans to start stormwater harvesting by directing it to either detention ponds or nearby wetlands in the case of SBL and Daveyton Campuses. The diversion of storm runoff to wetlands for storage rather than into storm drains will aid in managing both the stormwater quantity and quality as well as other associated amenity and biodiversity aspects, thereby addressing SDG 15. However, the water from storm detention ponds and wetlands may need further treatment before it can be used on campus. Besides saving municipal water, this is also addressing target 6.3 of SDG 6, which aims by 2030 to improve water quality by reducing pollution. The use of water conjunctively for different uses and at different times is part of a global movement known as water-sensitive urban design. Its vision is to manage the urban water cycle in its entirety and sustainably. This includes water supply, sanitation, stormwater and groundwater (Brown et al. 2008). It also aims to use the water in different areas of the urban water cycle in a fit-for-purpose fashion.

Apart from natural wetlands, UNISA has several artificial wetlands that require water and contain thriving biodiversity. For example, the main campus in Muckleneuk in Pretoria has three artificial wetlands (Fig. 13.13), and the SBL has one.

In terms of the way forward to accelerate progress on SDG 6 implementation, the UN-Water (2020) observed that most SDG targets are off-track and highlighted that action during the next few years will be critical to the 2030 Agenda's success. In light of this challenge, the United Nations Secretary-General launched a decade of action on water and sanitation. The experiences at UNISA as a living lab for the implementation of SDG 6 show the successes needed to be scaled up to other sections of society so that the benefits can be realised by the greater community. As such UNISA's example fits well into the SDG 6 Global Acceleration Framework being envisaged by the United Nations.

5 Conclusion

South Africa is increasingly becoming water stressed due to climate change, and variability, which combined with rapid population increase, may lead to negative impacts on water provision and sanitation. The University of South Africa has positively interacted with and localised SDG 6. This has seen the university developing an integrated water efficiency and rainwater harvesting plan which saw the installation of smart meters around the campus and on all the building to monitor flow and consumption as well as leakages and having efficient irrigation systems based on borehole and greywater, wetland restoration and rehabilitation projects, rainwater harvesting and stormwater detention and utilisation. UNISA interaction with SDG 6 and other interlinked ones such as SDGs 13 and 15 have, to a large extent, allowed the university to manage water resources more efficiently, reduce water use, enable alternative water supply to potable municipal supply and increase its participation in water efficiency and harvesting initiatives. Some of the benefits realised in localising SDG 6 include leg-

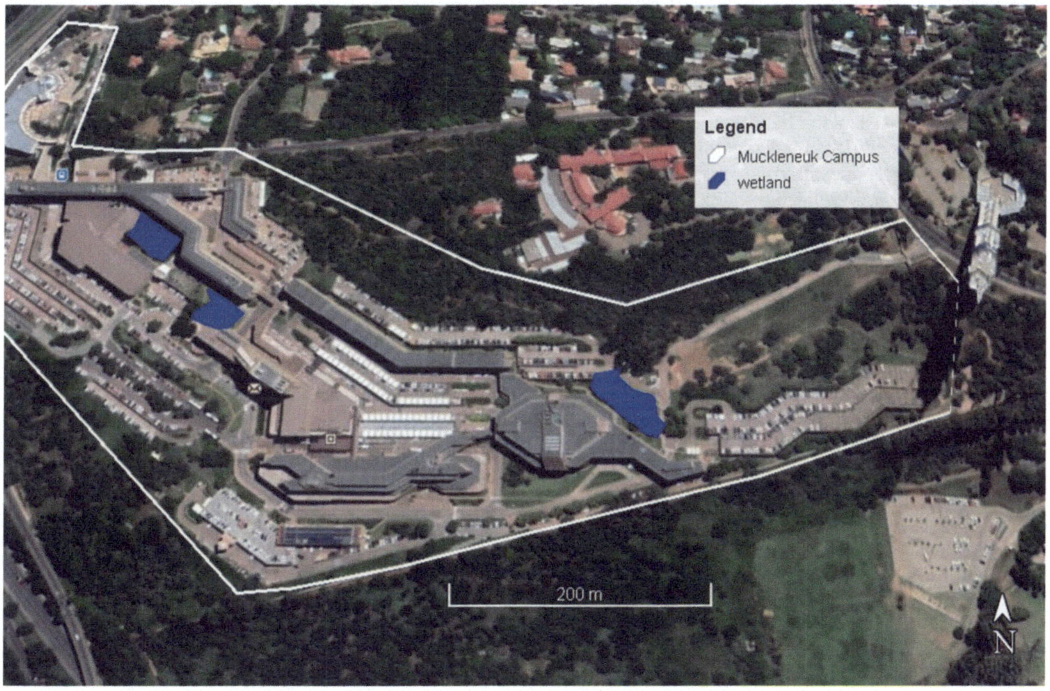

Fig. 13.13 Artificial ponds at UNISA's Muckleneuk Campus in Pretoria. (Source: Authors, image from on Google)

islative compliance, improved reputation, reduction of water bills due to the use of alternative sources of water, reduction of pollutant loads to nearby watercourses and mitigation of associated environmental issues and attenuation of flood peaks due to wetland rehabilitation.

This initiative can be adopted by other South African universities, especially those with residences within campuses as well as those from disadvantaged backgrounds. Big organisations with multiple buildings in both the private and public sectors can also localise SDG 6 using the UNISA example as a template to start from and improve on to suit their local conditions and realities. In the light of UNISA's successful utilisation of Advanced Metering Infrastructure, there is a high potential for country-wide embracing and adoption of similar technologies. Such disruptive innovation reduces costs through minimising the time of execution and improved capacity. This is a core feature of the Fourth Industrial Revolution which has the Industrial Internet of Things (IIoT) as its core operating system.

References

Augustine, E.E. & Hanafiah, M.M. (2019). Awareness level of water resource conservation of university students. Water Conservation and Management. 3(2): 18-21. https://doi.org/10.26480/wcm.02.2019.18.21.

Bai X, Surveyer A, Elmqvist T, Gatzweiler FW, Güneralp B & Parnell S. (2016). Defining and advancing a systems approach for sustainable cities. Curr Opin Environ Sustain 23:69–78.

Bath Spa University. (2018). Sustainability Strategy 2018. Bath Spa University.

Beynaghi, G., Trencher, F., Moztarzadeh, M., Mozafari, R., Maknoon, W., & Filho, L. (2016). Future sustainability scenarios for universities: Moving beyond the United Nations decade of education for sustainable development. Journal of Cleaner Production, 112, 1464–1474.

Biswas, J.K. and Dash, M.C. (2019). Building a green image: best sustainability practices in universities. Science and Culture. 1: 308-14. https://doi.org/10.36094/sc.v85.2019.

Bhowmik, J., Selim, S. A., & Huq, S. (2017). The role of universities in achieving the sustainable development goals. CSD-ULAB and ICCCAD policy brief. Dhaka:ULAB.

Brown, R., Keath, N., & Wong, T. (2008). Transitioning to water sensitive cities: historical, current and future transition states. In 11th International Conference on Urban Drainage. Edinburgh, Scotland, UK.

Edge Hill University. (2016). Report on Edge Hill University Water Efficiency May 2016. Lancashire: Edge Hill University.

Filho, W.L., Morgan, E.A., Godoy, E.S., Azeiteiro, U.M., Bacelar-Nicolau, P., Avila, L.V., Mac-Lean, C. & Huge, J. (2018). Implementing climate change research at universities: Barriers, potential and actions. Journal of cleaner Production. 170: 269-277. https://doi.org/10.1016/j.jclepro.2017.09.105.

Findler, F., Schönherr, N., Lozano, R., Reider, D. & Martinuzzi, A. (2019). "The impacts of higher education institutions on sustainable development", International Journal of Sustainability in Higher Education, Vol. 20 No. 1, pp. 23-38.

Ishii, S.K.L. & Boyer, T.H. (2016). Student support and perceptions of urine source separation in a university community. Water Research 100: 146-156. https://doi.org/10.1016/j.watres.2016.05.004.

Kestin, T., van den Belt, M., Denby, L., Ross, K., Thwaites, J., & Hawkes M. (2017) Getting started with the SDGs in universities: A guide for universities, higher education institutions, and the academic sector. Australia, New Zealand and Pacific Edition. Sustainable Development Solutions Network – Australia/Pacific, Melbourne.

Klopp, J.M. & Petretta, D.L. (2017). The urban sustainable development goal: indicators, complexity and the politics of measuring cities. Cities; 63:92–7.

Sharp, L. (2009). Higher education: the quest for the sustainable campus, Sustainability: Science, Practice and Policy, 5:1, 1-8, https://doi.org/10.1080/15487733.2009.11908023.

Sharp, L. (2013). Green campuses: The road from little victories to systemic transformation. Journal of Education for Sustainable Development, 7(1): 135-135.

South Africa Department of Water and Sanitation. (2017). Benchmarking of water loss, water use efficiency and nonrevenue water in South African municipalities (2004/05 to 2015/16). Retrieved from: https://africacheck.org/wp-content/uploads/2018/04/National-benchmark-2017-09-12-final.pdf (Accessed 19 August 2020).

Manchester Metropolitan University. (2016). Manchester Metropolitan University Environmental Sustainability Statement 2015-16. Manchester: Manchester Metropolitan University.

Nhamo, G & Mjimba V (eds.). (2020). Sustainable Development Goals and Institutions of Higher Education, Sustainable Development Goals Series, https://doi.org/10.1007/978-3-030-26157-3_1.

Nottingham Trent University. (2019). Nottingham Trent University Water Policy. Nottingham: Nottingham Trent University.

Nunes, B.T., Pollard, S.J.T., Burgess, P.J., Ellis, G., de los Rios, I.C. & Charnley, F. (2018). University Contributions to the Circular Economy: Professing the Hidden Curriculum. Sustainability 10, 2719; https://doi.org/10.3390/su10082719.

Owens, T. L. (2017). Higher education in the sustainable development goals framework. European Journal of Education, 52, 414–420. https://doi.org/10.1111/ejed.12237.

Patel, Z., Greyling, S., Simon, D., Arfvidsson, H., Moodley, N., Primo, N. (2017). Local responses to global sustainability agendas: learning from experimenting with the urban sustainable development goal in Cape Town. Sustain Sci;12(5):785–97.

Ramsar Convention Secretariat. (2016). An Introduction to the Ramsar Convention on Wetlands, 7th ed. (previously The Ramsar Convention Manual). Ramsar Convention Secretariat, Gland, Switzerland.

SANBI. (2014). A Framework for Investing in Ecological Infrastructure in South Africa. South African National Biodiversity Institute, Pretoria.

Tan D.T., Siri, J.G., Yi, G., Ong B, Lim, S.C., MacGillivray, B.H., & Marsden, T. (2019). Systems approaches for localising the SDGs: co-production of place-based case studies. Globalisation and Health 15:85 https://doi.org/10.1186/s12992-019-0527-1.

UNISA. (University of South Africa). (2016). Integrated water efficiency and rain water harvesting plan and waste management master plan. Pretoria: University of South Africa.

UNISA (University of South Africa). (2019). UNISA SDGs Localisation Declaration. Pretoria: University of South Africa.

UNISA (University of South Africa). (2020). Governance and Management. www.unisa.ac.za/sites/corporate/about/governance&management.

United Nations (2016) Roadmap for localising the SDGs: Implementation and monitoring at subnational level. The Global Taskforce of Local and Regional Governments. gtf2016.org.

United Nations. (2015). Transforming our world: the 2030 Agenda for Sustainable Development. United Nations New York.

University of Bristol. (2017). University of Bristol Carbon & Water Strategy – November 2017. Bristol: University of Bristol.

University of Oxford. (2011). University of Oxford Water Management Strategy January 2011. Oxford: University of Oxford.

University of Plymouth. (2018). University of Plymouth Sustainability Report 2018. Plymouth: University of Plymouth.

University of Surrey. (2019). University of Surrey Water Use Policy (2019-2021). Surrey: University of Surrey.

UN-Water. (2018). Sustainable Development Goal 6: Synthesis Report on Water and Sanitation 2018. New York: UN Water.

UN-Water. (2020). UN-Water Bulletin, Vol. 82, No. 40 International Institute for Sustainable Development http://enb.iisd.org/water/un/32/.

14 Climate Change in Zimbabwe's Vulnerable Communities: A Case Study of Supporting Enhanced Climate Action Project (SECA Project) in Bulilima District

Veronica Nonhlanhla Gundu-Jakarasi and Justice Nhidza

Abstract

There are noticeable changes in climatic conditions in the sub-Saharan region, characterised by unreliable rainfall which adversely affect agriculture and food production. Research on Zimbabwe reveals that the agricultural sector is already suffering from changing rainfall patterns, temperature increases and extreme weather events such as floods and droughts. This compromises the country's aim of achieving Sustainable Development Goals 1 and 2 which focus on eliminating poverty and hunger. Consequently, the Zimbabwean government is working with strategic partners to implement projects aimed at sustaining smallholder farmer livelihoods within vulnerable communities whilst also addressing the impacts of climate change. Thus, this study sought to map best practices and lessons learnt in the implementation of the Supporting Enhanced Climate Action (SECA) project (2016–2020). Through focus group discussions and key informant interviews, the study established that solar-powered and tailor-made multiple-use irrigation were crucial to enhancing community adaptation. Additionally, integrated planning was shown to have a high impact in addressing other sustainable development goals. The study provided lessons on strengthening governance systems and for improved adaptation and resilience building through engagement. Finally, the study recommends multi-stakeholder engagements to develop intra-adaptation and mitigation actions, tailor-made climate information and strong markets and sustainable value chains.

Keywords

Climate change · Poverty · Sustainable development goals · Adaptation · Bulilima district

1 Introduction

The world is experiencing an unprecedented climate change phenomenon that is likely to cause a human survival dilemma and difficulties in

V. N. Gundu-Jakarasi (✉)
Department of Environmental Sciences, University of South Africa, Pretoria, South Africa
e-mail: vjakarasi@idbz.co.zw; 61962074@mylife.unisa.ac.za

J. Nhidza
Young Volunteers for the Environment, Harare, Zimbabwe

G. Nhamo et al. (eds.), *Sustainable Development Goals for Society Vol. 2*, Sustainable Development Goals Series, https://doi.org/10.1007/978-3-030-70952-5_14

national development unless urgent steps are taken to curtail human behaviours driving climate change (Intergovernmental Panel on Climate Change (IPCC) 2013). There is already evidence that Africa is warming faster than the global average, and this is likely to continue (Government of Zimbabwe (GoZ) 2015a). Studies have revealed that climate change will have long-lasting negative effects for populations and their environment in most parts of the world. Climate change is said to threaten the achievement of a number of sustainable development goals (SDGs) such as ending poverty (SDG 1) and zero hunger (SDG 2). According to the United Nations, human health, food security, economic activities, physical infrastructure and natural resources will be greatly undermined by the climate change phenomenon (UNDP 2017).

In most parts of Africa, particularly in the sub-Saharan region, climate change has exerted enormous pressure on rural livelihoods, particularly smallholder farming which is the mainstay of rural livelihoods. This situation has arguably plunged populations into food insecurity, hunger, ill health, limited access to income and ultimately poverty (Madzwamuse 2010). Smallholder farming is important for food production and income generation for many in most parts of rural Africa. A study by the United Nations Food and Agriculture Organization (FAO 2008) found that about 75% of rural populations in sub-Saharan Africa rely on smallholder farming as a source of livelihood. In Southern Africa, impacts of climate change have resulted in declining water resources and water quality, reduced agricultural productivity, damaged infrastructure, loss of lives and ecosystem degradation, among other impacts (Gukurume 2013).Whilst issues related to access to clean water and poor climate change governance at national and local levels further add to the already existing challenges (Dodman and Mitlin 2015), the livelihoods of the people in most parts of sub-Saharan Africa are now at risk due to rapid increased intensity and frequency of extreme weather events such as floods and droughts (Mushita & Thompson 2013). This is also demonstrated by the 2019 Cyclone Idai that resulted in loss of life of over 1300 people in Mozambique, Malawi and Zimbabwe and contributed to about US$2.7 billion economic loss (AON 2020). As a result, extreme weather events are also expected to result in environmental degradation, increased water stress and a decrease in agricultural production which will result in livelihood decline and food insecurity.

This study, therefore, seeks to demonstrate that integrated programming is a strategic tool of addressing different SDGs with multiple outcomes. The chapter provides a review of how the Government of Zimbabwe has worked with development partners to address food security and poverty in different communities whilst also dealing with the impacts of climate change. Although such initiatives have the capacity to change lives, there is need to foster sustainability post-programme implementation by supporting coordinated stakeholder engagement and leveraging continued government support. The chapter starts by providing an overview of climate change and its impact on food security and how it aggravates poverty. This is followed by an outline of the goals and impacts of the Supporting Enhanced Climate Action (SECA) project including its achievements. The chapter concludes by explaining the challenges, outlining the lessons learnt and making recommendations.

2 Literature Review

2.1 Climate Change, Food Security and Poverty Alleviation

The world is facing consummate changes in the pattern and variability of climate phenomenon that are likely to lead into a catastrophe of human survival and sustainable development unless urgent steps are taken to curtail human behaviours that are causing climate change (Government of Zimbabwe (GoZ) 2016). There is overwhelming evidence that climate change is real and upon us. This has been seen through the rising sea levels and how some parts of the world are getting warmer (Adedeji et al. 2014). The sharp rainfall seasonal contrasts (Imada et al. 2017), noticeable increases in solar radiation (Ohunakin et al. 2015), increased variation in both intensity and frequency of rainfall (IPCC 2013), increase in soil salinity (Dasgupta et al. 2015), increase in both intensity and frequency of highly extreme wind events (Bloom et al. 2008), increased frequency of El Nino and cyclone events (e.g. Cyclone Katrina and Cyclone Idai), increase in intensity and frequency of storms (IPCC 2013) and consequently, increasing rates of fluctuation of water levels in some rivers (Ubeda et al. 2013) are some of the clear examples of climate change impacts. Climate change presents a significant threat to the environment, food security, human life and economic development. Davis and Hirji (2014) indicate that globally, sectors such as agriculture, health, ecosystems, forestry, water, disaster risk reduction, infrastructural development, transport and settlements are highly vulnerable to climate change. According to the IPCC Fourth Assessment Report (2007), Africa will be the most affected continent due to poverty and low adaptive capacity.

The increasing temperatures due to anthropogenic activities and decline in water resources in most parts of Southern Africa will affect food production. The increase in the frequency of droughts has made rainfed agriculture less sustainable. This has been further exacerbated by the intermittent rainfall and increased mid-season dry spells (GoZ 2015a). The inextricable links between agriculture and climate change have threatened food security globally (Arora 2019). The decline in agriculture production has also increased poverty among rural communities who are dependent on agriculture for their household food and extra income. Zimbabwe has not been an exception to the impacts of climate change where 70% of her population lives in rural areas and at least 80% of them rely on rainfed agriculture or climate-sensitive sectors (GoZ 2015b).

2.2 Overview of Climate Change and the Resultant Vulnerability in Zimbabwean Communities

Zimbabwe is among the most susceptible countries in Southern Africa due to its high exposure and low adaptive capacity (IPCC 2012). The vulnerability of Zimbabwe to climate change is driven by a range of factors that include over-dependency on rainfed agriculture, high reliance on ecosystem goods for livelihoods and less developed agricultural production systems (Manzungu and Moyo 2016). The risk of climate change on agricultural production, food security, water resources and ecosystem services will likely have increasingly severe consequences on lives and sustainable development prospects in Zimbabwe. Managing this risk requires integration of mitigation and adaptation strategies in the management of ecosystem goods and services and agricultural production systems in Zimbabwe.

Zimbabwe's economy is principally agro-based with over 70% of the population living in rural areas and dependent on climate-sensitive livelihoods such as arable farming and livestock (UNDP 2017). For that reason, the impacts of climate change are quite threatening. The Government of Zimbabwe (GoZ) regards climate change as a leading threat to the country and its people as demonstrated by its commitments to the 2015 Paris Agreement and its Nationally Determined Contribution obligations to address climate change. This is also supported by Jakarasi and Moyo (2018) who articulate that the GoZ established a full-fledged Climate Change Management Department to address the country's climate change challenges. GoZ (2015a) further iterates that climate change has the potential to undermine many of the positive developments made in meeting the country's sustainable development goals as well as constraining further progress. The impacts are most pronounced among young people, women and people with disabilities and others who may be marginalised or vulnerable. The most vulnerable parts of Zimbabwe are worsened by the problems of poverty, food insecurity and malnutrition exacerbated by the economic constraints in the country (ZimVAC 2016). These and other challenges caused by climate change need to be addressed urgently since they are likely to undermine the achievement of the Sustainable Development Goals (SDGs).

Climate change poses a significant threat to human security especially for women who represent 70% of the world's poor (UNDP 2017). In a country like Zimbabwe where women are more vulnerable than men, the highest percentage of the poorest and most disadvantaged are women. Women in Bulilima District are mostly involved in climate-sensitive tasks such as gathering and securing food, water and firewood, and these tasks ensure food security and household well-being. This, however, increases their vulnerability to climate change (Tshuma and Mathuthu 2014). Furthermore, the drought spells that took place in the years 2008/2009, 2012/2013 and 2018/2019 led to the failure of most crops and perishing of thousands of livestock. Also, drought-tolerant crops such as sorghum and rapoko could not thrive. Rising temperatures and increased rainfall variability have led to declining agriculture outputs, and this has further compromised the economic growth and stability of Bulilima District. This has also led to rampant unemployment levels, food insecurity and low productivity in the district, which have largely contributed to the outflux of the youth and economically active people into neighbouring countries of Botswana and South Africa (Swain et al. 2011).

Research shows that many districts in Zimbabwe are vulnerable to the effects of climate change and variability. These include low-lying areas such as Gokwe, Chiredzi, Wedza and Bulimia districts that receive low and unreliable rainfall. These regions are characterised by very high temperatures which are combined with erratic rainfall conditions, thereby making agricultural output relatively low. Zimbabwe has five agro-ecological zones which are divided based on the amount of rainfall received, mean annual temperatures and soil types, among other elements, and their potential for agriculture (Vincent and Thomas 1960).

2.3 SECA Project as a Resilience Initiative for Zimbabwe's Vulnerable Communities

Cognisant of these already existing challenges which were being exacerbated by climate change, in the year 2016, the Government of Zimbabwe in collaboration with the UNDP-Zimbabwe, undertook to implement a climate change adaptation and mitigation project in the four districts of Bulilima, Lupane, Gokwe and Buhera from 2016 to 2020. This partnership was emphasised by Dyer et al. (2013), who observe that "partnerships models" are increasingly being accepted as a vital tool for effective climate change adaptation in developing countries. The project's main objective was to help the targeted communities respond to the impacts of climate change on a day-to-day basis and to improve their livelihoods. The project sought to address problems related to perennial food shortages, high levels of environmental degradation and reduction in the productive capacity of agricultural land. The project aimed at scaling up adaptation and mitigation action in most vulnerable communities such as Bulilima, supporting climate and gender-sensitive disaster risk management (DRM) planning and implementation and supporting government institutions to meet obligations under DRM.

The project also aimed at enhancing community livelihoods and promoting more sustainable natural resource management practices. The SECA project implemented a number of initiatives which included climate-smart agriculture (CSA), climate awareness programmes, solar-powered irrigation schemes, biogas systems as well as helping the communities to draft DRM plans. Bulilima residents used to have different levels of knowledge on climate change and also varied perceptions on the phenomenon. According toLipper et al. (2014), CSA is an approach that seeks to address the interlinked challenges of food security and climate change and aims to (i) sustainably increase agricultural productivity to support equitable increases in farm incomes, food security and development; (ii) adapt and build resilience of agricultural and food systems to climate change at multiple levels; and (iii) reduce and/or remove greenhouse gas emissions, where possible. CSA techniques and initiatives introduced in the district incorporated research, development, promotion of indigenous knowledge systems (IKS) (Mugambiwa 2018), advocacy and training workshops on breeding, introduction and seed multiplication of drought-tolerant crops and animals.

2.4 SECA Project Linkage with SDGs

The core aim of the project sought to enhance community livelihoods and promote more sustainable natural resource management practices through Supporting Enhanced Climate Action (SECA) for low carbon and climate-resilient development pathway as a way of localising the SDGs. The SECA project also sought to build community resilience to the impacts of climate change by implementing tangible gender-sensitive interventions in line with sustainable eradication of hunger and poverty, as well as preventing the discrimination of women (SDG 5). This includes building resilience to the impacts of droughts and prolonged dry spells which are common in the southern and western regions of Zimbabwe. The project further desired to increase equitable and sustainable access to and use of safe water supply and improved sanitation in the district which is in line with mandate 6 of ensuring sustainable management of water and sanitation for all. This initiative addressed infrastructure imbalances between urban and rural services (MEWC 2016). The project established models where livelihood-centred approaches to disaster management are linked with wider institutional structures involved in disaster and development planning. The project endeavoured to be robust and inclusive, basing on the understanding of the cross-cutting nature of the SDGs. As a result, the project supported the implementation of SDGs 1, 2, 5, 6, 7, 13 and 17.[1]

In every project design, it has become imperative to ensure that a clear monitoring and evaluation (M&E) framework is in place. Hence, the project M&E was designed to capture lessons learnt from experiences in implementing different climate change approaches including potential lessons for scalability and replicability. An analysis of best practice in building consensus among stakeholders on how to link most effectively with and support communities' own disaster planning in a sustainable way was also considered. The project further recognised the need to influence policymakers at all levels involved in climate change management, disaster management and development planning to adopt a livelihood-centred approach to adaptation and disaster risk management. This is a critical initiative that builds synergies and bridges the gap between central governments and communities at the grassroots, playing an important role in fostering the inclusivity of everyone towards achieving the 2030 Agenda (Jakarasi and Moyo 2018).

3 Methodological Orientation

The study used a qualitative approach with a focus on a case study design. The case study was used to elaborate on how the SECA project implemented different strategies that supported the communities of Bulilima District to cope with climate change. The study area was Bulilima District (see Fig. 14.1), situated in Matabeleland South Province, a vulnerable district that lies in the driest agricultural regions IV and V.

The district consists of 15 wards, with 13 wards clearly displaying characteristics of high temperatures and low moisture, making the area unfavourable for rainfed agriculture. The district receives an average rainfall of 500 mm per annum whilst temperatures can reach up to 40 °C during summer (Practical Action 2014). It has a population of 90, 961 people and an area of 6439 km^2 with the average density of 14.06/km^2. The overall population declined by 0.41% per annum from 2002 (Zimstat 2012), and this has been attributed to migration by young adults most probably to urban areas in the country and neighbouring states such as South Africa and Botswana in search of greener pastures. The communities depend on cross-border trading, minor trade, limited crop production, livestock rearing and remittances (Bulilima Rural District Council (BRDC) 2020).

The study made use of both primary and secondary data. Primary data was obtained using focus group discussion (FGD) guide and semi-structured interview guide. Five FGDs were conducted with 30 randomly sampled farmers who were beneficiaries of the SECA project. Purposive

[1] SDG 1, end poverty; SDG 2, zero hunger; SDG 5, gender equality; SDG 6, clean water and sanitation; SDG 7, affordable and clean energy; SDG 13, climate action.

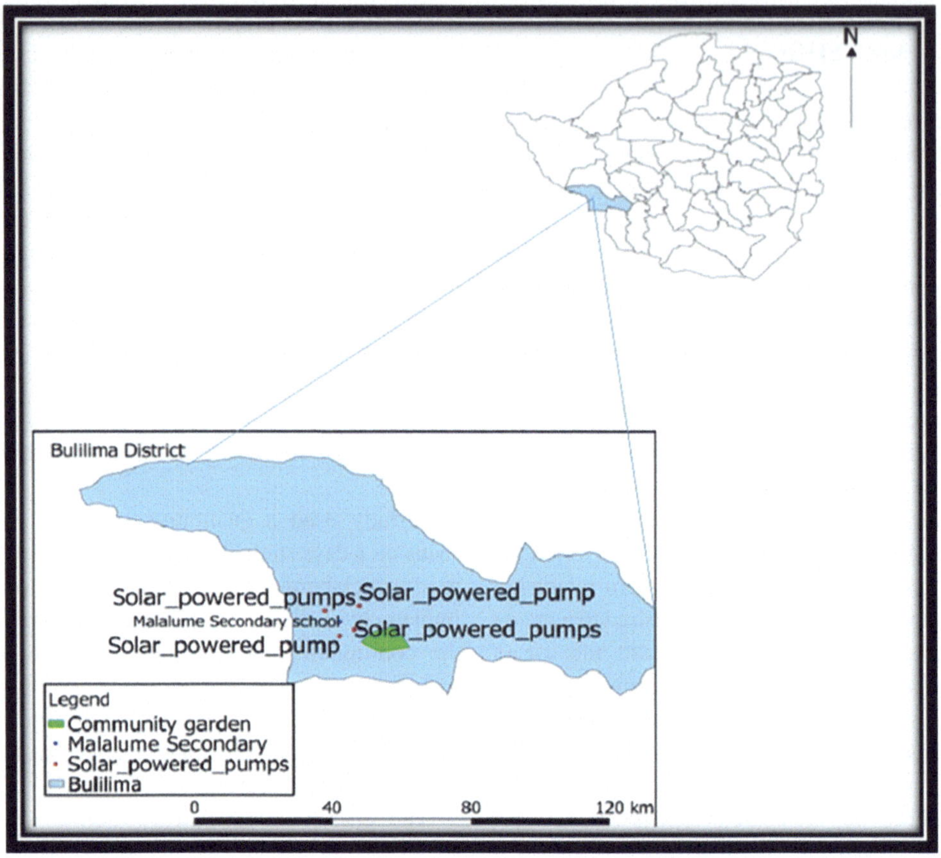

Fig. 14.1 Description of the case study area: Bulilima District. (Source: Authors (2020))

sampling, a non-probability sampling method, was conducted for the key interview informants. Twelve interviews were done with the District Administrator, Agriculture Extension Officer, United Nations Volunteer who was managing the project and the Chiefs, among others. The FGDs and the interviews sought to assess community perspectives on the impacts of climate change and the outcomes of the SECA project and to identify positive project benefits that can be scaled up or replicated in the future in similar areas. Secondary data was obtained from the Zimbabwe National Statistics Agency (Zimstat), Climate Change Management Department, UNDP and the Bulilima District Office in the form of reports, policies and project documents.

The findings of the interviews were transcribed into NVivo, a computer analysis program that allows for the discovery of trends and complex relationships in qualitative data. Each interviewee was assigned certain characteristics, including roles and activities in climate change management. The statements were examined on the basis of given characteristics by looking for commonalities, divergences and trends in the data collected. Responses from the FGDs were manually coded and themes and patterns were also identified.

4 Presentation and Discussion of Findings

4.1 Critical Review of SECA Project Case

The SECA project, through initiating climate change-related mitigation and adaptation interventions, played a crucial role in promoting climate resilience in Bulilima District. These actions enhanced community livelihoods and promoted sustainable natural resource management practices in that district through various ways. Some of the project benefits are outlined below.

A study was undertaken in October 2019, where community participants were engaged through the survey and focus group discussions, whilst key informants were engaged through interviews. The community participants pointed out that CSA techniques and initiatives such as diversification of crop and animal production and promotion of organic farming brought to the district through the SECA project were a major achievement in building climate resilience of the district. In one of the (FGDs), the following was said:

> We appreciate the significant role of (NGOs) in our district such as (UNDP-SECA) project are playing in meeting our needs in times of perennial erratic rainfall and ever-increasing temperature resulting from climate change like this. The organisation is supporting us in the best way they can, we have learnt a lot about the causes of droughts, and best ways of coping with the situation and how best to remain resilient. *(Malalume community member, Female, 45 years of age)*

Water harvesting and efficient irrigation (particularly drip) are key adaptation practices for a variety of crops, especially for winter crops, fruits and vegetables which can also have an impact on incomes and nutrition. Soil management-based practices such as precise fertiliser application, manure application, agroforestry, crop rotations and intercropping, along with soil conservation structures, are critical in enhancing and maintaining soil health (World Bank and CIAT 2018). CSA achieves food security under a changing climate and increasing food demand. It also enhances resilience and reduces greenhouse gases (GHGs) and requires planning to build synergies between the three pillars of productivity, adaptation and mitigation.

From the interviews, respondents pointed out that many households benefited from growing drought-tolerant crops and short-season varieties which are deemed more effective since yield possibility is higher and the crops have high nutritive value. Many households are benefitting since the crops are also important for food security during drought times. The community participants also highlighted that nutritional gardens played a pivotal role in providing community training and links to climate-proofing livelihood programmes. Farming of both staple foods and high-value crops ensured that households had food and nutrition security and income-generating opportunities. Additional benefits on health, education and poverty alleviation were also realised, and this was also attributed to mixed and diversified farming that reduced risk and supported a balanced food basket for the households. The results from focus group discussions also indicated that there is reduced manual work through agriculture machinery as well as artificial application of water to the crops throughout the year.

Improved dissemination mechanisms of seasonal climate forecasting to smallholder farmers in the district played a pivotal role in climate change mitigation and adaptation. The forecasts were tailored to the needs of local farmers and related end users by downscaling regional model outputs to sub-national level. The plots produced better predictions and easy interpretation to the final user.

The overall response from the interviews and FGDs showed that the project beneficiaries are now equipped with knowledge about climate change through the awareness campaigns facilitated by the SECA project. The respondents pointed to changes in climate conditions in rainfall and temperature patterns as the key indicators of climate change in their communities. The respondents outlined the drastic changes in rainfall patterns over the past decades as a strong case in point for a changing climate. A total of 80% (24 out of 30) of the farmers reported that over the years the amount of rainfall received has significantly reduced in Bulilima District. They further indicated that the frequency of droughts has increased in the past two decades. This is consistent with the Meteorological Services Department (MSD) (2011) statistics which show that the average rainfall in Bulilima District has decreased over the years and the findings by Dube et al. (2018) that the frequency of droughts in Bulilima District increased in the past 10–15 years.

Rural livelihood diversification has generally occurred as a result of an increased importance of various sources of income due to threatened rain-fed agriculture which increases uncertainty. Nutritional community gardens have been a bridging solution for the communities as they provide three farming seasons for the farmers/households, fresh produce of non-conventional marketable commodities and required day-to-day foods for the households. Other communities' participants now depend on apiculture (bee keeping), agroforestry and aquaculture projects which have high-income returns through CSA. As a result, this diversification has provided households with a basket of livelihood portfolios to augment household capacity to cope with the unexpected constraints of climate change.

The potential and participation of rural women in the district presented women as agents of change for climate mitigation and adaptation initiatives as compared to men. Women are regarded as more vulnerable than men, but the reality of the SECA project in Bulilima displayed energetic and enlightened women as better than some men. Their extensive theoretical and practical knowledge of the SECA initiatives in community gardens as well DRM campaigns showed a unique feature in the project. The domination of women in areas regarded as male dominated showed their vital role, knowledge and capabilities as strong change agents and key contributors to climate change mitigation and adaptation programmes at local, regional and international levels. Key Informant 1 said "As Bulilima District we applaud the government of Zimbabwe in partnership with UNDP for remembering most vulnerable districts in Zimbabwe through climate change initiatives that also have a goal of empowering women and youth as well as people living with disabilities. The SECA project came at the right time during hard times of prolonged drought which left the district dependent on government support for food".

Child-headed households were another vital group that brought in a new dimension in participating towards meeting adaptation, mitigation as well as DRM initiatives in the district. They can be attributed to the rapid increase in numbers of parental deaths, poverty and the HIV/AIDS pandemic. The orphaned children show vulnerability due to responsibilities as they care for their siblings and struggle to access basic needs. The households brought a unique feature through participation in the community gardens of fresh vegetable produce and the farming of new and drought-tolerant crops. This increased agricultural production leads to greater food security, to production and marketing of surpluses and, ultimately, to a source of income. The systemic issues of these households in the district project, such as gender inequality, played an important role, thereby making their stories vital and necessary to understand for the benefit of the project success.

The project's DRM programmes are integrating gender-sensitive approaches, as well as incorporating most vulnerable groups (women, people living with disabilities and aged community members) with the view of achieving SDG 5. The disaster risk reduction is a locally based disaster preparedness and reduction strategy using the livelihood-centred approach in building resilience against disasters in all facets, not excluding people living with disabilities. The programmes placed growing emphasis on working with women in communities through local hierarchy engagement in order to develop Community-Based Disaster Risk Reduction (CBDRR) since women are more vulnerable than men based on their productive and reproductive roles. The evidence produced from the case study areas provided the basis for advocacy strategies to influence policymakers at district, provincial and national levels, resulting in the official adoption of livelihood-centred approach to disaster risk reduction in Bulilima District. The project also demonstrated that whilst women were among the most vulnerable, in some instances they fared better than other male-headed households as they demonstrated the capacity to be agents of change and transformation in leading climate change actions and enhancing household food security.

Key informants at the Bulilima District Office confirmed that the district now had an effective disaster preparedness team whose major objective was to be proactive in managing disasters to ensure minimal damage at any given time when a disaster occurs. The disaster risk reduction capacity entails measures employed to deal with any hazards, which are most prevalent in the district. It also encompasses focusing on early warning systems and ways to improve the readiness of key actors to respond appropriately to hazards as they occur (Preventionweb 2015). The disaster management strategies that Bulilima has embraced include enforcement of environmental management laws, public education, climate change and disaster management campaigns and emergency drills. "The project managed to introduce adaptation and mitigation initiatives to help our communities address the effects of climate change. Bulilima District received equipment as well as information to respond during disasters through the SECA Disaster Risk Management Programme. We hope the project is going to be cascaded to other parts of the districts of our province" (Key Informant 3).

The SECA project introduced a climate-smart development initiative through the solar-powered system and water storage system in the district so as to alleviate the use of fossil fuel consumption in pumping water and also lower greenhouse gas emissions, hence contributing to a safer planet. The main aim of the intervention was to augment the existing water supply, which was seasonal, at the same time ensuring water efficiency and providing clean and affordable energy for the communities. This was confirmed by Key informant 3 who indicated that "the communities are now equipped to cope with climate change and build resilience for household food and nutrition security. The installation of the solar grid at Malalume school has improved lighting for the staff and learning environment for the students. However, some intended timeline activities are still incomplete which makes it difficult for the continuity of the initiative after project decommissioning".

The irrigation scheme implemented in Bulilima addressed the crucial water crisis in the district, and the community gardens cushioned smallholder farmers for food provision as well as selling fresh produce for income generation. This is also supported by FAO (2018) who aver that solar-powered irrigation systems are reliable and environmentally sustainable options in a growing number of contexts where solar-based irrigation systems can be scaled up to meet diverse energy demands and can contribute to a decoupling of growth in irrigated land areas from fossil fuel use whilst improving livelihoods. Building on the Bulilima District case study, results show that key socioeconomic and environmental benefits of solar-powered irrigation systems include supply of clean energy and improved access to water for irrigation, improved crop yields and increased incomes. Key informant 10 commented that "this clearly shows progress made through increasing food security and improved nutrition: 60 percent of households have reduced food gaps from 6 months to 2 months from their own crop production". Such initiatives have supported the GoZ rural electrification programme which has been slow in reaching the most marginalised communities and supporting the decentralisation of energy access.

Important interventions were also taken to prevent siltation through the construction of gabions for the sustainability of the irrigation scheme. The solar-powered power plant provides energy for a high-yielding borehole for community nutrition gardens and also for the communities at large. Climate change mitigation and adaptation initiatives supported by the project fostered agroforestry through Rural District Councils (RDCs) and consolidated gardens that reduced forest destruction, land decongestion, fire guard construction, gulley reclamation, water harvesting and sustainable wetland utilisation.

The project also addressed land degradation and natural resource management challenges through locally appropriate climate-smart agriculture practices and investments, which improve the natural resource base through agroforestry. CSA practices played a key role in ensuring food security in a declining natural resource base and a changing climate. The study area clearly elaborated the enhancement of the initiatives of CSA in agroforestry through establishing new plantations as well as gully reclamation in different parts of the district.

4.2 SECA Project Case Analysis: Challenges and Lessons Drawn

4.2.1 Challenges

Some of the challenges, obstacles and lessons facing adaptation and mitigation strategy implementation for the Bulilima case study for improving food security and climate resilience in Bulilima District include lack of assets, income and diversified means of livelihoods and sensitive economies, present low adaptive capacity among communities and enhance their vulnerability to climate shakes such as droughts (Mushore et al. 2013). Thus, the continuous economic constraints in the country bring a major hindrance in the continuity of the project after decommissioning. The majority of the population in the district are unemployed and this can result in survival challenges or coping strategies by individuals that endanger the success of the project where they may not afford the high cost of maintaining the project equipment such as the irrigation scheme. The income generation from the SECA project initiatives might be compromised by high inflation. The respondents mentioned that due to a lack of capital, they do not have access to inputs such as fertiliser, seeds and farming equipment. Hence, the sustainability of the project, post-completion, is compromised.

During FGDs, the respondents revealed that there are limited markets for their produce, especially for vegetables and drought-tolerant crops such as millet and rapoko so the majority cultivated maize for food production, which limits diversification of crop production and affects de-risking of their produce. A similar finding was made by Chazovachii et al. (2010) who state that "there are limited markets for drought-tolerant crops and people are only relying on the local market".

The geographical location and remoteness make the district vulnerable to the impacts of climate change and isolated from various agricultural markets. The poor road and communication networks make the area inaccessible as the roads are in poor conditions that donors and investors shun some wards in the district. This isolation due to lack of infrastructure may limit choices and coping strategies during times of stress and drought.

There is lack of integration and coordination among government departments, non-governmental organisations (NGOs) and other institutions in addressing climate change issues in vulnerable communities in the district. This issue created a serious problem of discontinuity in building resilience as there were repetition, competition and implementation of overlapping climate change initiatives in the same area. Respondents stated that many organisations were replicating the same climate initiatives which end up bringing confusion to the community beneficiaries. It is therefore imperative to have the District Administrator's Office playing a gatekeeping role that also facilitates the coordination of activities so that there is continuity and sustainability. Good district governance is important for facilitating economic growth and transformation of livelihoods.

The project execution faced challenges at one point which included political interferences within the district structures, and this prevented the systematic flow of project activities. The respondents stated that as a result, only a few people benefit from similar projects and programmes, with the poor and the less vocal being more disadvantaged.

4.2.2 Lessons Learnt

A common lesson learnt was the significance of local context in developing adaptation and mitigation strategies in rural communities, ensuring that projects and programmes are built on local knowledge and cultural norms, practices and value systems. In the case of Bulilima District, the traditional leaders of all wards were involved throughout the project cycle. Participatory methods were successful in engaging traditionally marginalised groups, instilling a sense of ownership of the project among participants, which was found to markedly increase the chances of project success. The engagement of traditional leaders is also supported by Kurebwa (2018) stating that they play important developmental, administrative and political roles in rural areas, despite modern state structures that exist.

The respondents in Bulilima communities also identified the potential for greater knowledge and information sharing between and/or among stakeholders, especially to strengthen early warning systems in disaster preparedness. The project also demonstrated the importance of co-ordination and facilitation roles among smallholder farmers in dialogue with the surrounding communities on climate-related issues that are intra-specific to the district. The use of the media for disseminating climate knowledge is important for capacity building. Chagutah (2010) highlights that climate information dissemination is an essential tool for effective climate risk management, and this was also confirmed by the project that more tailored climate information facilitated timeous and strategic response by both farmers and disaster management teams. The ability of the media to reach the community level presents an opportunity to disseminate accurate and useful climate information, including weather forecasts and early warning projections for the benefit of the district.

5 Conclusion and Recommendations

The study found that the SECA project resulted in a number of climate change adaptation strategies being adopted by rural communities where the project was being undertaken. The adaptation strategies allow the communities to build some form of climate resilience. Regardless of the gains made by the project, there are concrete actions and lessons that can be learnt from the Bulilima case study. To the extent of this study, the following recommendations and the general lessons arising from this study are relevant for a broader set of countries that are dealing with similar environmental, demographic and institutional challenges, particularly in sub-Saharan Africa.

- In future, more research is needed to improve regional model outputs by using more powerful models and better-quality data. In addition, more research is needed to improve skills to include more useful information such as the start and the end of the rainy season and the probability of dry spells.
- Cascade the National Climate Adaptation Plan to all the districts to develop intervention strategies and plans that hinge on matching and differentiated district adaptation and mitigation actions that are particular to specific agro-ecological zones in the country.
- Examine the potential of multi-level risk governance to support "linked-up" action between communities, civil society, the private sector and government at all levels. Facilitate for policy and governance frameworks that are inclusive, hence engaging marginalised groups in more participatory and inclusive decision-making processes and planning procedures in climate-sensitive and climate-smart agriculture initiatives across all districts of the country.
- Politics, as elaborated by (Leftwich 2008), is clearly essential for decision-making related to climate change. Governments and civil society actors are supposed to actively engage in working out both the specifics of climate change policies and programmes and more generally how societies might form binding agreements on what needs to be done. At the political level, it looks like it is a question of priorities. As in other low-income countries, Zimbabwe's developmental priorities are still placed on food security, health, job creation and education than on environmental protection. As a result, disaster mitigation and adaptation strategies receive comparatively lower priority in national policies and development plans. Hence, developing cross-cutting policies and initiatives that include addressing hunger, poverty and climate change can be a great leverage for a triple win.
- The government is supposed to promote institutional collaborations that mainstream climate change into rural planning, infrastructure, investments and service delivery. They should institutionalise climate change response into strategies, actions and budgets of climate-sensitive sectors such as agriculture, wildlife, forestry, water, environmental management, fisheries, settlements, infrastructure and

health, and support post-project sustainability and growth post implementation.

- There is need for research to continuously work together with communities, NGOs and academia to develop intra-adaptation and mitigation strategies that respond directly to the needs and vulnerabilities of children, youth, women and men, as well as people living with disabilities. The need to engage the private sector is also important to build stronger and sustainable markets, value chains and inter-linkages.

References

Adedeji, O., Okocha, R., & Olatoye, O. (2014). Global Climate Change. *Journal of Geoscience and Environment Protection,* 2(5), 114-122

AON. (2020). Weather, Climate & Catastrophe Insight: 2019 Annual Report. Accessed from: https://reliefweb.int/sites/reliefweb.int/files/resources/20200122-if-natcat2020.pdf

Arora, N. K. (2019). Impact of climate change on agriculture production and its sustainable solutions. *Environmental Sustainability,* 2, 95–96 https://doi.org/10.1007/s42398-019-00078-w

Bloom, A. A., Kotroni, V., & Lagouvardos, K. (2008). Climate change impact of wind energy availability in the Eastern Mediterranean using the regional climate model PRECIS. *Natural Hazards and Earth System Sciences,* 8(3), 1249-1257. https://doi.org/10.5194/nhess-8-1249-2008

Bulilima Rural District Council (BRDC). (2020). http://www.bulilimardc.co.zw/en/ about-us

Chagutah, T. (2010). *Climate change vulnerability and preparedness in Southern Africa: Zimbabwe country report.* Cape Town, Heinrich Boell Stiftung.

Chazovachii, B., Chigwenyu, A., & Mushuku, A. (2010). Adaptation of Climate Resilient Rural Livelihoods Through Growing of Small Grains in Munyaradzi Communal Area, Gutu District. *African Journal of Agricultural Research,* 8, 1335-1345.

Dasgupta, S., Hossain, M. M., Huq, M., & Wheeler, D. (2015). Climate change and soil salinity: The case of Coastal Bangladesh. *Ambio,* 44(6), 815-826. https://doi.org/10.1007/s13280-015-0681-5.

Davis, R. and Hirji, R. (2014). *Climate Change and Water Resources Planning, Development and Management in Zimbabwe.* An Issues Paper. Washington, D.C., World bank.

Dodman, D., & Mitlin, D. (2015). The national and local politics of climate change adaptation in Zimbabwe. Climate and Development, 7(3), 223–234. https://doi.org/10.1080/17565529.2014.934777

Dyer, J., Leventon, J., Stringer, L., Dougill, A., Syampungani, S., Nshimi, M., ... Kafwifwi, A. (2013). Partnership models for climate compatible development: Experiences from Zambia. *Resources,* 2, 1–25. https://doi.org/10.3390/resources2010001

Dube, N., Sithole, M., Ngwenya, T., Muhwati, T., Moyo, E., & Manyanga, K. (2018) Impact of climate change on sustainability in semiarid, rural Africa: Lessons from rural Zimbabwe, *Cogent Social Sciences*, 4(1), 1553327, https://doi.org/10.1080/23311886.2018.1553327

Food and Agriculture Organisation of the United Nations (FAO). (2008). *Seasonal Update Report, 2008. Agricultural Season Update*, Zimbabwe. Harare, Ministry of Agriculture

FAO. (2018). *The benefits and risks of solar powered irrigation – a global overview.* Rome, FAO

Government of Zimbabwe (GoZ). (2015a). *National Climate Change Response Strategy,* Ministry of Environment, Water and Climate (MEWC). Harare, Zimbabwe.

Government of Zimbabwe (GoZ). (2015b). *Zimbabwe's Intended Nationally Determined Contribution*, Ministry of Environment, Water and Climate (MEWC). Harare, Zimbabwe

Government of Zimbabwe (GoZ). (2016). *Third National Communication Strategy to the United Nations Framework Convention on Climate Change*, Ministry of Environment, Water and Climate (MEWC). Harare, Zimbabwe

Gukurume, S., (2013). Climate change, variability and sustainable agriculture in Zimbabwe's rural communities. Russian Journal of Agricultural and Socio-Economic Sciences, 14(2), 89-100.

Imada, Y. Maeda S., Watanabe, M., Shiogama, H., Mizuta, R., Ishii M., & Kimoto, M. (2017). Recent enhanced seasonal temperature contrast in Japan from large ensemble high-resolution climate simulations. *Atmosphere,* 8(57), 40-58. https://doi.org/10.3390/atmos8030057

Intergovernmental Panel on Climate Change (IPCC). (2007). Climate Change 2007: The Physical Science Basis. Contribution of Working Group I to the Fourth Assessment Report of the Intergovernmental Panel on Climate Change [Solomon, S., D. Qin, M. Manning, Z. Chen, M. Marquis, K.B. Averyt, M. Tignor and H.L. Miller (Eds.)]. Cambridge, Cambridge University Press, 996 pp.

IPCC. (2012). Managing the risks of extreme events and disasters to advance climate change adaptation, in C.B. Field, V. Barros, T.F. Stocker, D. Qin, D.J. Dokken, K.L. Ebi, et al. (eds.), A special report of working groups I and II of the intergovernmental panel on climate change, Cambridge University Press, Cambridge, New York, 582

IPCC (2013). *Climate Change 2013: The Physical Science Basis-Contribution of Working Group1 to the fifth Assessment of the Intergovernmental Panel on Climate Change,* London: Cambridge University Press.

Jakarasi, V. N., Moyo, E. N. & Manyena, B. (2018). The reality and rhetoric of integrating climate change adaptation into economic sectors in Zimbabwe: In Petrik, D., Ashburner, L. (2018). Conference Proceedings of Adaptation Futures 2018. Adaptation Futures 2018. University of Cape Town, Cape Town.

Kurebwa, J. (2018) The Institution of Traditional Leadership and Local Governance in Zimbabwe International Journal of Civic Engagement and Social Change 5(1),1-22, https://doi.org/10.4018/IJCESC.2018010101.

Leftwich, A. (2008). Developmental States, Effective States and Poverty Reduction: The Primacy of Politics, https://doi.org/10.1177/097370302011 0205

Lipper, L., Thornton, P., Campbell, B.M., Baedeker, T., Braimoh, A., Bwalya, M., Caron, P., Cattaneo, A., Garrity, D., Henry, K. & Hottle, R. (2014). *Climate-smart agriculture for food security. Nature climate change*, 4(12),1068-1072. https://doi.org/10.1038/nclimate2437

Madzwamuse, M. (2010). *Climate Governance in Africa: Adaptation Strategies and Institutions*. Heinrich Boll Stiftung. Accessed From: https://www.boell.de/en/ecology/africa-climate-governance-in-africa-adaptation-strategies-and-institutions-10914.html

Manzungu, E. & Moyo, S. (2016). *Climate Change Adaptation Options in Zimbabwe. World Bank Technical Paper to inform Zimbabwe's Climate Policy*, Harare: World Bank.

Ministry of Environment, Water and Climate. (2016). Supporting Enhanced Climate Action Project Document. Harare, Zimbabwe

Mugambiwa, S.S. (2018). Adaptation measures to sustain indigenous practices and the use of indigenous knowledge systems to adapt to climate change in Mutoko rural district of Zimbabwe. *Journal of Disaster Risk Studies*. 10 (1), a388, https://doi.org/10.4102/jamba.v10i1.388

Mushita, A., & Thompson, C. (2013). More ominous than climate change? Policy hazards to African food production. *African Studies Quarterly, 13*(4), 1–25.

Mushore, T. D., Mudavanhu, C. & Makovere, T. (2013). Effectiveness of drought mitigation strategies in Bikita District, Zimbabwe. *International Journal of Environmental Protection and Policy,* 1 (4) 101-107. https://doi.org/10.11648/j.ijepp.20130104.19

Ohunakin, O. S., Muyiwa S., Adaramola, M. S., Olanrewaju M., Oyewola,O. M., Olaniran J.Matthew O. J., & Fagbenle, R. O. (2015). The effect of climate change on solar radiation in Nigeria. *Solar Energy*, 116(14), 272-286. https://doi.org/10.1016/j.solener.2015.03.027

Practical Action. (2014). Coping with drought: Research findings from Bulilima and Mangwe districts, Matabeleland South, Zimbabwe. Harare, Practical Action.

Preventionweb (2015, 12 November). Disaster Risk: Capacity. Accessed from: https://www.preventionweb.net/risk/capacity

Swain, A., Swain, R. B., Themnér, A. & Krampe, F. (2011). Climate change and the risk of violent conflicts in Southern Africa. Global Crisis Solutions.

Tshuma, N. & Mathuthu, T. (2014). Climate change in Zimbabwe: Perceptions of smallholder farmers in Mangwe district. *The International Journal of Humanities & Social Studies*, 2(5), 318-325, ISSN 2321-9203.

Ubeda, B., Giacomo, A. S. D., Neiff, J. J., Loiselle, S. A., Poi, A. S. G., Gálvez, J. A., Casco, S., & Andrés Cózar, A. (2013). Potential effects of climate change on water level, flora and macro-fauna of a large neotropical wetland. *PloS one,* 8(7), 112-123. https://doi.org/10.1371/journal.pone.0067787

United Nations Development Programme (2017) *Human Development Report 2017, Zimbabwe*. Harare: UNDP.

Vincent, V. & Thomas, R.G. (1960). *An agricultural survey of Southern Rhodesia: Part I: agro-ecological survey.* pp 25-30. Salisbury, Government Printers.

World Bank & CIAT. (2018). *Climate-Smart Agriculture in Lesotho. CSA country profiles for Africa, Asia, and Latin America and the Caribbean series*. Washington D.C., The World Bank Group.

Zimbabwe Statistical Agency (Zimstat), (2012) *Census 2012 National Report*, Harare. Accessed From: from http://www.zimstat.co.zw/documents/census.

Zimbabwe Vulnerability Assessment Committee (ZimVAC). (2016). *Zimbabwe Vulnerability Assessment Committee Rapid Assessment Report*, Harare, Food and Nutrition Council.

15 Climate Resilience Strategies and Livelihood Development in Dry Regions of Zimbabwe

Leonard Chitongo

Abstract

The Sustainable Development Goals (SDGs) adopted by the United Nations member states in 2015 were aimed among other things at improving people's well-being. Despite goal 2 on zero hunger by 2030, climate change has adversely affected agriculture, threatening food security and rural livelihood security for many Zimbabweans. Extreme weather conditions induced by climate have led to food insecurity, precarity, poverty and vulnerability Against this background, the study assesses climate resilience strategies in three dry regions of Zimbabwe, namely, Nyanyadzi, Chiredzi and Gwanda. The study adopts a mixed-methods approach that combines quantitative and qualitative approaches. Data were collected through document reviews, key informant interviews (KIIs), focus group discussions (FGDs) and household survey. One thousand one hundred forty-one respondents were recruited through random sampling. This chapter argues that climate resilience and livelihood development in Zimbabwe are compromised by aridity. Furthermore, the paper illustrates ways in which the adopted resilience strategies improve food security, thus reducing poverty and deprivation. The study found that achieving SDGs requires a partnership between governments, the private sector, civil society and citizens in a participatory manner. This chapter recommends a shift towards a holistic integrated approach of all stakeholders' initiatives in order to achieve sustainable livelihood development for rural communities in arid regions.

Keywords

Climate change, Resilience · Livelihoods · Aridity · Vulnerability, Sustainability, SDGs

1 Background and Introduction

Global climate change has had a negative bearing on the socioeconomic and biophysical environment. This threatens the achievement of the Sustainable Development Goals (SDGs) adopted and ratified by the United Nations member states that signify a global blueprint for socioeconomic development by the year 2030. Climate change is featured in a number of SDG targets with SDG 13 calling for climate change action (United Nations 2015). This entails harnessing resilience

L. Chitongo (✉)
Postdoctoral Researcher SARChI Chair Sustainable Local (Rural) Livelihoods, School of Management, IT & Governance, University of KwaZulu-Natal, Durban, South Africa
e-mail: ChitongoL@ukzn.ac.za

G. Nhamo et al. (eds.), *Sustainable Development Goals for Society Vol. 2*, Sustainable Development Goals Series, https://doi.org/10.1007/978-3-030-70952-5_15

and adaptation capabilities of communities to reduce the adverse impacts of climate change.

Climate change is expected to exacerbate Africa's critical water situation, with Southern Africa being one of many water-stressed regions which could see a further decrease in the flow of streams and the ability of groundwater to recharge (Intergovernmental Panel on Climate Change 2012). Mutowo and Chikodzi (2014) noted that climate variability had been a key driver of change in rainfall patterns across Zimbabwe. There is significant evidence of changes in the spatial characterisation of agro-ecological regions, which were last zoned in 1965 by Vincent and Thomas based on biophysical characteristics (soil, vegetation and climate). Evidence based on spatial modelling of the agro-ecological regions suggests that more regions are becoming vulnerable to water scarcity because of climate change, hence crippling the ability to cope with or recover from shocks.

The Millennium Development Goals (MDGs) and, subsequently, the SDGs seek to improve the well-being of people and communities with SDG 1 calling for an end to poverty by 2030 (United Nations 2015). In the face of climate change, building climate resilience is a prerequisite for sustainable livelihood security and development. In sub-Saharan Africa poverty is predominantly rural, with 80% of the population being considered extremely poor. Rural and smallholder agriculture is the major source of livelihood security (Hajdu et al. 2020). Thus, any disruption in agricultural production systems will lead to precarity and vulnerability (UNICEF and World Bank 2016).

Given the above, the main focus of the paper is to highlight climate resilience strategies being adopted and executed by people living in Gwanda, Chiredzi, and Chimanimani classified under agro-ecological region IV. Resilience is defined as the capacity of a system to absorb disturbance and reorganise while undergoing change to still retain essentially the same function, structure, identity and feedbacks (Walker et al. 2004). Most livelihoods are sensitive to climate-induced vulnerability as a significant proportion of people are still locked in the deprivation trap such as weak institutions which interact with climate change to perpetuate the existence of fragile livelihoods.

2 Methodological Orientation

2.1 Description of Study Area

Zimbabwe was divided into five agro-ecological regions in the 1960s; however, the increased variability of rainfall has possibly affected the agro-ecological region boundaries. This chapter targets three districts which are confronted by the effects of climate change, namely, Gwanda, Chiredzi and Chimanimani. They belong to the agro-ecological region IV, which is characterised by low rainfall (450–650 mm per year) with severe dry spells during the rainy season and frequent seasonal droughts (TANGO International 2018). These districts are mainly suitable for livestock and drought-tolerant field crops such as sorghum, millet, cowpeas and groundnuts. Other crops are grown under irrigation and cattle ranching is practised.

The study was carried out in Zimbabwe, a country located between 15°37′S and 22°24′S and longitudes 25°14′E to 33°04′E and covering an area of 390,580 km^2. The study focused specifically on three districts: Gwanda, located at 20.9398° S, 29.0111° E, southwest of Bulawayo with a population of 114,250 people and 26,201 households; Chiredzi, located at 21.0333° S, 31.6796° E, with a population of 271,721 and 63,459 households; and Nyanyadzi which lies in Chimanimani District, a mountainous area in the province of Manicaland in eastern Zimbabwe, located at 19.8032° S, 32.8733° E, with a population of 132,014 and 32,291 households (ZIMSTAT 2012). These areas are located in agro-ecological region IV. The region is characterised by erratic rainfall events with predominantly drought-resistant crops being grown there. The regions were selected because they represent the north-south and east-west rainfall and temperature gradient in Zimbabwe and are characterised by erratic and unreliable rainfall. The biophysical environment interacts with climate change to amplify vulnerability which manifests in less desirable livelihood outcomes such as food insecurity as a function of water scarcity, poverty, human-wildlife conflict as well as post-harvest losses (Fig. 15.1).

2.2 Target Population and Sampling Strategy

In this research, the target population included 121,951 households which were based in Chiredzi, Chimanimani and Gwanda Districts. Population is regarded as a unit of analysis which possesses traits which are under consideration in a study (Bhattacherjee 2012). It was from these 121,951 households that a sample of 1141 was drawn for the questionnaire survey. Inferences were drawn from this sample about the characteristics of climate resilience strategies. The sampling strategy was based on Cochran (1963:75) due to the fact that a very large population provides proportionally more information than a smaller population. The formula is provided as:

$$n_0 = \frac{Z^2 pq}{e^2}$$

which is valid where n_0 is the sample size, Z^2 is the abscissa of the normal curve that cuts off an area α at the tails ($1 - \alpha$ equals the desired confidence level, e.g. 95%), e is the desired level of precision, p is the estimated proportion of an attribute that is present in the population and q is $1\text{-}p$. The value for Z was derived from statistical tables for normal distribution. Sampling is the process used in the selection of individuals/groups for use in a specific study, and the selected individuals represent the target population from which they have been selected so as to enable the researcher to fulfil the research questions (Creswell 2009).

A total of 1141 questionnaires which comprised open-ended, multiple response and dichotomous questions were administered by the

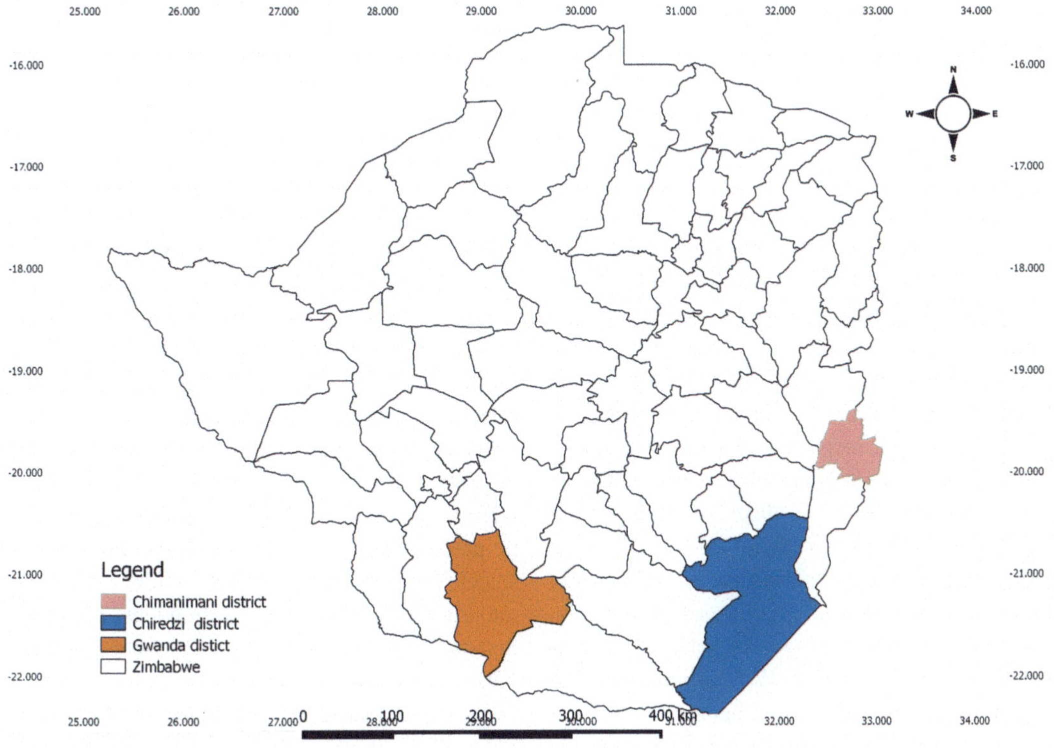

Fig. 15.1 Map of study area. (Source: Author Creation 2019)

researcher. Questionnaires involve a large amount of data that can be collected with greater accuracy (Harris and Brown 2010). According to Patton (2002), the use of face-to-face interviews with the purposively selected participants enables a researcher to acquire the respondents' mental transformations. It was from the established interpersonal nature of the interview context that participants responded in ways that they deemed socially desirable. Specifically, the questionnaires contained questions on the demographic profiles of the respondents, the on-farm and off-farm resilience strategies and the contribution of resilience strategies to livelihood development.

Key informant interviews provided the researcher with the ability to explore and probe participants so as to get in-depth responses about livelihood diversification strategies in the face of climate change. Employees from non-governmental organisations (NGOs), traditional leadership (chiefs and headman) and government departments formed the key informant group. Creswell (2009) indicates that from the use of interviews, one can determine the attitudes, interests, feelings, concerns and values of the respondents in coping and adapting to challenges. However, the interviews in the study had the disadvantage of being time consuming, given the extent and magnitude of the geographical area covered by the researcher.

Focus group discussions permit flexibility in the collection of data that is not usually achieved when applying an instrument individually. At the same time, they permit spontaneity of interaction among the participants (Krippendorff 2010). Focus group discussions allowed the respondents to unearth a 'basket' of diversification strategies employed in their respective districts. The application of the focus group discussion technique allows researchers to collect an appropriate amount of data in a short period of time (Bhattacherjee 2012). In this case, a total of three FGDs were carried out, one in each study area. Each FGD had 40 purposively selected participants, making a total of 120 participants. The researcher allowed the participants who did not take part in the questionnaire survey to provide information about the vulnerability context they are trapped in, as well as the livelihood resilience strategies adopted to minimise the multiple stresses of climate change.

2.3 Ethical Considerations

Based on the definition by Neuman (2000), research ethics are moral ways of conducting research. The research was conducted through written informed consent with participation being voluntary. The respondents in the study were given the opportunity to be anonymous. According to Christians (2000), the researcher has an obligation to keep the respondents' identities and responses private.

3 Results and Discussion

3.1 Demographic Characteristics of the Respondents

The demographic survey of respondents to determine the livelihood resilience strategies of the households of Nyanyadzi, Gwanda and Chiredzi in the face of climate change indicated that most of the respondents were aged between 26 and 30 years constituting 22% and those between 51 and 60 years constituting 28% of the study population. The elderly and younger generations contributed to the highest proportion of respondents. The issue of age was necessary so as to have an inference on their experience with diverse livelihoods and livelihood strategies that they have implemented as a result of climate change and how these have empowered them.

There were more males than females who took part in the study, and this suggests there were more male-headed households which participated in the study. 28% of the households selected were female-headed, while 72% were male-headed. Most of the respondents indicated that they were married (68%), 9% were divorced, 12% were widowed and 11% of the respondents were single, indicating that some of the households used in this study were child-headed. There was a need to determine the level of education of the respondents as this would have an impact on their knowledge of climate change and how it has affected their livelihoods and resultant resilience. Thus, it can be noted that 53% of the respondents were holders of ordinary level education. Moreover, 12% of the respondents had primary level education. Thirty-five per cent of the respondents had advanced level or professional qualifications. The professional qualifications comprised of diploma and certificate holders.

3.2 On-Farm Resilience Strategies

SDG 2 calls for an end of hunger and malnutrition by 2030. In order for this to be achieved there is a need to double up agricultural productivity and incomes of small-scale food producers, especially women and indigenous peoples. This can be done by ensuring sustainable food production systems and by progressively improving land and soil quality. The results which are presented in Fig. 15.2 show the on-farm resilience strategies which are being developed by the households in the study areas.

It is important to note that there are spatial variations in the strategies due to resource availability. This implies that some of the strategies can be more pronounced in one place and can be less pronounced or absent in the other place, depending on the livelihood assets. In the end, there are spatial variations in the resilience strategies which are adopted by the rural households to enhance livelihood sustainability. The most commonly used strategy was pen fattening which was dominant in Gwanda and Chiredzi. This is attributed to the realisation that cattle act as some form of assets which can be used as a livelihood strategy to buffer households against insecurities or contingencies precipitated by erratic rainfall synonymous with agro-ecological region IV. Irrigation agriculture is one of the most popular activities in the area, given the prevailing climatic conditions that allow for farming throughout the year. The deliberate government policy of establishing dams and irrigation projects were out of the realisation on the need to ensure livelihood security in the areas under study. Not everyone can participate in irrigation schemes; consequently, those who are not partakers in irrigation schemes depend on planting drought-resistant crops for subsistence and also surplus sales.

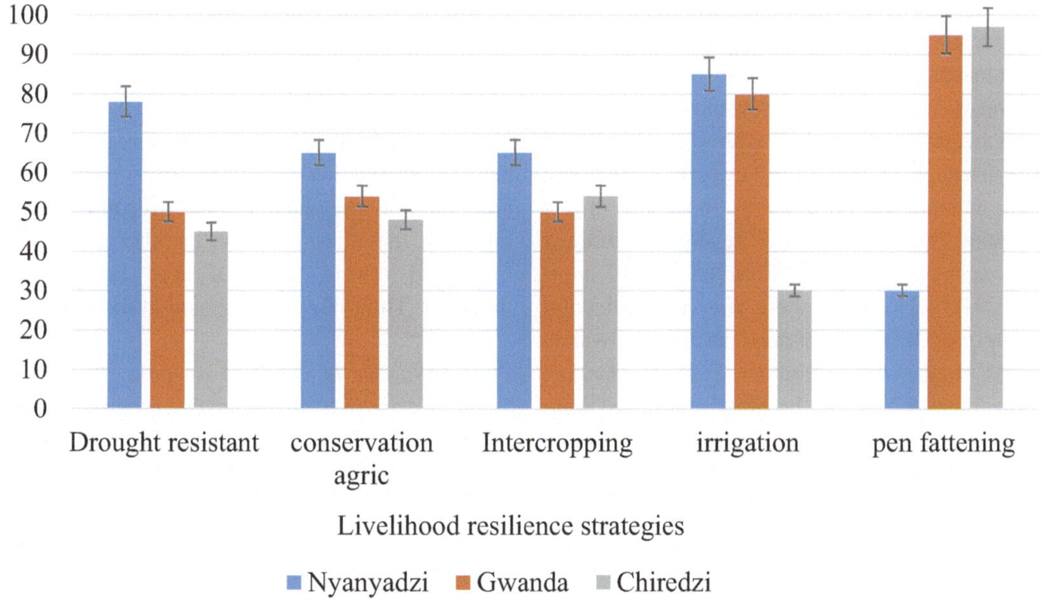

Fig. 15.2 On-farm strategies. (Source: Field Survey 2019)

From Fig. 15.2, overlapping error bars are used to show no significant differences and non-overlapping error bars are used to show significant differences in the variables between the three study sites used in the study. The results which are presented in Fig. 15.1 show that there are significant differences between the livelihood resilience strategy of using drought-resistant crops between Nyanyadzi, Gwanda and Chiredzi ($F(2, 3420) = 18{,}426$, $p = 0{,}0000$). Chitongo (2019) argues that drought-resistant crops such as millet and sorghum can thrive and yield relatively well in the face of water scarcity. However, it can be noted that there were no significant differences between Gwanda and Chiredzi ($p > 0.05$) though Nyanyadzi significantly used drought-resistant cropping than the other two districts. In terms of using conservation agriculture, there was an existence of significant differences between the districts, although there were no significant differences between Gwanda and Chiredzi. Nyanyadzi area tended to use more of conservation agriculture than the other study regions. In terms of intercropping, Nyanyadzi used it more than Chiredzi and Gwanda ($F(2, 3420) = 388{,}26$, $p = 0{,}0000$). The results of the study showed that there were no significant differences between Nyanyadzi and Gwanda ($p > 0.05$) in terms of using irrigation as a livelihood resilience strategy. Most of the respondents used irrigation as a livelihood resilience strategy. However, Nyanyadzi and Gwanda significantly used irrigation more than Chiredzi ($p < 0.05$). The findings of the study suggested that there were no significant differences in the use of pen fattening as a livelihood resilience strategy between Gwanda and Chiredzi study areas ($p > 0.05$). However, there was a significantly lower ($p < 0.05$) usage of pen fattening in Gwanda and Chiredzi when compared to Nyanyadzi. From Fig. 15.2, it can be noted that there were diverse on-farm resilience strategies which were used in the study sites. The variability shows that there are strategies which are area-specific than others, and there are other strategies which do well in one region and not in the other. More so, this implies that if one of the diversification strategies fails, there are safety nets to fall on as a way of building resilience within communities.

Data which were collected from the key informant interviews and focus group discussions also confirmed that there are a number of resilience strategies which are being used by the people in the three districts. Key informants indicated that they are reliant mostly on pen fattening as they have the necessary infrastructure for such. The following are excerpts from the interviews:

KII 1: *Here in Chiredzi, we are very prone to droughts due to climate change as we receive very little rainfall. So we resort to pen fattening as a resilience strategy whereby we can feed our cattle even when there is limited pasture.*

KII 2: *As for us here in Gwanda on the ground, we have pen fattening projects where we give feed supplements to our cattle and then sell in times of need. We are getting less and less rains, so the farmers have to rely on irrigation of crops as a strategy against the low rains.*

KII 3: *Here in Nyanyadzi we are not much into cattle fattening as you can see that we have our irrigation which has been recently rehabilitated, and as such, we can grow our crops even though we have less rain sometimes. There is conservation agriculture where farmers conserve moisture by not ploughing the whole farm but only single pits where they plant their seedlings.*

The same views were indicated in the FGDs in Gwanda and Chiredzi; the selected household heads indicated that the issue of climate change is real, and now they have resorted to cattle fattening such that they can have another source of livelihood during times of crop failure and droughts. The following was said in the FGDs:

FGD Gwanda: *In Gwanda we have diversified our livelihoods on the farm as we have now the Guyu-Chelesa irrigation scheme which was recently rehabilitated. This has now been our source of water, and we can irrigate our crops and use the same water for pen fattening, which has been lucrative for us for a while. So in the event of droughts which have been common here, the irrigation scheme has been helpful to the community.*

Different views were expressed in FGDs in Nyanyadzi, where it was noted that the most dominant form of response to climate change was the growing of drought-resistant crops and also the use of irrigation. The following was indicated by the FGD:

FGD Nyanyadzi: *In this place, the climate is becoming increasingly dry. The irrigation scheme was no longer functional, but thanks to NGOs, it has been rehabilitated. Due to failure of crops such as maize, we have resorted to growing drought-resistant crops such as sorghum (mapfunde), pearl millet (mhunga) and finger millet (rukweza or njera). These can withstand heatwaves and droughts.*

The findings concur with Whingwiri et al. (1992) who noted that millet is adapted to warm and dry climates and good yields can be obtained from the crop with as little as 250 mm of rainfall provided it is well distributed. The Nyanyadzi irrigation scheme helps farmers to successfully grow crops (Mandizvidza 2006). Thus, farmers in the area can grow a variety of crops despite the dry conditions in the area.

3.3 Off-Farm Resilience Strategies

The results which are presented in Fig. 15.3 show the off-farm strategies which are used by the rural households in Gwanda, Chiredzi and Nyanyadzi in the face of climate change. For Gwanda specifically, there is a dominance of gold panning. This has its own attendant problems of environmental degradation where rivers are silting and drying up because of wanton digging and poor environmental management systems (EMS). However, it can also be noted that the rural households in all the study areas resorted to the use of non-timber forest products (NTFPs) as a strategy to mitigate the impacts of climate variability and change. In Nyanyadzi, the community is engaged in honey production. From the study, it can be noted that rural households also resorted to 'vending'. The vending involved the buying and selling of food items, clothes and hardware materials from the surrounding countries so as to find a way of survival in the harsh conditions. Due to the presence of some non-governmental organisations (NGOs) in the areas, some of the households, especially women, have been trained in Internal Saving and Lending Club (ISLC) where there is a generation of income with minimal interests. Food-for-work projects have been introduced whereby the beneficiaries work on a project and get paid by the government and/or implementing partner through food handouts.

There were no significant differences in terms of the off-farm activities for the people in Nyanyadzi, Gwanda and Chiredzi with regard to food for work ($F(2, 3420) = 872$, $p = 0{,}3021$). This implies that the three study sites used similar forms of food for

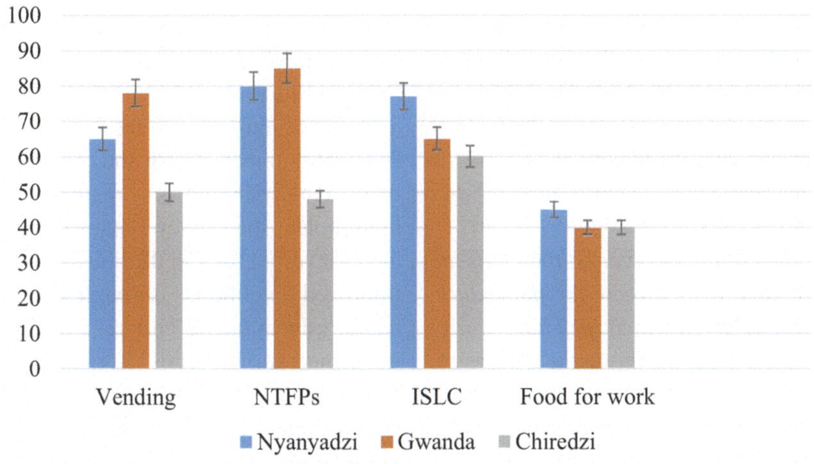

Fig. 15.3 Off-farm strategies (Field Survey 2019)

work as an off-farm resilience strategy. In terms of Internal Savings and Lending Schemes (ISALs), it was noted that there were no significant differences ($p > 0.05$) between Gwanda and Chiredzi. However, ISALs were dominantly used in Nyanyadzi as compared to Chiredzi and Gwanda, and there were significant differences between the three study sites ($p < 0.05$). There were no significant differences in the use of NTFPs as an off-farm resilience strategy between Nyanyadzi and Gwanda ($p > 0.05$) although significant differences existed when these two study areas are compared with Chiredzi ($p < 0.05$). Vending significantly differed between the three study sites ($F(2, 3420) = 211, p = 0{,}000$). However, it was noted that the off-farm strategy of vending was significantly used in Gwanda than in all the other study sites.

Results from the KII and FGDs concur with the findings from the household heads who responded to questionnaires. The major issues to emanate from off-farm activities are mining, NTFPs and vending. The following was indicated by the KII informants:

KII Nyanyadzi: *In this area, people rely on a number of activities, and some are vending mats which are made from baobab bark, there is a honey production project that we have here, and so people sell the honey to shops and roadsides. In short, NTFPs help them survive the challenges of climate change that are being felt in this place.*

KII Gwanda: *Here we have been victims of the climate which is changing, and as such we have some people who are engaged in gold panning, and they are actually able to provide food for their families through that. There are also non-timber forest products in the form of mopane worms which are found and harvested locally; they are both a source of food and income.*

KII Chiredzi: *The situation is very different as the households here also rely, other than cattle fattening, on ISALs in which they have a saving and lending scheme among the club members and they exchange with one another with little interest. This helps them to generate income for food and even school fees.*

The same results were also indicated from the household heads who participated in the FGDs. Off-farm strategies such as mining, ISLC, NTFPs and vending were commonly used strategies. The following was noted from the FGDs:

FGD Gwanda: *Here almost everyone wants to be engaged in artisanal gold mining which has quick returns in terms of cash as there is also selling of the gold in foreign currency which is a plus for us. The other natural resource which we can use here in Gwanda is the mopane worm which we can sell and get some income.*

FGD Nyanyadzi. *In our situation, we are involved in beekeeping and honey production which is a crucial economic activity. Produced honey is sold to various companies and individuals in and around Chimanimani. Other people produce baobab mats and conduct vending activities. . There are also Internal Savings and Lending Schemes that we have which were introduced to us by NGOs. They help in the generation of income.*

FGD Chiredzi: *In Chiredzi, we rely on vending some of the farms produce that we get and also on food-for-work programmes from government and NGOs. They help us have extra income for use as a safety net in time of need, especially in terms of food.*

The study showed that households, mainly from Gwanda, rely on mining as an off-farm activity and this confirms the findings by the Civil Protection Organisation in Zimbabwe (2009) which concluded that agriculture was no longer viable, and this has forced many people living in the marginal regions of Zimbabwe into artisanal mining despite the fact that the strategy is harmful to their health and the natural environment. Gold panning is sometimes beneficial to households; the money obtained from the strategy is invested back in the household, resulting in improved livelihoods through the purchase of cattle, education of children, health and other goods and commodities (Bernsterin et al. 1991). Bernsterin et al. (1991) further argue that gold panning and sieving can also increase people's vulnerability instead of reducing it because it leads to diversion of labour from agriculture, the major livelihood activity in rural areas, leading to reduced agricultural yields. In addition, it is a 'hit and miss' activity that is not reliable and easy to monitor like agriculture.

The study also indicated that the Gwanda community uses mopane worms as a strategy, and this is supported by Mogotsi et al. (2011) who indicated that being exposed to harsh climatic conditions, rural people are forced to harvest mopane worms within their residential areas or migrate to areas with mopane trees to harvest these invertebrates that rescue many people in drylands from poverty, although the respite is usually on a temporary basis. These worms are used to meet their dietary requirements and/or sold to generate cash to buy household essentials during drought.

More so, in this study, some of the respondents use craftwork as a source of livelihood. According to Bernsterin et al. (1991), livelihood strategies such as pottery, basketry, metalwork, woodwork and mat-making serve as safety nets that do not only help people meet their needs but also reduce rural people's dependence on agriculture. Chimhowu et al. (2011) reveal that in Biriwiri, most households are into craftwork, particularly women. This has provided them with financial capital to buy food, educate their children, buy livestock and improve their standards of living, thereby enhancing sustainability. However, craftwork in most rural areas is a low productivity work with little economic resources generated from the strategy; hence, it has limited economic viability (Bernsterin et al. 1991).

4 Contribution of Resilience Strategies to Livelihood Development

Results from the KII informants from Gwanda, Nyanyadzi and Chiredzi showed that the strategies which were used by the local people and the NGOs for the purposes of adaptation to and mitigation of the effects of climate change had helped them to some extent. There are issues which have to do with Internal Savings and Lending Schemes (ISALs) which were applauded by the people as they have empowered them significantly. Honey production has also empowered them, although they are affected by markets. The other KIIs from Chiredzi and Gwanda indicated that cattle fattening is a good project, but the challenge they face is of diseases and declining prices. The following are excerpts from each of the KIIs from the three areas:

KII Gwanda: *The NGOs have empowered us through the rehabilitation of the dam, and we can grow our own crops through irrigation and also our livestock can drink the same water. Off-farm activities such as mining are empowering people, although they are affected because of lack of environmental compliance as encouraged by organisations such as EMA. As for the mopane worm, it's ok with us; we only need to have value addition for us to get more.*

KII Chiredzi: *The people of Chiredzi have been empowered through the use of ISALs, and they can now generate income of their own; the only challenge that they have are the prevailing economic conditions which affect their savings. Livestock project has empowered most of the communities as they can use the livestock for other purposes such as purchasing of food. However, the challenge, for now, is that there is a serious disease outbreak; as such, most of the livestock is lost.*

KII Nyanyadzi: *The use of baobab as a livelihood has been empowering communities for ages, but now the challenge is that some of the trees are dying due to poor methods of bark harvesting. As for the honey, initially, the community was empowered when there was assistance from the NGO; now the big challenge is that of the market for the honey and as such not much is being realised by the households. The ISALs schemes are ok; they have empowered the people into financial management.*

The private sector can also play the role of value addition. This will go a long way in creating employment, thus leading to local development through the multiplier effect. These efforts can be supported by the provision of both financial services and climate information services (Carabine and Simonet 2018).

5 Conclusion

Climate change has greatly affected livelihood strategies in dry regions. Since most rural areas depend on agriculture, this has resulted in livelihood diversification strategies. The most common on-farm resilience strategy adopted is the growing of drought-resistant crops. The prospects of developing irrigation schemes are high as this will go a long way in increasing agricultural yields. Off-farm strategies include internal savings and lending schemes, food for work and harvesting of non-timber forest products. A myriad of challenges limited climate change resilience strategies. Poor communication networks have affected the marketing of products. This is exacerbated by poor farming methods which have affected the quality of both crops and livestock. To enhance livelihoods, there is a need for integrated resource management. Climate change effects, like any other hazards, can be reduced by coming up with sustainable disaster management plans. The adoption of renewable sources of energy such as solar can go a long way in the quest to achieve sustainable communities.

6 Recommendations

6.1 Livelihood Resilience Strategies

Within the context of prevailing and anticipated harsh climatic conditions, efforts must be stepped up to improve the viability of livelihood resilience strategies that have been adopted. Efforts to reconfigure the existing strategies and/or adequately design alternative strategies must be made by various stakeholders involved in rural development such as non-governmental organisations, the government, some quasi-government institutions and the rural people themselves.

6.2 Investment in Water Harvesting

Aridity is the greatest hindrance to food security in sub-Saharan Africa. Therefore, there is a need to adopt a holistic, integrated approach, in the investment of water harvesting techniques in order to optimise agricultural yields. Thus, various stakeholders involved in rural development should combine resources and channel them towards the construction of more dams and the desiltation of existing ones.

6.3 Integrating Indigenous Knowledge Systems with Biotechnology

The chapter noted the adoption of drought-resistant crops, which is influenced by indigenous knowledge systems. In order to achieve sustainability, efficiency and effectiveness of farming systems, there is a need to incorporate a biotechnological approach guided by the already available local knowledge. The indigenous knowledge from local communities should be harmonised with modern science in coming up with sustainable resilience strategies.

6.4 Fostering Internal Savings and Lending Schemes

This chapter documented the use of internal savings and lending schemes as an off-farm livelihood diversification strategy. Thus, the researcher recommends that self-generated schemes such as the ISALs should be supported because of the

Table 15.1 Population and sample size

District	The total population of households	Sample size
Chiredzi	63,459	382
Chimanimani	32,291	380
Gwanda	26,201	379

Source: Field Survey 2019

organic nature of their evolution out of society's own means. To this extent, this article corroborates the notion that home-grown solutions are more appropriate and suited for local conditions compared to externally imposed prescriptions.

Acknowledgement Special mention goes to my postdoctoral supervisor, Professor Betty C Mubangizi, for the technical support. I greatly appreciate the financial support from her SARChI Chair Sustainable Local (Rural) Livelihoods throughout the research.

References

Bernsterin, H., Crow, B. and Johnson, H. (1991). *Rural Livelihoods: Cries and Responses*. New York: Oxford.

Bhattacherjee, A. (2012). *Social Science Research: Principles, Methods, and Practices*. Florida: University of South Florida.

Carabine, E. and Simonet, C. (2018). Value Chain Analysis for Resilience in Drylands (VC-ARID): Identification of adaptation options in key sectors. VC-ARID synthesis report. [online] Available at: http://prise.odi.org/research/value-chain-analysis-for-resilience-in-drylands-vc-arid-identification-of-adaptation-options-in-key-sectors-2/.

Chimhowu, A, Manjengwa, J. and Feresu, S. (2011). *Moving Forward in Zimbabwe: Reducing Poverty and Promoting Growth*. Harare: Institution of Environmental Studies.

Chitongo, L. (2019). Rural livelihood resilience strategies in the face of harsh climatic conditions. The case of ward 11 Gwanda, South, Zimbabwe. *Cogent Social Sciences*, 5(1) 1–19, https://doi.org/10.1080/23311886.2019.1617090

Christians, C. G. (2000). Ethics and politics in qualitative research. In: *The Sage Handbook of Qualitative Research*. London: Sage.

Civil Protection Organisation of Zimbabwe. (2009). *Disaster Risk Management*. Harare: Civil Protection Organisation of Zimbabwe.

Cochran, W.G. (1963). *Sampling Techniques*. Wiley, New York.

Creswell, J. (2009). *Research Design: Qualitative, Quantitative and Mixed Methods Approaches*. London: Sage.

Godfrey Mutowo, David Chikodzi, (2014),Remote sensing based drought monitoring in Zimbabwe, Disaster Prevention and Management, Vol. 23 Iss 5 pp. 649 - 659 Permanent link to this document: https://doi.org/10.1108/DPM-10-2013-0181

Hajdu, F., Neves, D., & Granlund, S. (2020). Changing Livelihoods in rural eastern cape, South Africa (2002–2016): diminishing employment and expanding social protection. *Journal of Southern African Studies*, *46*(4), 743-772.

Harris, L., and Brown, G. (2010). Mixing interview and questionnaire methods: Practical problems in aligning data. Research and Evaluation, 15(1), 230-32.

Intergovernmental Panel on Climate Change (2012). Managing the risks of extreme events and disasters to advance climate change adaptation. In C. B. Field, V. Barros, T. F. Stocker, D. Qin, D. J. Dokken, K. L. Ebi, … P. M. Midgley (Eds.), A special report of working groups I and II of the intergovernmental panel on climate change. Cambridge, UK, and New York, USA: Cambridge University Press.

Krippendorff, K. (2010) *Content analysis: an introduction to its methodology*. The Sage CommText Series.

Mandizvidza, M. (2006). Dynamics of agriculture. College Press, Harare

Mogotsi, K. Nyangito, M. M. and Nyariki, D. M. (2011). Drought Management Strategies Among Agro-Pastoral Communities in Non-Equilibrium Kalahari Ecosystems. *Environmental Research Journal*. Available at: http://www.medwelljournals.com/fulltesxt/?doi=erj.2011.156.162.

Neuman, W. (2000). Social Research Methods: Qualitative and Quantitative Approaches (4th Edition ed.). Boston: Allyn and Bacon.

Patton, M. Q. (2002). Qualitative evaluation methods. Beverly Hills: Sage.

TANGO International. (2018). Zimbabwe Resilience Research Report. Produced as part of the Resilience Evaluation, Analysis and Learning (REAL) Associate Award. Photo Credit: Colin Crowley/Save the Children

UNICEF and the World Bank Group (2016) Ending Extreme Poverty: A Focus on Children, October 2016. Available at: https://www.unicef.org/publications/index_92826.html accessed 14 August 2020

United Nations, (2015) Transforming Our World, 2015: The 2030 Agenda for Sustainable Development. Geneva: United Nations Goal 13. Accessed 20 January 2020 at: https://sustainabledevelopment.un.org/content/documents/21252030%20Agenda%20for%20Sustainable%20Development%20web.pdf

Walker, B., C. S. Holling, S. Carpenter, and A. Kinzig. 2004. Resilience, adaptability and transformability in social-ecological systems. *Ecology and Society* 9 (2) 5. http://www.ecologyandsociety.org/vol9/iss2/art5.

Whingwiri, E. E., Mashingaidze, K. and Rukuni, M. (1992). *Small-Scale Agriculture in Zimbabwe*. Harare: Rockwood Publishers.

ZIMSTAT (2012). *Poverty and Poverty Datum Line Analysis in Zimbabwe 2011/12*. ZIMSTAT, Harare

Climate Action at International Airports: An Analysis of the Airport Carbon Accreditation Programme

16

Kaitano Dube

Abstract

The aviation industry is often criticised over its handling of carbon emissions and consequently contributes to climate change. There has been increasing pressure for the industry to re-align its carbon management strategies with the sustainable development pathway. The industry was widely discussed during the COP25 in Madrid, Spain, in 2019 given its influence on climate change. This study is aimed at examining the handling of climate change action under Sustainable Development Goal 13 (climate action) by airports under the voluntary mechanisms championed by the Airport Carbon Accreditation (ACA) under Airports Council International (ACI). The study makes use of a critical document analysis of the 10-year annual reports prepared by the ACA and a review of sustainability reports. The study found that airports had made some significant efforts in curbing carbon emissions with the participation of 312 airports in 71 countries that cover 44.7% of the global air passenger traffic in less than 11 years of the programme coming into effect through a raft of measures. The study recommends policy framework support to scale up the efforts by airports to ensure sustainability in the transport and travel industry.

Keywords

Climate change · SDG 13 · Aviation · Airports · Green aviation · Sustainability

1 Introduction

The sustainability of airports has been at the centre of debate amongst academics and airport stakeholders for some time now. This is given the fact that airports have drastically expanded extensively over the years in both scope and functions. While the significant growth in airports has been characterised and matched by an equal urban expansion, they have also contributed to global environmental challenges (Boons et al. 2010; Postorino and Mantecchini 2014). This growth brought about pain and suffering to the host communities who often complain of various challenges posed by air traffic, particularly noise pollution from landing and departing flights (Raimi and Adindu 2019). Communities close to airports have also been complaining about respiratory diseases, amongst others, attributed to their proximity to the airports (Black et al. 2007;

K. Dube, PhD (✉)
Department of Hospitality, Tourism and PR, Vaal University of Technology, Vanderbijlpark, South Africa
e-mail: kaitanod@vut.ac.za

G. Nhamo et al. (eds.), *Sustainable Development Goals for Society Vol. 2*, Sustainable Development Goals Series, https://doi.org/10.1007/978-3-030-70952-5_16

Hsu et al. 2012; Penn et al. 2017). Other environmental challenges have also been brought to the fore, such as the impact of airports on bird populations (Hauptfleisch and Avenant 2016). Environmentalists and conservationist have been quite critical of the impact of airports on bird species, which are often victims of departing and landing aircraft but also posing a threat to aircraft (Lopez-Lago et al. 2017; Zhao et al. 2019; Metz et al. 2020).

Notwithstanding the above, airports as large entities where millions of people move across daily contribute to significant levels of air pollution (Barrett et al. 2013). Airports are major contributors to other sources of land and atmospheric pollution caused by departing and landing aircraft (Ison et al. 2014; Sarbassov et al. 2020). Also, in recent debates over climate change; airports have been pointed out for their immense contribution for Scope 1, 2 and 3 carbon emissions. Additionally, airports have been regarded as gateways for smuggling illicit drugs, contrabands and other prohibited substances such as drugs (Steinberg 2005; Kirman et al. 2012).

The year 2015 was monumental in dealing with carbon emissions and other environmental challenges. The world witnessed several international protocols and treaties that were aimed at addressing a host of environmental challenges (Dube and Nhamo 2020a). Chief amongst these was the Paris Agreement (United Nations Climate Change 2015) and Agenda 2030 on Sustainable Development, which promulgated the Sustainable Development Goals (SDGs) (United Nations 2015). These global developments added more pressure and momentum for the global airports to invest more and focus on sustainability, which also led to the adoption of the ambitious aviation plan by the International Civil Aviation Organization (ICAO) titled "On Board a Sustainable Future". The document was aimed at ensuring that the aviation sector becomes more sustainable through the adoption of global SDGs (ICAO 2016).

Key amongst such aspirations by the aviation sector is the quest for the sector to reduce its carbon footprint in line with SDG 13 on climate action, which directly speaks about the Paris Agreement. Chapters 4 and 7 of the "On Board a Sustainable Future" policy document talk directly about climate change initiatives within the aviation sector (ICAO 2016). Addressing climate change has particularly been an issue of interest to the sector, given the costs and taxes of carbon in some countries where carbon pricing has been adopted (Poll 2017; Gratton et al. 2020). Consequently, various aviation subsectors have been adopting several green initiatives. One of the initiatives is the airport carbon accreditation programme aimed at achieving carbon neutrality at various airports. Against that background, this study is aimed at examining the Airport Carbon Accreditation under Airports Company International's role in addressing SDG 13 on climate action. Furthermore, the study also seeks to examine how the adoption of the SDG agenda shaped the green initiative at various global airports.

2 Literature Review

The increased costs from extreme weather events and pressure from environmentalists and academics have put the environmental and climate change agenda at the focal point of global discourse (Dube and Nhamo 2020b). This agenda is building up from the momentum that was generated from the commissioning of the Brundtland Report (1987) and the subsequent global environmental initiatives (Nhamo et al. 2020). The quest for a sustainable future building is now a key issue in global agenda setting. The aviation sector, amongst others, has come under intense scrutiny for its carbon emissions that contribute to climate change (Lenzen et al. 2018).

The increase in carbon emissions is predicted to influence an increase in the frequency and severity of extreme weather events. *The Special Report on Global Warming of 1.5 °C* bears testament to the dangers of tipping beyond 1.5 °C temperature as carbon emissions continue to soar across the world (Intergovernmental Panel on Climate Change- IPCC 2018). Since aviation is a significant contributor to climate change, it is, on the other hand, a victim given the multiple vulnerabilities of various subsectors of the aviation industry. Airports in particular which are a focus of this study have been predominantly vulnerable to extreme weather events which have forced in some cases changes to design protocols as a measure to reduce climate change vulnerability (Burbidge 2016; Burbidge 2018). The International Civil Aviation Organization (2016) proposes new designs and upgrades for new airports to factor in the impacts of climate change and ensure that the airports are climate-smart.

Pejovic et al. (2009) observe, for example, that at London Heathrow Airport, as a consequence of climate change weather-related challenges, aircraft delays had gone up 25% due to changes in parameters such as fog, thunderstorms, wind speed and fog. Coffel et al. (2017) argue that the increase in temperature at various airports particularly those at higher latitude is likely going to result in challenges of weight restrictions to allow for a safer take off with the demand in other instances for airport runway extensions at several airports. On the other hand, Poo et al. (2018) highlight that airports in coastal areas are under threat from rising sea level, flooding damages and increased storm activities.

Conversely, Zhou and Chen (2020) report that most airports remain vulnerable to extreme weather events as they have not built enough capacity for adaptation. Addressing climate change under the auspices of SDG 13 is, therefore, in the interest of the airport industry as the sector stands to lose from inaction on climate change. SDG 13 target 13.3 calls for an improvement on education, awareness raising and human and institutional capacity on climate change mitigation, adaptation, impact reduction and early warning (United Nations 2015). Tackling climate change at airports under SDG 13 requires that airports make substantial cuts to carbon emissions throughout their value chain. The airports have various activities where carbon emissions can be cut or offset to align airport activities to ensure carbon reduction. Fig. 16.1 highlights the various points where carbon emissions are produced at airports. It is at those points that are shown which have been the focus of climate change action.

3 Research Design

The study made use of a critical document analysis, which utilised archival data and secondary data analysis. The process involved the analysis of all the annual reports from 2009 to 2019, starting with the first annual report, as shown in Table 16.1. Other reports that were reviewed include airport sustainability reports and a few selected airport websites, amongst others.

The Airport Carbon Accreditation (ACA) produces annual reports, outlining all the key figures for each year, which summarised the number of airports accredited, the collective reduction achieved and case studies detailing some of the innovative and original ways that airports have achieved their accreditation. This provided a rich source of data for this study. The ACA is a brainchild of Airports Council International (ACI) and was adopted to fight climate change by airports. The programme is voluntary, was launched in 2009 and is administered by WSP, an international consultancy appointed by ACI EUROPE to enforce the accreditation criteria for airports on an annual basis. Airports applying for accreditation must have their carbon footprints independently verified per ISO14064 (Greenhouse Gas Accounting). The definitions of emission footprints used by the ACA follows the principles of the World Business Council for Sustainable Development (WBCSD) and the World Resources Institute (WRI) "Greenhouse Gas Protocol" Corporate Accounting and Reporting Standard, and as such the programme meets the best international standards whose accreditation system can be relied upon.

Then, content and thematic analysis were conducted to gain insights into carbon-related initiatives that were covered by the reports. Prescripts and guidance provided by Bowen (2009) were followed through data gathering and analysis process. This approach was preferred as it was found to be cost-effective as it would have been

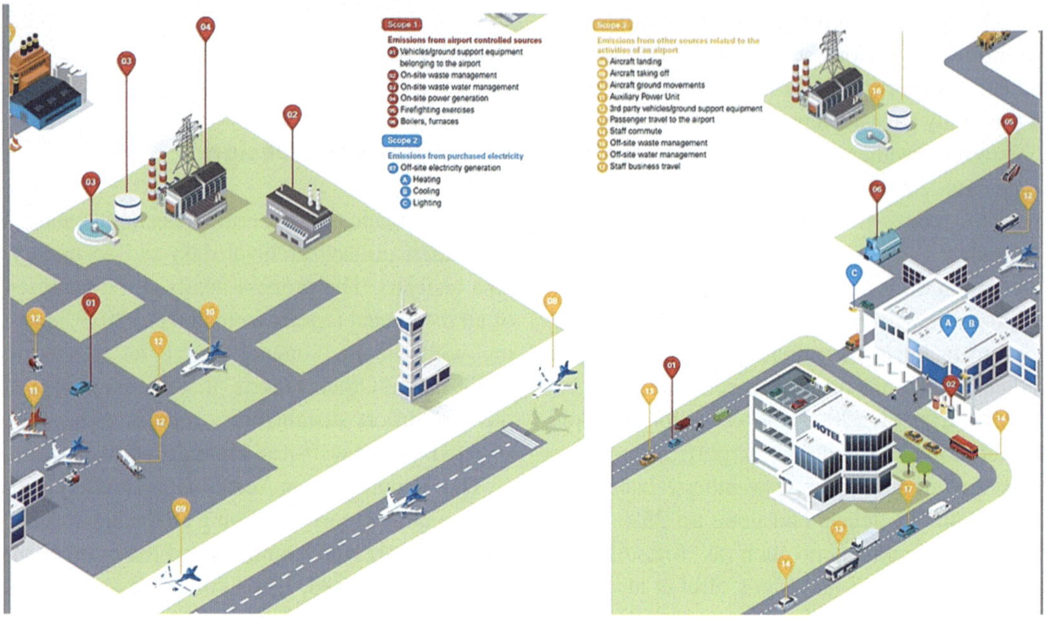

Fig. 16.1 Types of carbon emissions produced at an airport. (Source: Author's Adaptation from Airport Carbon Accreditation (2018:11–12))

Table 16.1 Breakdown of reports that were reviewed

Annual report	Number of pages
2009–2010	21
2010–2011	34
2011–2012	31
2012–2013	39
2013–2014	48
2014–2015	66
2015–2016	84
2016–2017	96
2017–2018	72
2018–2019	64

Source: Author

ordinarily cumbersome to visit and gather data from all the nation-states, which constitutes the Airport Carbon Accreditation programme. All the data that were needed for this study were documented in the reports in a summarised manner which made it easy to scrutinise and analyse it promptly. A critical document analysis is also an acceptable research approach that is growing in sage and is also used by other scholars in similar studies such as Nhamo et al. (2020).

4 Results and Discussion

The study found that the ACA programme is an innovative programme that provides a platform for airports across the globe to aspire to achieve carbon neutrality where airports can be accredited under four levels of certification as outlined in Fig. 16.2. It is under those accreditation levels that 312 airports across the world have joined since the inception of the programme in 2009 to July 2020.

The voluntary project programme was designed to ensure that airports across the world can participate in reducing their carbon footprint, which has been of huge concern in the era of climate change. An airport can start at the mapping, which is the lowest level of accreditation. At that stage, airports are expected to conduct a carbon audit annually within their operational parameter. The carbon audit is done by the airport and verified by an independent third-party verifier which takes into consideration direct (Scope 1) and indirect (Scope 2) emissions emanating from the airport's operational boundaries. This audit must meet ISO14064 and accreditation requirements. For an airport to be considered and accredited at

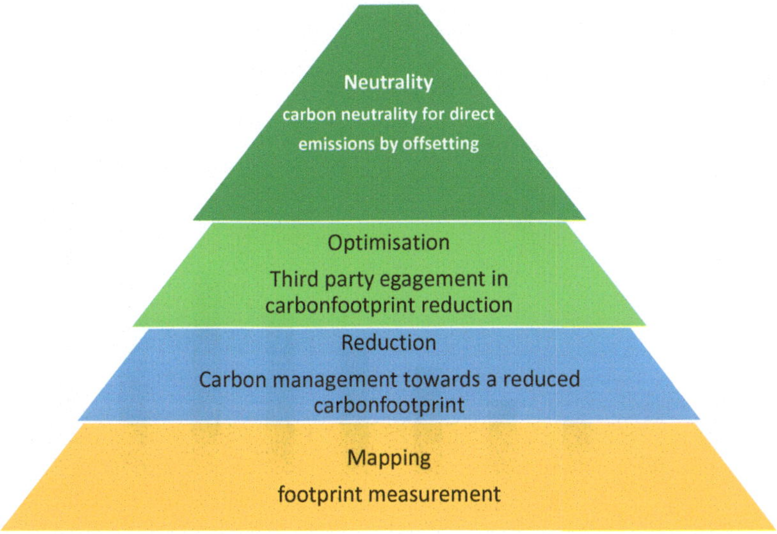

Fig. 16.2 The four levels of carbon accreditation under the Airport Carbon Accreditation. (Source: Authors, Data obtained from ACA (2020),

Stage 2, which is carbon reduction, it has to meet all conditions set out in Stage 1 (carbon mapping) and adopt several measures aimed at reducing carbon emissions. Amongst such measures is the need to have a low carbon policy, carbon or energy reduction targets, embarking on employee training on ways to reduce carbon emissions and also putting in place measures to monitor and track adherence to carbon reduction.

Under the ACA programme Stage 3, carbon optimisation requires airports to engage with third parties in carbon footprint reduction. Such third parties include airlines, independent ground handlers, air traffic controllers, caterers and all others such as service providers on the airport site. Also, this engagement is extended to surface access modes (road, rail) with authorities and users in a bid to ensure that every layer cuts their carbon emissions. This entails a reduction of scopes of carbon emissions as set out in the GHG Protocol, which covers Scope 1, 2 and 3 emissions. Scope 3 emissions cover indirect emissions from outsourced services and activities, including purchased fuels and waste disposal. Stage 3+ carbon neutrality entails that airports fulfil all requirements of "Mapping", "Reduction" and "Optimisation" and in addition to offsetting its remaining Scope 1 and 2 carbon emissions using internationally recognised carbon offset mechanisms. Market-based measures are part of the basket of measures that were agreed in 2016 as the aviation industry strategy to reduce the amount of GHG by the aviation industry (ICAO 2016).

Exponentially, the number of airports joining the airport carbon accreditation has been increasing yearly. When the programme started, it had the full participation of 17 airports (Fig. 16.3). While the number might look smaller, this was a significant start as the number of airports that were in the programme accounted for 20.5% of European passenger traffic. The airports were utilised by about 310 million passengers annually. As a consequence of the programme, airports at various levels of accreditation had a reduction of carbon emissions equivalent to 411,390 tCO_2. The four categories of airports that had achieved a carbon-neutral airport status managed to offset 13,129 tonnes of CO_2 in the first year alone (Airport Carbon Accreditation 2010).

By the time the world adopted the all-inclusive SDGs in 2015, at least about 92 airports were participating in the programme. This shows that there was an increase of 75 airports within 5 years from the onset of the programme. Post-adoption of the SDGs in 2016, the programme seems to

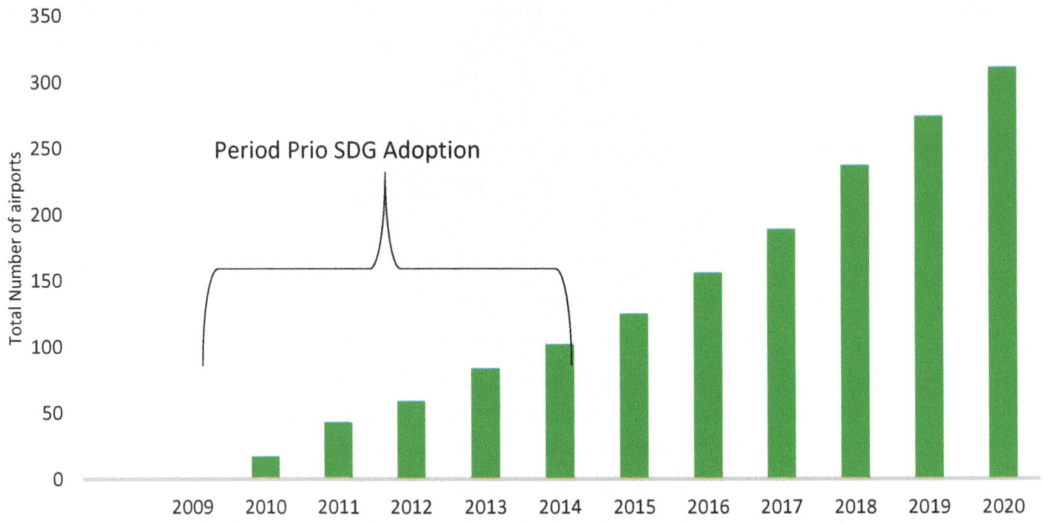

Fig. 16.3 Trends in the number of accredited airports 2009–2020. (Source: Author Data from Airport Carbon Accreditation (2020))

have witnessed more pronounced growth momentum as the number of airports that joined the programme grew from 92 airports to 311 by mid-2020. During the period the number of airports in the programme tripled over the same period. By the end of 2019, 274 accreditations were marking a 16% growth from the previous year. As a result of this phenomenal growth, about 43% of global passengers made use of accredited airports, and the number of countries participating in the programme grew to 68 (Airport Carbon Accreditation 2019).

Nonetheless, Europe had the largest share of accredited airports globally. As a consequence of the airport accreditation programme, the participating airports reported a carbon reduction under Scope 1 and 2 of −322,297 CO_{2e} tonnes. On the other hand, –710,673 tonnes of CO_{2e} was offset by airports that had achieved the carbon-neutral status. The figures represent a significant leap from the first year of the inception of the programme and great momentum. This demonstrates the popularity of the programme amongst major airports across the world. The results suggest that there was an intention on the part of the airports' community to respond to global pressure for the aviation sector to embrace sustainability and combat climate change within their businesses.

It was critical for the study to examine where growth was happening within the various levels of accreditation. To that extent, the study found that, even from the inception of the carbon accreditation programme, some airports were already implementing some measures aimed at ensuring carbon management and reduction. In the first year of the programme, nine airports had an accreditation level 2 and above, meaning that they were actively engaged in carbon reduction initiatives through various measures. Out of the airports that had carbon reduction initiatives, four of these were accredited at a carbon-neutral level. These airports are Trondheim Airport in Norway which handles about 4.5 million passengers in 2018 (Statista 2020); Oslo Airport in Oslo, Norway, which handles five million passengers (Oslo Airport 2020); Stockholm Bromma Airport which handles 26 million passengers (Swedavia Airports 2019); and Stockholm Arlanda Airport. Europe was the pioneer region for the ACI programme, and the strict regulation of emissions and relevant legislation assisted in nudging the airports towards adopting a greener path.

By 2015, there were 125 airports at various levels of accreditation as depicted in Figs. 16.3 and 16.4. It is important to note that the number of level 1 airports was surpassed by airports in

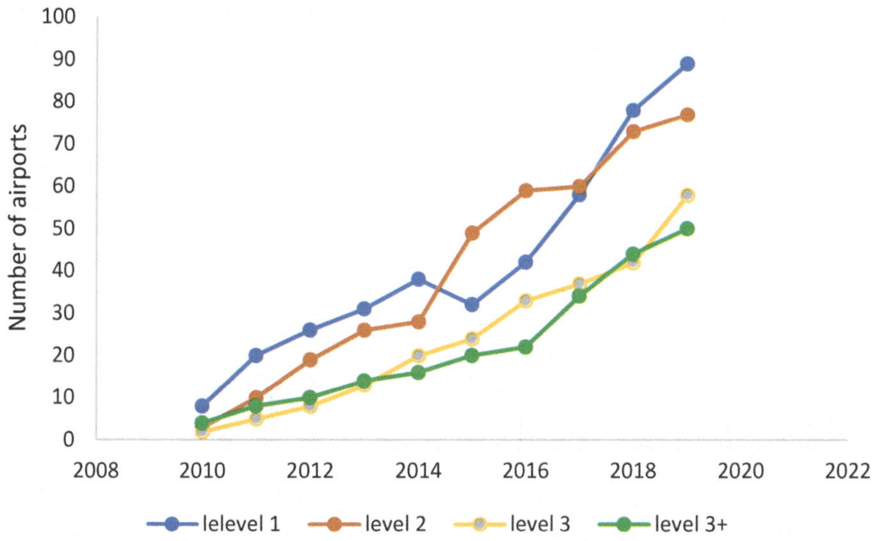

Fig. 16.4 Trend in accredited airports at various stages 2009–2020. (Source: Author Data from ACA (2020))

level 2 as more airports sought to address their waste and also carbon emissions as they moved from the first stage of merely auditing the emissions. Between 2016 and 2020 the number of airports participating in the carbon-neutral programme grew although the number of airports in the higher category remains lower than those in the first two categories pointing to increased carbon reductions. While this is not surprising, it will be useful to have far more airports in the higher band of accreditation. The reason is that, in the higher accreditation band, it is where the highest carbon emission reduction is likely to be significant for a sector that was expanding until the advent of COVID-19.

It would seem that the advent and proclamations made in the Paris Agreement and the 2030 Agenda for Sustainable Development (2015) combined and exerted the ethical and moral pressure for more airports to join the airport accreditation programme as the figure rose from 125 airports in 2015 to 311 airports across the world. Figure 16.4 shows how the accreditation of airports has taken place between the first year (2009), in 2015 and in the tenth year in 2019. The figures show that there have been great strides at all levels with the number of airports participating in high levels of accreditation. This has led to a significant reduction in carbon emission reduction at airports since the programme started in 2009/2010.

Inasmuch as there has been a growth of the number of accreditations across the world, there has not been uniformity about the spatial growth of accredited airports across the world. As can be seen in Fig. 16.5 and already discussed earlier, the European region is leading other regions in the number of accredited airports with 82% of carbon-neutral airports located in Europe, about 40% of level 3 airports, 44% of level 2 airports and 52% of level 1 airports.

While the European region entered the programme at its inception, other aviation regions entered the programme later with the number of accredited airports having grown from 4 in 2013 to 57 by mid-year 2020. There is a predominance of level 2 and above with airports accredited at level 3 in dominance (Fig. 16.6). Notably, some of the most popular airports such as Frankfurt, Heathrow and Paris Charles de Gaulle Airport are accredited at level 3, while some other airports such as Amsterdam Airport Schiphol and London Gatwick Airport are accredited as carbon-neutral airports. This means a significant number of travellers are making use of carbon accredited airports which are taking concrete and measurable steps to reduce their impact on the environment to ensure sustainability. This is commendable as

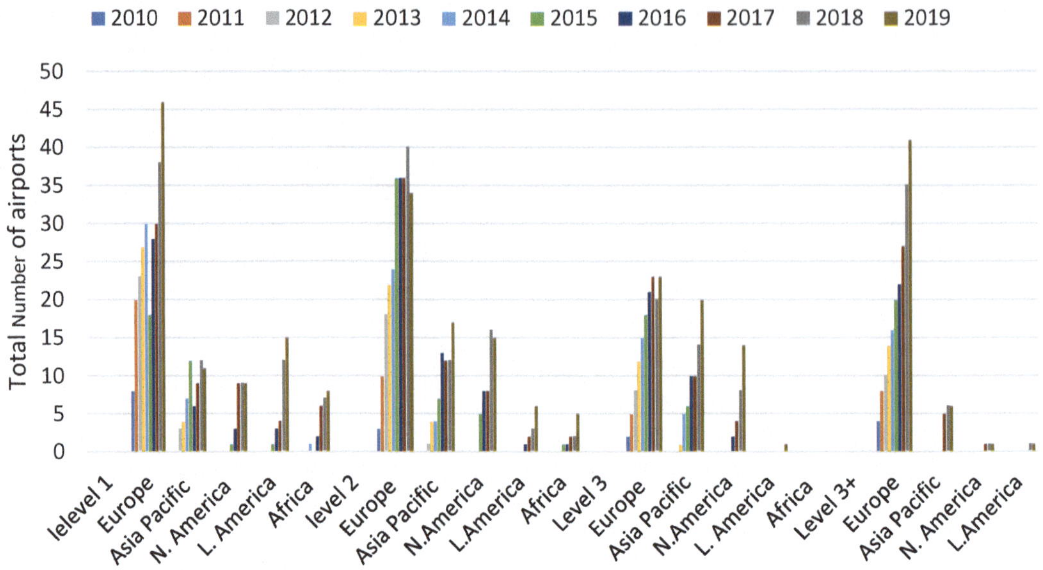

Fig. 16.5 The distribution of accredited airports across the world 2010–2019. (Source: Author)

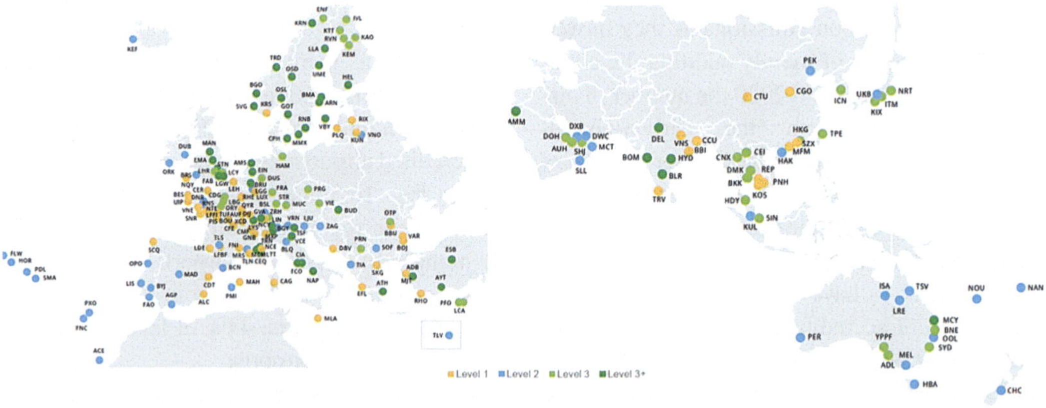

Fig. 16.6 Distribution of carbon accredited airports across Europe and Pacific Asia. (Source: Author Adapted from Airport Carbon Accreditation (2019))

this means many participating airports are making a difference in combating climate change as prescribed under SDG 13 on climate change action and helping in the contribution of achieving the aspiration of green travels.

North America has also recorded some progress in the uptake of the carbon accreditation programme with 47 accredited airports as of mid-year 2020. South America and Africa remain underrepresented (Fig. 16.7) with very few airports having achieved carbon-neutral status. In North America, several large international airports such as JFK International, New York International Airport and Dallas/Fort Worth International Airport also participate in this voluntary programme at high levels with the first two participating at level 3 and the Dallas/Fort Worth International Airport accredited at level 3+, which is a neutral stage. This means that many passengers that make use of the mentioned airports are assured of carbon-neutral business at those airports. The three airports facilitated the movement of about 273 million passengers in 2019 from hundreds of thousands of aircraft that

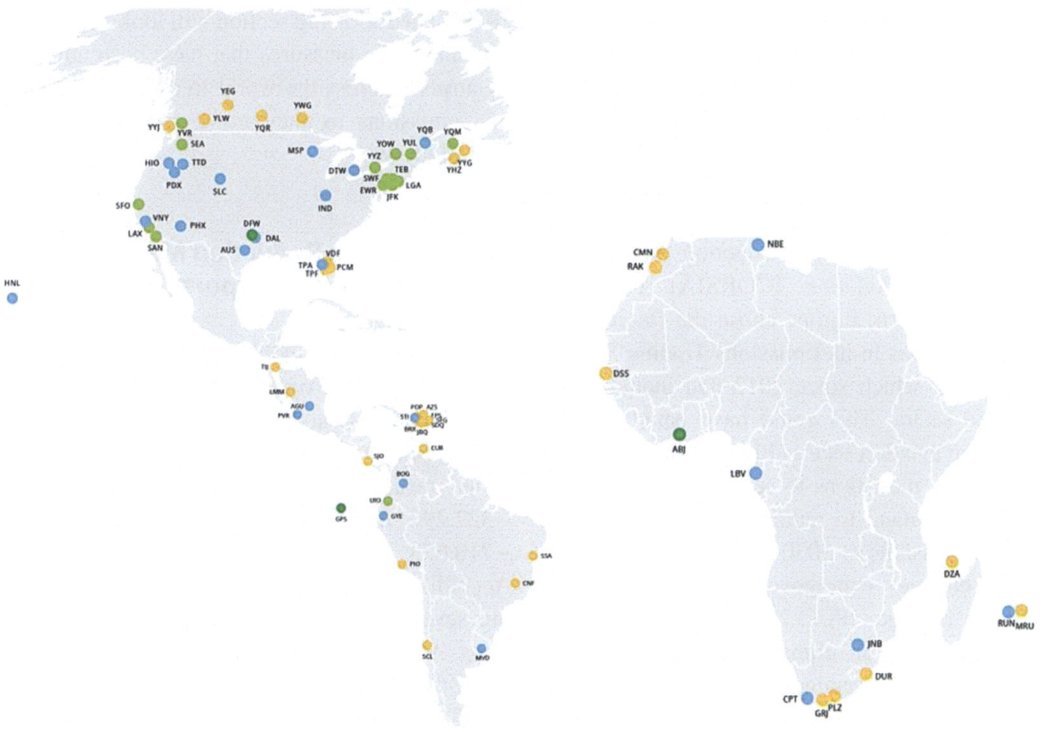

Fig. 16.7 Distribution of accredited airports in North and South America and the African region. (Source: Author Adapted from Airport Carbon Accreditation 2019) (*Map not to scale)

utilises these airports annually, which can contribute immensely to carbon emissions if not well managed.

Only Seymour Airport in South America and Félix Houphouet-Boigny International Airport in Africa have achieved a carbon-neutral airport status. In Africa, airports in South Africa are well represented, although these are only at level 2 and level 1 in the main. Johannesburg International Airport or OR Tambo International Airport and Cape Town International Airport are some of the largest airport hubs serving the African continent. Other important airport hubs in the region of Africa, which include Bole International Airport in Ethiopia and Jomo Kenyatta International Airport in Kenya, have not yet joined the programme. Given the volume and traffic processed by these airports, it will be important to learn how these airports are making strides to ensure climate change mitigation from airport carbon accreditation position.

A study by Frömming et al. (2011) on the contrail effect from aviation found that the greatest intensity was experienced on the route path between North America and the European region and some strips in the Asia Pacific. Predictably, these regions have the densest aviation traffic. These findings were also confirmed by Kärcher (2018), who bemoaned the impact of contrail effect on altering climate at the behest of the aviation sector. Therefore, it is encouraging to note that the European region, in particular, has taken a central lead in addressing its carbon emissions in a move that can assist in avoiding the dangerous climate change and aberrations envisaged by IPCC if the world does not address its carbon emissions (Intergovernmental Panel on Climate Change- IPCC 2018).

The momentum in aviation sustainability also has to be understood in the context of consumer pressure as they demand better environmental action from the sector. Gössling et al. (2020)

noted that flight shame, a spirited movement by passengers to force aviation to go green, had resulted in declining passengers in Germany, a move that could be forcing the aviation to take concrete measures to address climate change. According to Larsson et al. (2019), while the rest of the world found impetus from the ICAO's Carbon Offsetting and Reduction Scheme for International Aviation (CORSIA) that was adopted in 2016, Europe chose the faster emission reductions in its Emissions Trading System (EU ETS), which since 2012 had included the aviation sector. This could have benefitted the European aviation sector although there is a need to quantify reductions to see if they are in line with the demand to reduce the ambitious 45% carbon emissions by 2030 to avoid dangerous climate change (Intergovernmental Panel on Climate Change- IPCC 2018). The presence of concrete policy in climate change could have assisted greatly to nudge the industry in the right direction.

Ritchie et al. (2020) confirmed the crucial role played by the policy in driving climate change action by pointing out that they certainly had a positive impact on carbon offsetting initiatives within the Australian market. As already witnessed, Australia has more than 12 accredited airports. Out of the 12, 4 airports are at level 3 and 1 carbon-neutral airport, the Sunshine Coast Airport. The pursuit of carbon neutrality within the airports, therefore, requires a robust and clear policy that is consistent. In countries where there is rhetoric on climate change, the growth in the carbon-neutral airport had been rather moderated such as in the USA. The withdrawal of the USA government from the Paris Agreement could have had sent a negative signal to the global and local community affecting and slowing down progress towards green investment (Zhang et al. 2017). According to Jotzo et al. (2018), Donald Trump's presidency undermined climate action in several ways as it led to the questioning of climate change science and several enabling policies.

Airports across the world have been active in fostering climate change which has resulted in about 43% of global passengers utilising accredited airports (Airport Carbon Accreditation 2019). The following section will look at some of the innovative measures that have been adopted by airports across the world to reduce their carbon emissions as they move towards carbon-neutral growth.

4.1 Carbon Reduction Measures at Selected Airports

In the tenth year of celebrating the carbon-neutral status under the Airport Carbon Accreditation programme, 274 airports across the world were taking concrete steps to reduce their carbon emissions resulting in emissions reductions of −322,297 tonnes of CO_{2e} and offsetting of −710,673 tonnes of CO_{2e} (Airport Carbon Accreditation 2019). This was achieved by adopting several diversified measures at various airports. According to the ACA report in the tenth year, the 50 carbon-neutral airports tended to show a preference for Certified Emissions Reductions (CER), and through that way at least 54% of emissions was offset. Another approach was the utilisation of the Verified Carbon Standard (VCS), which accounted for 27%.

To offset carbon emissions, airports have been involved in various projects that include biogas projects and hydroelectric projects being the most popular projects to be adopted by airports used in offsetting their carbon emissions (Fig. 16.8). Additionally, airports have also been ramping up efforts to ensure energy efficiency in a bid to reduce their carbon footprint drastically.

Several airports were investing in a massive solar project to run their airports from clean energy. In Latin America, for example, Carrasco International Airport/MVD was the first airport in that region to have a photovoltaic solar plant. The plant which had 0.5 MW capacity from 1540 photovoltaic panels was part of the initiative to ensure energy efficiency which saw the airport switching to LED lights, heat pumps and free cooling for thermal conditioning at the airport (Corporacion American Airports 2018). In the Nordic region, one of the airports that took to solar is the Helsinki Airport which built one of the biggest solar plants in the Nordic region. In

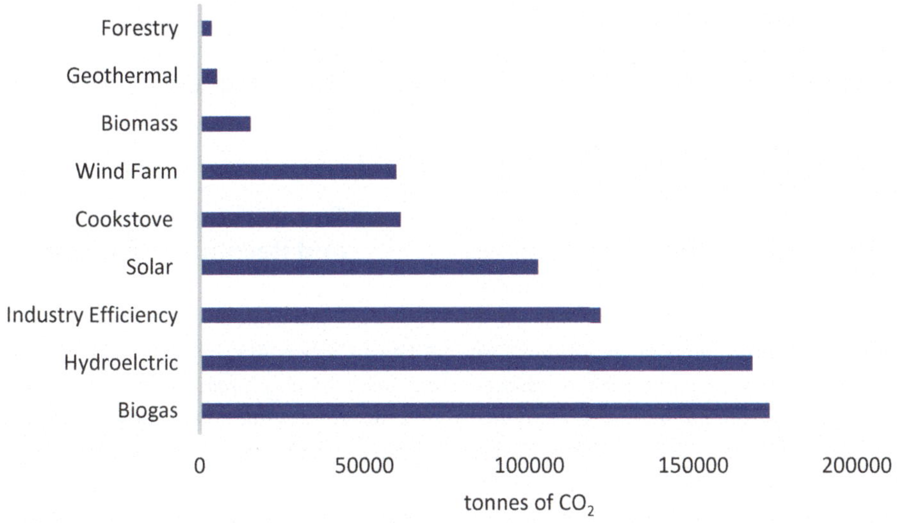

Fig. 16.8 Most popular carbon offsetting projects adopted by airports. (Source: Authors Data from Airport Carbon Accreditation (2019))

2019, it had an installed capacity of 330 kWp energy plant and this augmented wind energy with solar energy powering 5% of the airport's needs (Finavia 2019). There were plans to upscale production to reach 660 kWp which greatly assisted the airport to reach its carbon-neutral ambition. Malta Airport had also turned to solar as a measure of reducing its carbon emissions and invested €1.2 million in the procurement and installation of PV panels that produce 602,000 kWh of clean energy. As a result, between 2016 and 2018, the airport reported a carbon emission reduction of 12% of its carbon emissions in 2018 (Malta International Airport 2019).

Another major airport that stood out in the USA is the Dallas/Fort Worth International Airport. This airport served approximately 69 million passengers to 62 international destinations and 182 domestic destinations in 2019. The airport is a carbon-neutral airport (Dallas Fort Worth International Airport – DFW 2018). It supports 228,000 jobs and has cut its carbon emissions by 83% per passenger and runs on 100% renewable energy. DFW is also actively involved in ensuring the supply of sustainable aviation fuels and renewable diesel and propane. The airport has installed 25 electric vehicle stations to promote the use of less carbon-intensive modes of transport. Throughout the airport, there were several other initiatives aimed at increasing energy efficiency such as upgrading the airport to increase operational efficiency which resulted in cumulative emission reductions of 18% between 2015 and 2018, which could result in close to 40% emission reductions if the pace is maintained. The airport purchased renewable energy from Texas wind farms and had also moved a step further in establishing a geothermal and a solar energy plant at the airport. It also makes use of renewable fuel gas for its fleet, and by 2018 about 36% of its fleet's fuel consumption was from renewable natural gas (RNG) against a set target of 10% (Dallas Fort Worth International Airport 2018). It had also invested in LED lighting on terminal rumps which is expected to save more than 4 million kilowatt-hours of electricity per annum. The installation of dynamic glass for the airport was expected to reduce up to 67% of solar heat gain and ensure passenger and employee comfort. The airport also contributes to other SDGs through its Leadership in Energy and Environmental Design (LEED) Gold building and certified by the Sustainable SITES Initiative. The adoption of green building design in aviation has been a growing phenomenon in the quest to achieve sustainability within the sector with even aircraft manufacturing companies adopting the same (Dube and Nhamo 2019; Kaitano and Nhamo 2020).

The use of hydroelectricity and investment into hydroelectricity is also popular amongst carbon-neutral airports. One airport that adopted such a project is Aeroporto di Napoli. Hydroelectricity projects are interesting in the sense they have a heavy dominance in airports regardless of the declining global investment in such projects. However, inasmuch as they use renewable energy, they are vulnerable to climate change, particularly from the increased incidence and frequency of droughts. In Southern Africa, for example, the increased incidences of droughts had led to rolling blackouts and energy insecurity, e.g. Lake Kariba affected Zimbabwe and Zambia energy supply (Dube and Nhamo 2020c). In Italy, for example, where Aeroporto di Napoli airport is located, climate change is expected to result in a decline in hydro energy reduction due to a fall in rainfall and a decline in snowfall (Bombelli et al. 2019). Long-lasting solutions are, therefore, needed to address energy demands at such airports.

Another project that is also being used by airports is the cookstoves. Several airports globally are investing in cookstoves to offset their carbon footprint by investing in affordable and reliable energy services for cooking which is essential for developing countries. One of the largest Sweden state-owned airport company, for example, managed to offset 3000 tonnes of CO2e through investment into the purchase of 48,000 sets of solar cookers for usage by the rural populace in China totalling 170,000 individuals (Airport Carbon Accreditation 2019). Such projects conducted by several other airports across the world such as Copenhagen Airport and Geneva Airport amongst others are of great help in ensuring that rural communities reduce their dependence on other biomass forms of energy which is detrimental to both the environment and people's health. Wood fuel usage has several health effects on the several millions of people living in rural areas, affecting life expectancy, particularly that of women (Dilger et al. 2016; Liu et al. 2017). Also, such interventions are critical, more so, in the light of expected declining biomass anticipated under increasing climate change to ensure energy sufficiency. A replacement of firewood also assists in the maintenance of trees which acts as carbon sinks.

5 Conclusions and Recommendations

The study was aimed at examining the role of the Airport Carbon Accreditation (ACA) in addressing SDG 13 on climate action. The study shows that the ACA is one of the most coordinated programmes that has seen several airports across the world participating in the programme to climate change and other environmental concerns. Since the programme has gained considerable traction in the European region, which started with the programme, the growth in other regions is encouraging and has allowed airports across the world to learn from each other to build a sustainable airport future. The exponential growth in the number of participating airports particularly in the higher levels of accreditation such as level 3 and carbon-neutral stage 3+ has also benefited non-aviation actors through offset projects that has resulted in the airports investing in clean energy for rural people. This investment has helped in improving the standard and quality of life of people who are very vulnerable from the impacts of climate change to which the aviation industry is a contributor.

Inasmuch as this voluntary programme has gained traction within the aviation sector, there is a need for providing a supportive mechanism by policymakers to nudge non-participating airports to join the programme, which has seen a reduction of carbon emissions from the aviation sector. In regions where this programme has not been as popular, there is a need to find ways of including these regions to get everyone on board. This can be done by putting in place supporting regulation frameworks and providing incentives where possible. It is an excellent start to have airports at a lower level, but there have to be time limitations imposed for an airport to stay at the same level to increase higher-level participation where much of the emissions are being reduced and/or offset. There is almost a deathly silence on this programme in academic literature, and more researches and publications are needed to understand how the airport industry is addressing climate change, mainly focusing on mitigation strategies. A lot of other sectors can learn much about how this sector is coordinating efforts to combat climate change.

References

Airport Carbon Accreditation. (2010). *Annual Report 2009-2010.* Airport Carbon Accreditation. Retrieved from: https://www.airportcarbonaccreditation.org/component/downloads/downloads/29.html. (Accessed 26 July 2020).

Airport Carbon Accreditation. (2018). *Annual Report.* Airport Carbon Accreditation. Retrieved from: https://www.airportcarbonaccreditation.org/component/downloads/downloads/132.html. (Accessed 12 July 2020).

Airport Carbon Accreditation. (2019). *10 Years.* Airport Carbon Accreditation. Retrieved from: https://www.airportcarbonaccreditation.org/component/downloads/downloads/149.html.html. (Accessed 26 July 2020).

Airport Carbon Accreditation. (2020). *Annual Report 2019-2020.* Airport Carbon Accreditation. Retrieved from: https://www.airportcarbonaccreditation.org/component/attachments/?task=download&id=168. (Accessed 26 July 2020).

Barrett, S. R., Britter, R. E., & Waitz, I. A. (2013). Impact of aircraft plume dynamics on airport local air quality. *Atmospheric Environment, 74*, 247-258.

Black, D. A., Black, J. A., Issarayangyun, T., & Samuels, S. E. (2007). Aircraft noise exposure and resident's stress and hypertension: A public health perspective for airport environmental management. *Journal of air transport management, 13*(5), 264-276.

Bombelli, G. M., Soncini, A., Bianchi, A., & Bocchiola, D. (2019). Potentially modified hydropower production under climate change in the Italian Alps. *Hydrological Processes, 33*(17), 2355-2372.

Boons, F., Van Buuren, A., & Teisman, G. (2010). Governance of sustainability at airports: Moving beyond the debate between growth and noise. *Natural resources forum, 34*(4), 303-313.

Bowen, G. A. (2009). Document analysis as a qualitative research method. *Qualitative research journal, 9*(2), 27.

Burbidge, R. (2016). Adapting European airports to a changing climate. *Transportation Research Procedia, 14*, 14-23.

Burbidge, R. (2018). Adapting aviation to a changing climate: Key priorities for action. *Journal of Air Transport Management , 71*, 167-174.

Coffel, E. D., Thompson, T. R., & Horton, R. M. (2017). The impacts of rising temperatures on aircraft takeoff performance. *Climatic change, 144*(2), 381-388.

Corporacion American Airports . (2018). *Carrasco Airport is the first in Latin America to have a photovoltaic solar plant.* Retrieved from: http://caap.aero/n.php?id=100. (Accessed 27 July 2020).

Dallas Fort Worth International Airport. (20198). *2018 Environmental Social Governance Report Growing Responsibly.* Dallas Fort Worth International Airport. Retrieved from https://www.dfwairport.com/cs/groups/webcontent/documents/webasset/p2_950004.pdf. (Accessed 27 July 2020)

Dilger, M., Orasche, J., Zimmermann, R., Paur, H. R., Diabaté, S., & Weiss, C. (2016). Toxicity of wood smoke particles in human A549 lung epithelial cells: the role of PAHs, soot and zinc. *Archives of toxicology, 90*(12), 3029-3044.

Dube, K., & Nhamo, G. (2019). Climate change and the aviation sector: A focus on the Victoria Falls tourism route. *Environmental Development, 29*, 5-15.

Dube, K., & Nhamo, G. (2020a). Greenhouse Gas Emissions and Sustainability in Victoria Falls: Focus on Hotels, Tour Operators and Related Attractions. *African Geographical Review*, 1-16. doi:https://doi.org/10.1080/19376812.2020.1777437

Dube, K., & Nhamo, G. (2020b). Tourism business operators' perceptions, knowledge and attitudes towards climate change at Victoria Falls. *The Journal for Transdisciplinary Research in Southern Africa, 16*(1), 1-1016.

Dube, K., & Nhamo, G. (2020c). Vulnerability of nature-based tourism to climate variability and change: Case of Kariba resort town, Zimbabwe. *Journal of Outdoor Recreation and Tourism, 29*, 100281.

Finavia. (2019). *Helsinki Airport's solar power plant is now the largest airport-based solar power generation facility in the Nordics.* Retrieved from https://www.finavia.fi/en/newsroom/2019/helsinki-airports-solar-power-plant-now-largest-airport-based-solar-power-generation. (Accessed 27 July 2020).

Frömming, C., Ponater, M., Burkhardt, U., Stenke, A., Pechtl, S., & Sausen, R. (2011). Sensitivity of contrail coverage and contrail radiative forcing to selected key parameters. *Atmospheric Environment, 45*(7), 1483-1490.

Gössling, S., Humpe, A., & Bausch, T. (2020). Does 'flight shame' affect social norms? Changing perspectives on the desirability of air travel in Germany. *Journal of Cleaner Production, 266*, 122015. doi:https://doi.org/10.1016/j.jclepro.2020.122015

Gratton, G., Padhra, A., Rapsomanikis, S., & Williams, P. D. (2020). The impacts of climate change on Greek airports. *Climatic Change, 160*, 219-231.

Hauptfleisch, M. L., & Avenant, N. L. (2016). Actual and perceived collision risk for bird strikes at Namibian airports. *Ostrich, 87*(2), 161-171.

Hsu, H. H., Adamkiewicz, G., Houseman, E. A., Vallarino, J., Melly, S. J., Wayson, R. L., & Levy,

J. I. (2012). The relationship between aviation activities and ultrafine particulate matter concentrations near a mid-sized airport. *Atmospheric Environment, 50*, 328-337.

Intergovernmental Panel on Climate Change- IPCC. (2018). *Special Report: Global Warming of 1.5°C.* Retrieved from https://www.ipcc.ch/sr15/download/. (Accessed 12 July 2020)

International Civil Aviation Organization -ICAO. (2016). *2016 Environmental Report* . Retrieved from https://www.icao.int/environmental-protection/Documents/ICAO%20Environmental%20Report%202016.pdf. (Accessed 11 July 2020).

Ison, S., Merkert, R., & Mulley, C. (2014). Policy approaches to public transport at airports—Some diverging evidence from the UK and Australia. *Transport policy, 35*, 265-274.

Jotzo, F., Depledge, J., & Winkler, H. (2018). US and international climate policy under President Trump. *Climate Policy , 18*(7), 813-817.

Kaitano, D., & Nhamo, G. (2020). Major global aircraft manufacturers and emerging responses to the SDGs Agenda. In G. Nhamo, G. Odularu, & V. Mjimba (Eds.), *Scaling up SDGs Implementation* (pp. 99-113). Springer, Cham. https://doi.org/10.1007/978-3-030-33216-7_7

Kärcher, B. (2018). Formation and radiative forcing of contrail cirrus. *Nature communications, 9*(1), 1-17.

Kirman, B., Linehan, C., & Lawson, S. (2012). Blowtooth: a provocative pervasive game for smuggling virtual drugs through real airport security. *Personal and Ubiquitous Computing, 16*(6), 767-775.

Larsson, J., Elofsson, A., Sterner, T., & Åkerman, J. (2019). International and national climate policies for aviation: a review. *Climate Policy, 19*(6), 787-799.

Lenzen, M., Sun, Y. Y., Faturay, F., Ting, Y. P., Geschke, A., & Malik, A. (2018). The carbon footprint of global tourism. *Nature Climate Change, 8*(6), 522-528.

Liu, L. K., Zhang, Y., Wang, Y., Xu, L., Yan, Q., & Shao, L. (2017). Morphology, composition, and mixing state of primary particles from combustion sources—crop residue, wood, and solid waste. *Scientific reports, 7*(1), 1-15.

Lopez-Lago, M., Casado, R., Bermudez, A., & Serna, J. (2017). A predictive model for risk assessment on imminent bird strikes on airport areas. *Aerospace science and technology, 62*, 19-30.

Malta International Airport . (2019). *Sustainability Report* . Retrieved from https://www.maltairport.com/corporate/corporate-responsibility/sustainability-report/. (Accessed 27 July 2020).

Metz, I. C., Ellerbroek, J., Mühlhausen, T., Kügler, D., & Hoekstra, J. M. (2020). The Bird Strike Challenge. Aerospace. *Aerospace, 7*(3), 26.

Nhamo, G., Dube, K., & Chikodzi, D. (2020). Sustainable Development Goals: Concept and Challenges of Global Development Goal Setting. In H. R, K. I, G. D, & M. M (Eds.), *Handbook of Global Health* (pp. 1-40). Springer, Cham. https://doi.org/10.1007/978-3-030-05325-3_79-1

Oslo Airport. (2020). *Oslo Airport Passenger Numbers.* Retrieved July 28, 2020, from https://www.osloairport.net/passenger-statistics.shtml

Pejovic, T., Williams, V. A., Noland, R. B., & Toumi, R. (2009). Factors affecting the frequency and severity of airport weather delays and the implications of climate change for future delays. *Transportation research record, 2139*(1), 97-106.

Penn, S. L., Boone, S. T., Harvey, B. C., Heiger-Bernays, W., Tripodis, Y., Arunachalam, S., & Levy, J. I. (2017). Modeling variability in air pollution-related health damages from individual airport emissions . *Environmental research, 156*, 791-800.

Poll, D. I. (2017). 21st-Century civil aviation: Is it on course or is it over-confident and complacent?-thoughts on the conundrum of aviation and the environment. *The Aeronautical Journal, 121*(1236), 115.

Poo, M. C., Yang, Z., Dimitriu, D., & Qu, Z. (2018). Review on seaport and airport adaptation to climate change: a case on sea level rise and flooding. *Marine Technology Society Journal , 52*(2), 23-33.

Postorino, M. N., & Mantecchini, L. (2014). A transport carbon footprint methodology to assess airport carbon emissions. *Journal of Air Transport Management, 37*, 76-86.

Raimi, M., & Adindu, I. B. (2019). Impact of Airport Noise on the Health Situation of Host Communities: A Case Study of Obong Victor Attah International Airport, Akwa Ibom State, Nigeria. *SM J Public Health Epidemiol, 5*(1), 1052.

Ritchie, B. W., Sie, L., Gössling, S., & Dwyer, L. (2020). Effects of climate change policies on aviation carbon offsetting: a three-year panel study. *Journal of Sustainable Tourism, 28*(2), 337-360.

Sarbassov, Y., Venetis, C., Aiymbetov, B., Abylkhani, B., Yagofarova, A., Tokmurzin, D., & Inglezakis, V. J. (2020). Municipal solid waste management and greenhouse gas emissions at international airports: A case study of Astana International Airport. *Journal of Air Transport Management, 85*, 101789.

Statista. (2020). *Number of terminal passengers at Trondheim Airport in Norway 2008-2018.* Retrieved from: https://www.statista.com/statistics/803393/number-of-terminal-passengers-at-trondheim-airport-in-norway/. (Accessed 28 January 2020).

Steinberg, J. (2005). Drug smuggling and border control at Johannesburg International Airport and Durban Harbour . *Institute for Security Studies Papers, 104*(2005), 16.

Swedavia Airports. (2019). *Annual and Sustainability Report* . Retrieved from: https://www.swedavia.com/about-swedavia/financial-information/. (Accessed 28 July 2020).

United Nations. (2015). *Agenda 2030 on Sustainable Development* . Retrieved from: https://sustainabledevelopment.un.org/content/documents/21252030%20Agenda%20for%20Sustainable%20Development%20web.pdf. (Accessed 11 July 2020).

United Nations Climate Change . (2015). *The Paris Agreement.* Retrieved from: https://unfccc.int/process-and-meetings/the-paris-agreement/the-paris-agreement. (Accessed 11 July 2020).

Zhang, Y. X., Chao, Q. C., Zheng, Q. H., & Huang, L. (2017). The withdrawal of the US from the Paris Agreement and its impact on global climate change governance. *Advances in Climate Change Research, 8*(4), 213-219.

Zhao, B., Wang, N., Fu, Q., Yan, H. K., & Wu, N. (2019). Searching a site for a civil airport based on bird ecological conservation: An expert-based selection (Dalian, China). *Global Ecology and Conservation, 20*, e00729.

Zhou, L., & Chen, Z. (2020). Measuring the performance of airport resilience to severe weather events. *Transportation Research Part D: Transport and Environment, 83*, 102362.

Part IV

Health, Water and Biodiversity Engagements

17 Protected Areas Interventions and SDGs: The Case of Bolsa Floresta Programme in the Brazilian Amazon

Anne-Elisabeth Laques, Ana I. R. Cabral, Romero Gomes Pereira da Silva, and Carlos Hiroo Saito

Abstract

The Brazilian Amazon forest area has many isolated communities practising small-scale agriculture in a primary forest matrix. Grant programmes, such as the Bolsa Floresta Programme (BFP), aimed to protect the forest and reduce social injustice by offering financial support for family agricultural operations as a mitigation measure for deforestation. Once there seem to be convergences between the BFP and the Sustainable Development Goals (SDGs); the objective of this work is to assess the synergies and conflicts between SDGs and the BFP. The study relies on primary data from fieldwork that was conducted between 2017 and 2018. The fieldwork for this case study research was conducted at Uatumã Sustainable Development Reserve. The study sought to better understand the efficiency of BFP and its contribution to improving local communities' livelihoods. In addition, secondary data and remote sensed data were used to better understand the impact of swiddens (land cleared for cultivation by slashing and burning vegetation) on the Amazon forest. The study demonstrated how the BFP is connected to the most SDGs and that some of the BFP initiatives presented conflict with some SDGs at its target level. A concept map was produced to synthesize the web of interdependences. The results showed the complex interdependences between the BFP and the SDGs and the existing gaps and trade-offs at ground level.

Keywords

Primary forest · Slash-and-burn agriculture · Sustainability · SDG interactions · Concept map · Uatumã

A.-E. Laques
IRD - Institut de Recherche pour le Développement, France, Montpellier, France
e-mail: anne-elisabeth.laques@ird.fr

A. I. R. Cabral
Forest Research Centre, School of Agriculture, University of Lisbon, Tapada da Ajuda, Lisbon, Portugal, Lisbon, Portugal

R. G. P. da Silva · C. H. Saito (✉)
Center for Sustainable Development, University of Brasilia, Asa Norte, Brasília-DF, Brazil
e-mail: saito@unb.br

1 Introduction

Brazil made some progress in meeting some of the goals established under the Millennium Development Goals (MDGs), which ended in 2015. The target of reducing extreme poverty rates by half since 1990 was globally met 5 years ahead of the deadline and Brazil could present in

G. Nhamo et al. (eds.), *Sustainable Development Goals for Society Vol. 2*, Sustainable Development Goals Series, https://doi.org/10.1007/978-3-030-70952-5_17

2012 a stronger reduction corresponding to one-seventh of that existing in 1990. Regarding the fight against hunger, which was also included in MDG 1, the prevalence of child malnutrition (in children under 5 years old) in Brazil was reduced from 7.1% in 1989 to 1.8% in 2006, corresponding to a quarter of the initial value (Ipea 2014). According to Barros et al. (2010) and Victora (2011), there were signs of improvements with regard to maternal health care. While such signs of progress have been noted, other scholars such as Roma (2019) called for attention to the existence of regional inequalities around these indices by analysing the indicators of underweight children in reference to groups of healthy and well-nourished children.

Additionally, the Sustainable Development Goals (SDGs) came after the MDGs that aimed at maintaining the same social concerns but encompassing a range of themes as well as broader objectives and goals seeking engagement and commitment from the developed countries of the Global North (UN 2014). For countries that still had not covered much ground in making progress towards meeting the MDG goals, it can be said that the SDGs also represent the renewal of opportunities from a more integrated perspective that combines economic development with an end to social inequalities and ecological impacts.

At the dawn of SDGs, the northern parts of Brazil had not reached the goals in this regard since 3.2% of the children were underweight, which increased to 3.7% of children in the poorest 20% of the population (Ipea 2014). The North region of Brazil remains a challenge from both a social and environmental point of view: it is predominantly under the domain of the Amazon biome, with extensive humid tropical forest, large volumes of surface and underground water and areas with no populations. Some protected areas were created to safeguard forest domain, imposing to the traditional and riverine communities rules of sustainable practices, which create additional difficulties to ensure social and economic inclusion. To overcome these challenges, public policies such as Bolsa Floresta Programme (BFP) try to reconcile forest protection and social justice.

The BFP, which was created in 2007, represents a method of income transfer, and it rewards traditional communities for their commitment to stop deforestation by income transfer for ecosystem services to families, communities, and family associations (the payments were initiated in 2008). This is why the BFP is also known as the Forest Conservation Grant Fund.

The BFP and SDG present convergence and mutual reinforcement to mitigate social inequalities and promote environmental protection based on the development of sustainable practice and awareness. In this context, the present study aims to analyse the implementation of the BFP at the Uatumã Sustainable Development Reserve as a case study located in the state of Amazonas (Brazil) and to assess the possibility of linking the SDGs and the BFP by looking into its synergies and conflicts.

This chapter is presented in the following structure: after this introductory section, there is a background information section supported on literature review to present the general context of work followed by methodological orientation including a characterization of the study area. The main findings/results are presented in a separate section from the discussion. At the end, final remarks are brought.

2 Background Information

The BFP is offered to traditional communities living in a Sustainable Development Reserve (RDS, Portuguese acronym), which is a protected area category corresponding to IUCN category VI (Dudley 2008). It comprised four components: income, association, family and social (FAS 2009). The income component of the BFP is directed towards the development of sustainable production arrangements based on support for small enterprises that promote diversification in production. The association component of the BFP seeks to empower residents' associations and training for the management of collective enterprises. The family component of the BFP is responsible for the payment of the Bolsa Floresta grant, which is a direct monthly payment of R$

50.00 to the women of riverside families residing in the RDS. As a counterpart of this payment, families must commit themselves to zero deforestation in primary forests (which means not opening new areas to clearing in primary forest areas), in addition to keeping their children in school and participating in associations and training workshops on climate change and environmental services. Finally, the social component of the BFP is responsible for the strategy of the participatory definition of investments in the social area, such as education, health, communication and transportation, based on the demands expressed by the local community. In this last component, according to Viana et al. (2012), the BFP budget for improving education, health, communication and transportation was approximately US$ 35000/year/protected area.

The dilemma experienced by the community in maintaining the primary forest as part of the commitment to renouncing deforestation resulting from receiving the Bolsa Floresta grant is described in Laques et al. (2018) who noted that in as much as farmers understand the rationality of forest protection commitment and agree with the necessity to protect forest ecosystem, there is also pressing demand for farmers to open new fields (called as swiddens) in more fertile areas to increase family income from the planting of cassava and the commercialization of its flour. The opening of these new swidden areas is problematic because it is based on slash-and-burn practice which releases carbon emissions (van Vliet et al. 2013).

The greatest dissatisfaction of the community was the low remuneration of the Bolsa Floresta grant, which did not align with the increased cost of production and the low productivity of cassava fields. In general, the production cost of the swiddens in the capoeira area is higher than in newly opened areas by deforestation due to the increased effort required in weeding to control invasive plants (Jakovac et al. 2015).

The low amount paid by the BFP (R$ 50.00 monthly per family to waive the right to clearing primary forest areas) and the low productivity of the swiddens made in capoeira areas meant that this programme may have difficulties to fully combat poverty. Additionally, a productive swidden has a limited life span of up to 3 years (cycles of two to three annual plantings/harvests), post which they will be needed to open up of a new area for production (Laques et al. 2018).

These difficulties in the RDS-Uatumã are the same as those faced in the city of St George (French Guiana, another field site of the SINBIOSE project), where restricting the area dedicated to the cassava fields limits the fallow time of the *capoeiras* (Tchansia 2012). In fact, the lack of available space limits the possibility of respecting the cycle of slash-and-burn agriculture, which needs sufficient rest periods for the soils to fully recover to their initial fertility status (Jakovac et al. 2015; Lintemani et al. 2020). To that end therefore better planning of the carrying capacity and recovery of soil fertility is necessary for the continued success of the BFP and the achievement of the SDG goals.

3 Research Design

3.1 Characterization of the Study Area

The Uatumã Sustainable Development Reserve (RDS-Uatumã) was created in 2004 by State Decree No. 24,295 of Amazonas state, northern Brazil, and aims to conserve ecosystems and habitats and the associated cultural values and traditional natural resource management systems (Börner et al. 2013). The RDS-Uatumã was created in a territory with an extent of 4244.30 km^2, which is located 200 km from Manaus in the municipalities of São Sebastião do Uatumã and Itapiranga (Fig. 17.1). It was named after the river that intersects the protected area: the Uatumã River is its main water resource, and it is characterized as a dark water river, which is characteristic of rivers on the left bank of the Amazon River (Amazonas 2009). Its low soil fertilization capacity is due to flood and ebb pulses that are poor in suspended sediments (Junk et al. 2015) which are currently controlled by the Balbina Dam upstream of the protected area. On the banks of the main river and its tributaries, the

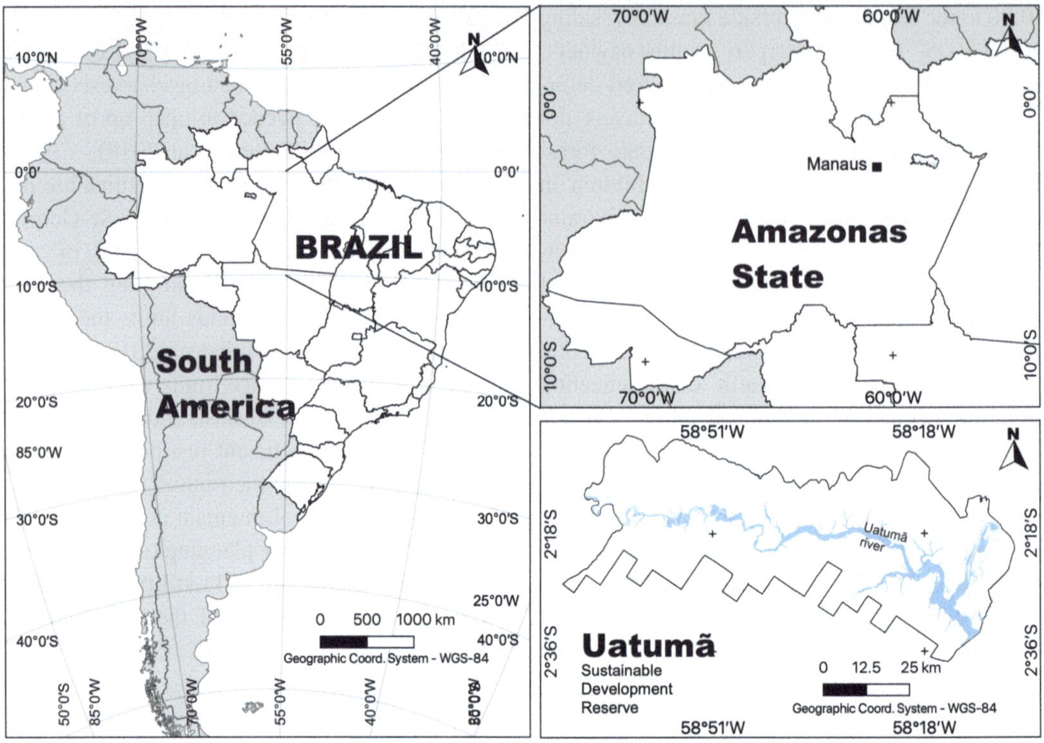

Fig. 17.1 Location of the Uatumã Sustainable Development Reserve – Brazil

RDS-Uatumã has 20 communities with fewer than 400 families that survive by cultivating cassava plantations (*Manihot esculenta*), extracting fruit and fishing on the river (Jakovak et al. 2017). Federal income transfer programmes, such as the Bolsa Família programme, together with state socioenvironmental initiatives, contribute to the composition of income in the community.

The communities are characteristic of the Amazon riverside; they are isolated from major economic circuits, living mainly according to the rhythms of nature, and are influenced by the rainy season (February to April) and the drought period (July to October), which influence river levels and its tributaries (Laques et al. 2018).

3.2 Methodological Orientation

Within a French-Brazilian cooperative effort, two field expeditions (2017–2018) were executed at Uatumã Sustainable Development Reserve by the SINBIOSE research project "Système d'INdicateurs de BIOdiversité à l'uSage des actEurs: Biodiversité terrestre et aquatique (Amazone & Oyapock)".

Fifteen local families were interviewed and questioned about the subsistence methods adopted and their compliance with the BFP, where their swiddens are located and how long they lasted. Each fieldwork had a duration of 1 week, implied to navigate along the Uatumã River and visited eight communities: (1) Nossa Senhora do Perpétuo Socorro Community (Maracaranã), (2) Manaim Community, (3) Santa Luzia do Caracarana Community, (4) Ebenezer Community, (5) São Francisco do Caribi Community, (6) Nossa Senhora do Livramento, (7) Santa Luzia do Jacarequara and (8) Nova Jerusalém do Amaro Community.

The fieldwork results were complemented with remote sensing data to assess the presence of swiddens inside the forest and to understand whether the financial compensation offered by the BFP can guarantee an acceptable level of well-being for households. To understand swidden dynamics, mapping using different Landsat imagery was used. The Thematic Mapper (TM) and Operational Land Imager (OLI) (Orbits 23/ points 61 and 62, with respective dates: June 5, 2011; August 8, 2015; and August 14, 2019) and the methodology based on the Normalized Difference Moisture Index (NDMI) were used after recommendations from Vogelmann and Rock (1988) and also Silva et al. (2019). The next section presents the results.

4 Results

The study found that the BFP project responds to several SDGs and targets. It emerged that the BFP projects were directly aligned with SDG-15 (Life on Land). The BFP seeks to fulfil SDG target 15.1 by proposing zero deforestation in primary forests and opening of swiddens only in *capoeira* areas. It also meets SDG target 15.2 of promoting the implementation of sustainable forest management, and in doing so, it meets SDG target 6.6. In addition to halting deforestation, the BFP specifically requires compliance with the RDS-Uatumã management plan, which, among other rules, establishes a logging ban for several protected tree species in the Amazon biome, thereby complying with SDG target 15.5.

It was also found that the projects have other direct links to other SDGs, namely: SDG-1 (No Poverty), which considers the financial remuneration of families engaged in the programme; SDG-2 (Zero Hunger), which considers that the programme is concerned with maintaining agricultural production activity within the protected areas and allows new swiddens to be opened in areas of old *capoeiras*; SDG-4 (Quality Education), which considers that the programme requires participating families to keep the school-age child attending school; SDG-5 (Gender Equality), because the BFP's remuneration is paid to women; and SDG-12 (Responsible Consumption and Production), which considers carrying out proper fire management and complying with the RDS-Uatumã management plan, which states that each family is allowed to use up to 3 hectares per year for swidden areas and requires approval by the deliberative council and SEMA's consent for opening swiddens above 3 hectares, which must be located in the intensive use zone according to the management plan of the protected area. The extensive use zone and the preservation zone have more limitations and do not allow agricultural activities, even on a small scale (Amazonas 2009; Viana et al. 2012).

Table 17.1 presents both those just positive interactions, indicating the BFP's compliance with the SDG targets, and also those complex and conflicted situations, in which the BFP acts in favour of some SDG targets and at the same time acts against other SDG targets.

Table 17.1 shows the interaction between the BFP and various SDGs as they relate to the Uatumã Sustainable Development Reserve and Amazon state, Brazil. Green arrows facing upwards indicate compliance with the SDG target. Green arrows facing upwards accompanied by red arrow facing downwards indicates a dual impact. The dashed horizontal arrow indicates a possible and desired connection with the SDG target that currently does not exist.

Figure 17.2 provides a summary of the SDGs and their relationships with the BFP, according to Table 17.1. It is a concept map showing the richness of the BFP's interactions with the SDG and illustrates how the 2030 Agenda can be implemented at the local scale. The concept map was build following Heemskerk et al. (2003), Saito (2017) and Novak & Cañas (2007). Table 17.1 and Fig. 17.2 are complementary: the first one presents the BFP-SDG relationship focused on each SDG target, and the later data shows an integrated view of all relationships and how BFP is immersed on the set of SDG targets and their interdependencies. The set of BFP-SDG positive or contradictory relationships are discussed in the next section.

About the main rule of BFP, the zero deforestation in primary forest (green box, Fig. 17.2), the results provided by remote sensing data showed that a sharp reduction in deforestation rates was observed after the implantation of the BFP, from 437 ha in 2011 to 190 ha in 2019 (Fig. 17.3). After intersecting the polygons of the swiddens year on year, we also found out that 60% of the

Table 17.1 BFP impacts on SDGs

Goal	Target	Brief target description (UN 2015)	Effect
1	1.1	Eradicate extreme poverty for all people everywhere, currently measured as people living on less than $1.25 a day	↑↓
1	1.2	Reduce at least by half the proportion of men, women and children of all ages living in poverty in all its dimensions according to national definitions	↑↓
2	2.1	End hunger and ensure access by all people, in particular the poor and people in vulnerable situations, including infants, to safe, nutritious and sufficient food all year round	↑↓
2	2.2	End all forms of malnutrition, including achieving the internationally agreed targets on stunting and wasting in children under 5 years of age, and address the nutritional needs of adolescent girls, pregnant and lactating women and older persons	↑↓
2	2.3	Double the agricultural productivity and incomes of small-scale food producers, in particular women, indigenous peoples, family farmers, pastoralists and fishers, including through secure and equal access to land, other productive resources and inputs, knowledge, financial services, markets and opportunities for value	↑↓
2	2.4	Ensure sustainable food production systems and implement resilient agricultural practices that increase productivity and production; that help maintain ecosystems; that strengthen capacity for adaptation to climate change, extreme weather, drought, flooding and other disasters; and that progressively improve land and soil quality	↑↓
3	3.3	End the epidemics of AIDS, tuberculosis, malaria and neglected tropical diseases and combat hepatitis, waterborne diseases and other communicable diseases	↑
3	3.d	Strengthen the capacity of all countries, in particular developing countries, for early warning, risk reduction and management of national and global health risks	↑
4	4.1	Ensure that all girls and boys complete free, equitable and quality primary and secondary education leading to relevant and effective learning outcomes	↑
4	4.2	Ensure that all girls and boys have access to quality early childhood development, care and pre-primary education so that they are ready for primary education	↑
4	4.7	Ensure that all learners acquire the knowledge and skills needed to promote sustainable development, including, among others, through education for sustainable development and sustainable lifestyles, human rights, gender equality, promotion of a culture of peace and non-violence, global citizenship and appreciation of cultural diversity and of culture's contribution to sustainable development	↑
5	5.5	Ensure women's full and effective participation and equal opportunities for leadership at all levels of decision-making in political, economic and public life	↑
5	5.a	Undertake reforms to give women equal rights to economic resources, as well as access to ownership and control over land and other forms of property, financial services, inheritance and natural resources, in accordance with national laws	↑
6	6.6	Protect and restore water-related ecosystems, including mountains, forests, wetlands, rivers, aquifers and lakes	↑
7	7.1	Ensure universal access to affordable, reliable and modern energy services	--▶
8	8.2	Achieve higher levels of economic productivity through diversification, technological upgrading and innovation, including through a focus on high value-added and labour-intensive sectors	↑↓
8	8.3	Promote development-oriented policies that support productive activities, decent job creation, entrepreneurship, creativity and innovation, and encourage the formalization and growth of micro-, small- and medium-sized enterprises, including through access to financial services	↑
8	8.4	Improve progressively global resource efficiency in consumption and production and endeavour to decouple economic growth from environmental degradation, with developed countries taking the lead	↑

(continued)

Table 17.1 (continued)

Goal	Target	Brief target description (UN 2015)	Effect
8	8.9	Devise and implement policies to promote sustainable tourism that creates jobs and promotes local culture and products	↑
9	9.1	Develop quality, reliable, sustainable and resilient infrastructure, including regional and transborder infrastructure, to support economic development and human well-being, with a focus on affordable and equitable access for all	↑
10	10.2	Empower and promote the social, economic and political inclusion of all, irrespective of age, sex, disability, race, ethnicity, origin, religion or economic or other status	↑
10	10.3	Ensure equal opportunity and reduce inequalities of outcome, including by eliminating discriminatory laws, policies and practices and promoting appropriate legislation, policies and action in this regard	↑
11	11.4	Strengthen efforts to protect and safeguard the world's cultural and natural heritage	↑
12	12.2	Achieve the sustainable management and efficient use of natural resources	↑
12	12.8	Ensure that people everywhere have the relevant information and awareness for sustainable development and lifestyles in harmony with nature	↑
13	13.2	Integrate climate change measures into national policies, strategies and planning	↑
13	13.3	Improve education, awareness raising and human and institutional capacity on climate change mitigation, adaptation, impact reduction and early warning	↑
15	15.1	Ensure the conservation, restoration and sustainable use of terrestrial and inland freshwater ecosystems and their services, in particular forests, wetlands, mountains and drylands, in line with obligations under international agreements	↑
15	15.2	Promote the implementation of sustainable management of all types of forests, halt deforestation, restore degraded forests and substantially increase afforestation and reforestation globally	↑↓
15	15.5	Take urgent and significant action to reduce the degradation of natural habitats, halt the loss of biodiversity and protect and prevent the extinction of threatened species	↑↓
15	15.7	Take urgent action to end poaching and trafficking of protected species of flora and fauna and address both demand and supply of illegal wildlife products	↑↓
15	15.8	Introduce measures to prevent the introduction and significantly reduce the impact of invasive alien species on land and water ecosystems and control or eradicate the priority species	↑
15	15.9	Integrate ecosystem and biodiversity values into national and local planning, development processes, poverty reduction strategies and accounts	↑↓
17	17.1	Strengthen domestic resource mobilization, including through international support to developing countries, to improve domestic capacity for tax and other revenue collection	↑

Source: Authors

mapped fields were located in old fields after the implantation of the BFP.

Figure 17.3 shows that there is a decline in the opening of primary forest over the period of the study by more than half the size between 2011 and 2019.

5 Discussion

The overall results did reinforce the literature review and the background information to raise awareness about the importance of deeper investigation about the synergies and conflicts between the SDGs and

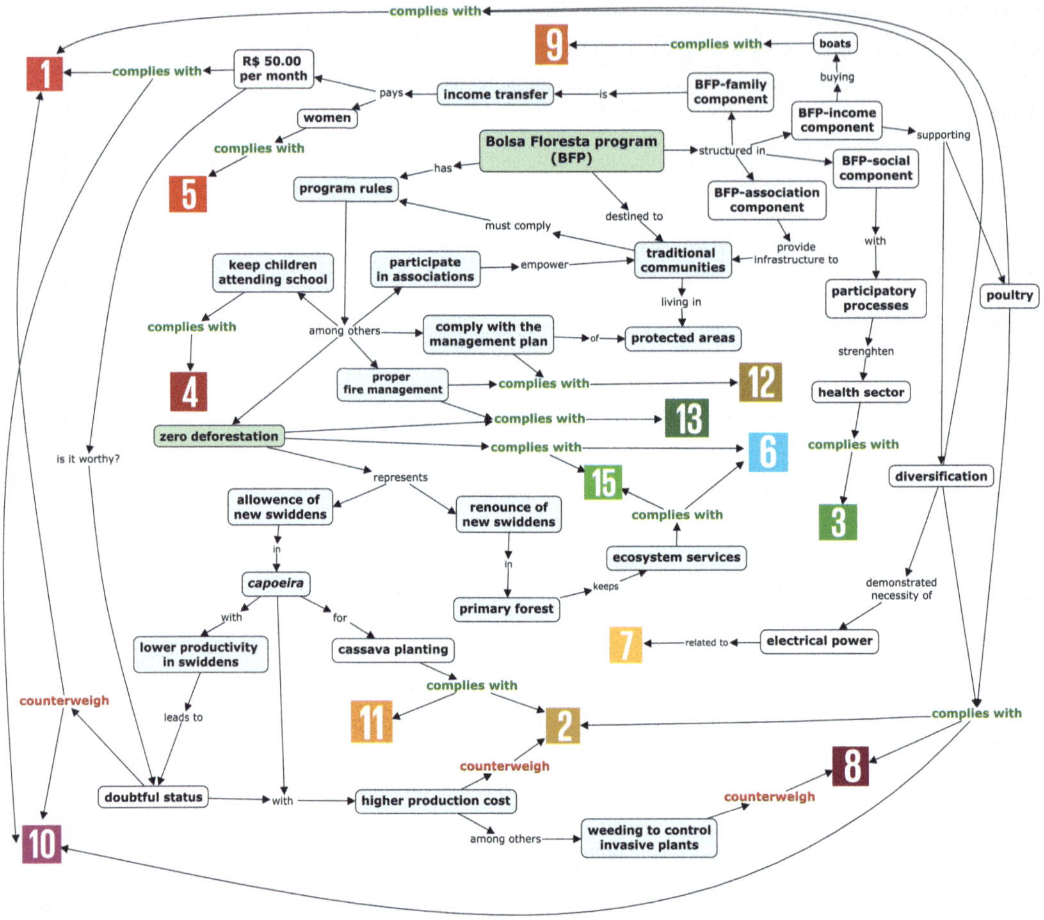

Fig. 17.2 Concept map summarizing the relationship between the BFP and the SDGs at the Uatumã Sustainable Development Reserve, Amazon state, Brazil. (Source: Authors)

the BFP. This horizontal ability of the BFP to connect with multiple SDGs must be analysed with caution to avoid being prideful and making assumptions. If the BFP has several interfaces converging with the SDGs, it also has contradictory interactions, some of which act as a counterweight against the objectives and goals of the same SDG with which they signalled compliance.

5.1 The BFP and the SDGs: Conflicting Goals or Lack of Synergy

One of the contradictory BFP-SDG relationships due to conflicting goals can be exemplified by the fact that the BFP stipulates that deforestation for the opening of swiddens occurs only in *capoeira* areas (secondary vegetation), and this leads to low productivity of the manioc swiddens, reducing income and therefore reducing families' food. This presents challenges for families and can potentially threaten the long-term project sustainability if a balance is not found.

Remote sensing mapping indicated an important reduction of primary forest deforestation (437.51 ha in 2011 to 190.95 ha in 2019). Such a decline is significant in as much as it ensures the protection of biodiversity and most importantly it also assist in the climate change battle. The Amazon forest is considered as one of the most important global carbon sinks so a reduction in

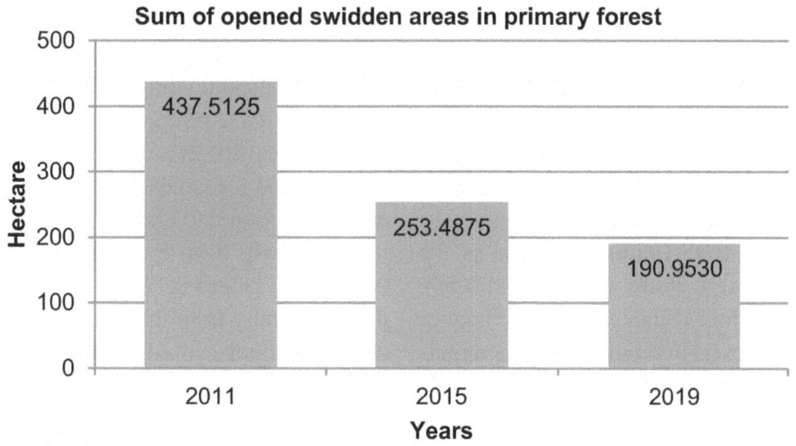

Fig. 17.3 Swidden areas opened in primary forest. (Source: Authors)

deforestation is a positive development in the protection of biodiversity and the climate change battle. On the other hand, it is critical to note that such a development comes at the expense of farmers who are unable to enjoy maximum profits due to a reduction of profits as they cannot open new forests. It is therefore important for BFP managers to have strategies so that reaching the SDG-15 objective does not increase poverty and hunger.

Finally, the interviewees' indication of the programme's low remuneration (R$ 50.00), added to the low price of cassava flour, warns of the vulnerability of families in the RDS. The drop in the income of BFP beneficiaries needs to be analysed in the search for solutions since it negatively impacts the secular cultural habits of planting and using cassava, which include the food base of traditional peoples.

Thus, SDG-1 conflicts with the actions of the BFP. Considering the BFP as a cash transfer programme for poor and vulnerable families, it can be said that it constitutes a step towards the fulfilment of SDG targets 1.1 and 1.2. However, because families complain that it is a very low amount (R$ 50.00 per family per month) for renouncing the opening of fields in the more fertile primary forest that would therefore produce more economic gains, both SDG targets could be compromised in the medium and long term when successive swiddens in *capoeira* areas show signs of falling productivity and profitability. Our findings showed that 60% of the mapped fields were located in old fields after the implantation of the BFP reinforce the tendency of an intensified use and reduced productivity.

The decreased deforestation trends in Fig. 17.3 were confirmed by the families interviewed during the fieldwork. The interviews indicated that, in addition to the zero-deforestation rule, another factor contributes to the reported decrease: the drop in the manioc flour price, indicating the importance of the benefit paid by the BFP as a supplement in the income to farmers. The interviewees also stated that zero deforestation is a difficult goal to be achieved, not only individually but also collectively, due to the constitution of new families (descendants from the already established families) associated with the lack of fertile areas to open swiddens. So, the results indicated the positive outcomes of the efforts of the families to fulfil the compliance rules of the BFP and thus achieve those SDGs goals and targets which are connected to it in Fig. 17.2.

Nevertheless, considering that the cost of production in the *capoeira* areas was reported by the interviewees as being higher than that in the primary forest because it requires greater effort to control invasive plants, the BFP, based on this aspect of the development of cassava crops, could negatively affect SDG target 8.2 which states the necessity of achieving higher levels of economic productivity through diversification, technological upgrading and innovation, including through a focus on high value-added and labour-intensive sectors. Furthermore, because of the additional efforts of weeding to control invasive plants in the *capoeira* areas, the production cost of the swiddens in these *capoeira* areas becomes higher; the BFP also ends up acting against SDG-8 by worsening the working conditions.

Regarding the RDS-Uatumã, cassava is the main subsistence agriculture item and is usually transformed into flour (Silva et al. 2017; Laques et al. 2018). However, in relation to SDG targets 2.1 and 2.2, it must be questioned whether simply the maintenance of the cassava plantation and its consumption guarantees the fight against malnutrition, since the challenge of ending hunger through the search for broadly considered food security includes nutritional security. Food security corresponds to a situation in which all people at all times have physical and economic access to sufficient, safe and nutritious food to meet their dietary needs and the food is preferable for active and healthy life (FAO 1996). For Quisumbing et al. (1995) the adequate availability of food in households is fundamental but insufficient to acquire nutritional security, which would be the nutritional status of having adequate protein, energy, micronutrients and minerals for all household members. Nutritional security, which is related to SDG target 2.2, thus would be one of the pillars of food security. The fact that these families live in a rural area is not a guarantee to ensure that the food resources necessary for nutritional security are available in the environment, which makes it possible for children and adolescents to become malnourished, as has been reported in relation to the Brazilian Cerrado (Graebner et al. 2007). Additionally, when eating habits are changed (e.g. introduction of rice and sugar) and families have access to a new income (Bolsa Família grants, retirement, etc.), agricultural practices change because it is easier to buy food than to produce it locally (Doyen et al. 2017; Laques et al. 2013). This situation also causes changes in food security from a qualitative point of view.

In line with the arguments that food insecurity must be combated by dietary changes rather than the enhanced ability of the household to obtain more food (Pinstrup-Andersen 2009), the income component of the BFP sought production and food diversification through agroforestry systems (AFS), which converges with SDG target 8.4, which states that we need to improve progressively global resource efficiency in consumption and production and endeavour to decouple economic growth from environmental degradation. AFS would indeed decouple economic growth in the community from the increase in environmental degradation caused by deforestation in primary forest. However, according to local respondents, the AFS initiative was inhibited by the quality of the land (the so-called fertile black soil was available in only a few places) and the availability of water for plants, especially in the summer (Laques et al. 2018). If these conditions are not resolved, the zero hunger goal will suffer a setback in SDG target 2.4.

In the specific case of water supply for plants (fruit trees, as indicated by the local residents), the main obstacle would be the absence of infrastructure to supply electricity capable of running the pumps to draw water from the river. This problem has not yet been addressed by the BFP, and although the RDS-Uatumã was impacted by the upstream construction of the Balbina hydroelectric plant, its communities do not have electricity, and the school operates at night by small generators powered by fuel engines. Bringing electrical power to rural and remote areas, as was done in large part of the Brazilian territory by the *Luz para Todos* programme, could also strengthen SDG-2 and SDG-3 in the RDS-Uatumã. Access to electricity would ensure methods for preserving food in refrigerators and freezers, increasing food and nutritional security and having direct and indirect benefits for family health (Bezerra et al. 2017). Thus, SDG-7, which has not been valued until now, should be included in the list of objectives to be developed within the RDS-Uatumã because it represents one of the challenges in all protected areas served by the BFP, in the view of local communities (Viana et al. 2012).

When AFS were introduced in areas during the first cycle of greater fertility, they can cause a discontinuity in cassava planting, which can consequently result in the opening of new areas to make other swiddens through forest clearing (Laques et al. 2018), which is contradictory to SDG targets 15.2 and 15.5. Likewise, SDG target 2.3 can be harmed by reducing the productivity of the swiddens that have been opened successively and only in *capoeira* areas. In other words, SDG-2 will find itself in conflicting situation, with actions converging towards and moving away from compliance.

On tourism in the reserve, although it complements the income of some families, this activity increases the pressure on natural resources, such as fish, when sport fishing develops without control and can negatively pressure SDG targets 15.7. They also introduce social inequalities when certain families receive an additional income through the accommodation of tourists, for example, that is much higher than the sale of flour, potentially conflicting with SDG target 10.3, which would otherwise be benefited.

5.2 The BFP and the SDG: Common Goals

The fact that the BFP was created in 2007 and the RDS-Uatumã management plan was created in 2009, both of which were prior to the creation of the Sustainable Development Goals, shows that the country was already seeking a greater commitment to ecological and sustainability ideas.

The targets most correlated with the context of the BFP at the RDS-Uatumã are SDG targets 12.2 and 12.8. The families' commitment to zero deforestation, proper fire management and compliance with the RDS-Uatumã management plan represent both steps towards achieving sustainable management and the efficient use of natural resources, and raising awareness about the plan itself corresponds to an appreciation of lifestyles in harmony with nature.

The BFP remunerates families through payments to the women of the families, which clearly indicates a link with SDG targets 5.5 and 5.a. Gender privilege issues in public policies have been a recent practice in Brazil, where this strategy also appears strongly in the 1 million cistern programme and in the Bolsa Família programme (Moraes and Rocha 2013; Bartholo 2016). Despite being a step forward as a policy for the protection and strengthening of women in society, it is necessary to address the issue of gender within a larger framework of seeking justice and social inclusion. The fieldwork detected situations in which the wife in a couple had died and the widower, without having sought a new marriage, was excluded from receiving the BFP. This anomaly could also be resolved if there was continuous monitoring by the institution managing the BFP of the families benefited to detect these situations and make the necessary corrections in the policy with the aim of maximizing social well-being. However, the same fieldwork detected a shortage of personnel to inspect and monitor the implementation of the BFP.

The BFP is also linked to SDG-4 when it requires family members to keep their children in school to be entitled to receive the benefit. In this way, the programme is able to act in favour of SDG targets 4.1 and 4.2. Additionally, an important initiative within the scope of the BFP is Conservation and Sustainability Centres (NCS, Portuguese acronym), which are education centres in the forests that seek to educate and generate knowledge for traditional peoples and communities. The NCS was already implemented in six Amazonian protected areas (Juma, Uatumã, Rio Negro, Mamirauá, Poranga Conquista and Cujubim RDS), strengthening SDG target 4.7.

Viana (2008) and Rival (2012) indicated that in the social component of the BFP, the planned health actions included training community agents and nurses and participatory planning of priority health initiatives with the presence of doctors, researchers, nurses and community health workers. The resultant plan included strategies for monitoring and evaluating results. We can say that these initiatives converge with SDG targets 3.3 and 3.d. The BFP's adherence to SDG target 3.3 is even more important if we consider that malaria, diarrhoea, flu and worms are responsible for 95% of the diseases found in all the protected areas covered by the BFP, even though these diseases vary among the localities (Viana 2008).

The income component of the BFP was also an important vector for strengthening the infrastructure for marketing community products. According to Viana et al. (2012), for BFP beneficiaries in the RDS-Uatumã, the most important investment made to improve production in the community was a boat (18.1% of the indications). This action converges with the goals of building resilient infrastructure, especially as stated in SDG target 9.1.

In the context of production diversification, as part of the income component of the BFP, the Amazonas Sustainable Foundation (FAS) implemented six poultry breeding projects and one sheep breeding project in the RDS-Uatumã (FAS, 2009, 2015). The projects benefited 61 families, which would have improved food diversification and gain a complement for school meals and a new income alternative. The communities of Manaim, Flexal, Emanuel, Ebenezer, Nossa Senhora de Fátima do Caioé Grande, and São Francisco do Caribí were considered by this project, and each received 400 free-range broiler chicken birds, two tons of feed, equipment and 250 m^2 aviaries in November 2011. Thus, it can be argued that although the renunciation of deforestation in primary forest apparently indicates an economic loss, the BFP as a whole would be working in favour of both SDG-1 and SDG-2 in addition to SDG targets 8.3 and 10.2. In the RDS Uatumã, from 2011 to 2015, there was an increase from 5.3% to 13.6% in the participation of activities related to the sustainable use of forest in the composition of the families' income (Brito et al. 2019).

Additionally, it can be said that the association component of the BFP allows for strengthening of the local organization, bringing not only collaborative and participatory processes but also critical thinking and transformative praxis. The associative strengthening led the community of Maracaranã to build a communitarian hostel in opposition to the model of individual tourist exploitation in another community of the same RDS-Uatumã. The process of political maturation also led this same community to make blockades and hold vigils on streams and lakes and to approach to boats that reached the demarcated areas for the protection of the turtles, especially during the breeding period (Laques et al. 2018). These initiatives make it possible to connect the BFP with both SDG targets 15.7 and 8.9.

The BFP meets the SDG target 15.8 because the RDS-Uatumã management plan guides the marketing of bee honey only by residents through projects authorized by the responsible environmental agency and through the promotion of native bees (Amazonas 2009). This means that they are opposing *Apis* sp., an exotic invasive species, by encouraging meliponiculture based on native bees. In the same way, the BFP is articulated with the SDG target 15.9, which considers that the management plan of the RDS-Uatumã establishes rules that seek to align economic production with the sustainability of the forest: for the harvesting of indigenous fruits (among others, *Euterpe oleracea*, which is locally known as *açai*), it is only done by climbing trees; the extraction of copaiba oil must obey good management practices, including following a prohibition of new extraction for 3 years (Amazonas 2009).

It should also be noted that the BFP can be recognized as a payment programme for environmental services (Pereira 2010; Fearnside 2012) with voluntary adherence. In this context, the BFP's zero deforestation in primary forests, in fact, represents the deforestation avoided by payment for environmental services, and thus, the BFP also shows convergence with the climate action goal, mainly in regard to SDG targets 13.2 and 13.3.

The BFP can represent an effort to reverse the trend of population decline in rural areas since there are an increasing number of people moving to urban areas in search of better access to schools and jobs (Schmink and Gómez-García 2015). This can even strengthen the original idea behind the creation of the RDS-Uatumã, which sought to preserve nature and ensure the necessary conditions for the reproduction and improvement of the ways of life of local populations and sustainable exploitation of natural resources (Laques et al. 2013).

It is worth mentioning that the actions within the BFP also work towards SDG target 11.4 to protect culture. In addition to this aspect, the production of cassava itself has strong cultural components: the collective process of peeling cassava includes the formation of a circle of children, the elderly and friends, who sing together; they eat freshly caught fish and exchange experiences while working. The production of cassava flour in flour houses also presents strong community collaboration.

According to the evaluation report of the effectiveness of the BFP, the family income of those participating in the BFP was significantly higher than the families that did not join the programme in the RDS-Uatumã: the mean difference between the two groups increased from R$ 55 in 2011 to R$ 182 in 2015 (Brito et al. 2019). This result also converges with SDG target 10.2.

The FAS, executor of the BFP, points out the link between this programme and the Millennium Development Goals (MDGs), specifically referencing MDG 1 (ending hunger and poverty), MDG 3 (gender equality), MDG 7 (quality of life and respect for the environment) and MDG 8 (everybody working for development). According to Brito et al. (2019), after the establishment of Agenda 2030 and the Sustainable Development Goals (SDGs), the BFP would have automatically embraced the SDGs, especially SDG-2, SDG-5, SDG-8, SDG-10, SDG-15 and SDG-17. To these, we add here SDG-1, SDG-3, SDG-4, SDG-6, SDG-9, SDG-11, SDG-12 and SDG-13, with the indication of the need to satisfy the community in relation to SDG-7. Although it is not presented in Fig. 17.2, actions such as the BFP can be framed within the strategies of SDG-17, especially in SDG target 17.1, hence the suggestion of Brito et al. (2019).

The FAS thus won second place in the "SDG Brazil Award" in December 2018 in the category of nonprofit organizations.

6 Conclusions

The BFP, which was implemented in several protected areas in the state of Amazonas before the establishment of the 2030 Agenda, largely complements the SDGs and acts in favour of its implementation at the local scale. This shows that policymakers had already been conscious about sustainable development issues in the country since the re-democratization process, which has been facilitating adherence to the objectives of Agenda 2030 and its implementation.

The analysis of the case study RDS-Uatumã allows concluding that the SDGs are important references to guide actions within the scope of the BFP to seek greater robustness and effectiveness with regard to the general objective of promoting sustainable development in the protected areas. We could demonstrate how the BFP is connected to all the SDGs in general, save for SDG16, with the indication on the need to satisfy the community in relation to SDG-7. At target level, some of the BFP initiatives present conflicts with these SDGs: examples could be found in relation to SDG target 1.1, SDG target 1.2, SDG target 2.1, SDG target 2.2, SDG target 2.3, SDG target 2.4, SDG target 8.2, SDG target 15.2, SDG target 15.5, SDG target 15.7 and SDG target 15.9.

It was also verified that a more accurate study of the interactions between the initiatives within the scope of the BFP and the goals of the different SDGs would reveal the trade-offs and the initiatives that need to be developed to give greater completeness to the actions. Despite the still-existing limitations of the BFP and the several required improvements, it seems to contribute to the effective implementation of the SDGs. Therefore, an expansion of this programme to other states could bring positive benefits to local populations and a higher commitment to SDGs at the local level.

Acknowledgements The authors are grateful to INCT-ODISSEIA Project, "Observatory of Socio-environmental Dynamics" at Brasília University (UnB) for their technical support. This work was also supported by the research project SINBIOSE "Système d'INdicateurs de BIOdiversité à l'uSage des actEurs: Biodiversité terrestre et aquatique (Amazone & Oyapock)", under the International Cooperation Programme GuyAmazon-Edital n.022/2014 (IRD/UFAM). The authors are also grateful to the Coordination for the Improvement of Higher Education Personnel (Capes), National Council for Scientific and Technological Development (CNPq) and Research Support Foundation of the Federal District (FAP-DF) for their financial support. Additional thanks to CEF, a research unit funded by Fundação para a Ciência e a Tecnologia I.P. (FCT), Portugal (grant UIDB/00239/2020).

References

Amazonas (2009). Plano de Gestão da Reserva de Desenvolvimento Sustentável do Uatumã, volumes 1, 2. Manaus: Secretaria de Estado do Meio Ambiente e Desenvolvimento Sustentável do Amazonas - SDS. https://documentacao.socioambiental.org/ato_normativo/UC/2107_20160311_173117.pdf Accessed 17 September 2020.

Bartholo, L. (2016). Bolsa Família and gender relations: national survey results. *The International Policy Centre for Inclusive Growth, Research Brief 55*. https://www.ipc-undp.org/pub/eng/PRB55EN_Bolsa_Familia_gender_relation.pdf. Accessed 03 August 2020.

Barros, F.C., Matijasevich, A., Requejo, J.H., Giugliani, E., Maranhão, A.G., Monteiro, C.A., et al. (2010). Recent trends in maternal, newborn, and child health in Brazil: progress toward Millennium Development Goals 4 and 5. *American Journal of Public Health, 100*(10), 1877-1889. https://doi.org/10.2105/AJPH.2010.196816

Bezerra, P.B.S., Callegari, C.L., Ribas, A., Lucena, A.F.P., Portugal-Pereira, J., Koberle, A., et al. (2017). The power of light: socio-economic and environmental implications of a rural electrification program in Brazil. *Environmental Research Letters, 12*(9), 095004. https://iopscience.iop.org/article/10.1088/1748-9326/aa7bdd/pdf

Börner, J., Wunder, S., Reimer, F., Bakkegaard, R.K., Viana, V., Tezza, J., Pinto, T., et al. (2013). Promoting Forest Stewardship in the Bolsa Floresta Programme: Local Livelihood Strategies and Preliminary Impacts. Rio de Janeiro, Brazil: Center for International Forestry Research (CIFOR). Manaus, Brazil: Fundação Amazonas Sustentável(FAS). Bonn, Germany: Zentrum für Entwicklungsforschung (ZEF), University of Bonn.

Brito, A., Ferreira, G.L., Budi, J., Rodekirchen, M., Sá, P. (2019). *Projeto Bolsa Floresta: Relatório Final de Avaliação de efetividade*. Rio de Janeiro: Banco Nacional de Desenvolvimento Econômico e Social (BNDES)/Deutsche Gesellschaft für Internationale Zusammenarbeit (GIZ) GmbH. http://www.fundoamazonia.gov.br/export/sites/default/pt/.galleries/documentos/monitoramento-avaliacao/avaliacoes-externas/FAS-Bolsa-Floresta-Relatorio.pdf Accessed 17 September 2020.

Doyen, M.-F., Laques, A.-E., Gurgel, H., Garcia, L. (2017). Systèmes agricoles et système d'indicateurs: évaluation de l'impact du changement climatique sur la sécurité alimentaire dans un bassin d'inondation amazonien. *Confins, 30*. https://doi.org/10.4000/confins.11828

Dudley, N. (2008). *Guidelines for Applying Protected Area Management Categories*. Gland, Switzerland: IUCN. https://repository.oceanbestpractices.org/bitstream/handle/11329/1177/PAPS-016.pdf?sequence=1 Accessed 17 September 2020.

FAO (1996). *Rome Declaration on World Food Security*. World Food Summit. 13-17 November 1996, Rome, Italy. http://www.fao.org/3/w3613e/w3613e00.htm. Accessed 03 August 2020.

FAS (2009). *Relatório de Gestão 2008*. Manaus: Fundação Amazonas Sustentável. https://fas-amazonas.org/wp-content/uploads/2020/03/Projeto-Relatório-2008-FAS.pdf Accessed 03 August 2020.

FAS (2015). Relatório de Atividades. Fundação Amazonas Sustentável, Manaus, 120p. https://fas-amazonas.org/wp-content/uploads/2020/03/Projeto-Relatório-2015-FAS.pdf Accessed 03 August 2020.

Heemskerk, M., K. Wilson, M., Pavao-Zuckerman (2003). Conceptual models as tools for communication across disciplines. *Conservation Ecology*, 7(3). http://www.consecol.org/vol7/iss3/art8/. Accessed 03 August 2020.

Ipea (2014). Objetivos de Desenvolvimento do Milênio - Relatório Nacional de Acompanhamento. Brasília: Instituto de Pesquisa Económica Aplicada. http://www.ipea.gov.br/portal/images/stories/PDFs/140523_relatorioodm.pdf. Accessed 03 August 2020.

Fearnside, P.M. (2012). Brazil's Amazon forest in mitigating global warming: unresolved controversies. *Climate Policy, 12*(1), 70-81. https://doi.org/10.1080/14693062.2011.581571

Graebner, I.T., Saito, C.H., de Souza, E.M. (2007). Biochemical assessment of vitamin A in schoolchildren from a rural community. *Jornal de Pediatria, 83*(3), 247-252. https://doi.org/10.1590/S0021-75572007000400010

Jakovac, C.C., Peña-Claros, M., Kuyer, T.W., Bongers, F. (2015). Loss of secondary-forest resilience by land-use intensification in the Amazon. *Journal of* Ecology, 103(1), 67-77. https://doi.org/10.1111/1365-2745.12298

Jakovak, C.C., Dutrieux, L.P., Siti, L., Peña-Claros, M., Bongers, F. (2017). Spatial and temporal dynamics of shifting cultivation in the middle-Amazonas river: Expansion and intensification. *PloS one, 12*(7), e0181092. https://doi.org/10.1371/journal.pone.0181092

Junk, W.J., Wittmann, F., Schöngart, J., Pidade, M.T.F. (2015). A classification of the major habitats of Amazonian black-water river floodplains and a comparison with their white-water counterparts. *Wetlands Ecology and Management, 23*(4), 677-693. https://doi.org/10.1007/s11273-015-9412-8

Laques A.-E., Léna P., Silva A.I.C., Martins A.L.U., Arvor D., Dessay N., et al. (2013). As políticas públicas e os efeitos sobre as estratégias de gestão de recursos: o caso do Alto Solimões, Amazonas, Brasil. In H. Noda, P. Lena, A.-E. Laques, S. N. Noda (Eds.), *Dinâmicas sócio ambientais na agricultura familiar na Amazônia* (pp.7-32). Manaus: INPA/UFAM.

Laques, A-E., Cabral, A.I.R., Silva, S.C.P., Pereira, H.S., Saito, C.H. (2018). Water and Forest in Uatumã Sustainable Development Reserve. *Sustentabilidade em Debate, 9*(2), 164-186. https://doi.org/10.18472/SustDeb.v9n2.2018.16792.

Lintemani, M.G., Loss, A., Mendes, C.S., Fantini, A.C. (2020). Long fallows allow soil regeneration in slash-and-burn agriculture. *Journal of the Science of Food and Agriculture, 100*(3), 1142-1154.

Moraes, A.F.J., & Rocha, C. (2013). Gendered waters: the participation of women in the 'One Million Cisterns' rainwater harvesting program in the Brazilian Semi-Arid region. *Journal of Cleaner Production 60*, 163-169. https://doi.org/10.1016/j.jclepro.2013.03.015

Novak, J.D. & Cañas, A.J. (2007). Theoretical origins of concept maps, how to construct them and uses in education. *Reflecting Education*, 3(1), 29–42. http://www.reflectingeducation.net/index.php/reflecting/article/view/41

Pereira, S.N.C. (2010). Payment for Environmental Services in the Amazon Forest: How Can Conservation and Development be Reconciled?. *The Journal of Environment & Development, 19(*2), 171-190. https://www.jstor.org/stable/10.2307/26199354

Pinstrup-Andersen, P. (2009). Food security: definition and measurement. *Food Security, 1*(1), 5-7. https://doi.org/10.1007/s12571-008-0002-y

Quisumbing, A.R., Brown, L.R., Feldstein, H.S., Haddad, L., Christine Peña, C. (1995). *Women: The key to food security*. Washington DC: IFPRI Food Policy Report. https://core.ac.uk/download/pdf/6289051.pdf. Accessed 03 August 2020.

Rival, L. (2012). *Sustainable Development through Policy Integration in Latin America: A Comparative Approach. Occasional Paper Seven*, Geneva: United Nations Research Institute for Social Development (UNRISD)/Friedrich-Ebert-Stiftung (FES). https://www.files.ethz.ch/isn/142048/7%20Rival-web.pdf. Accessed 03 August 2020.

Roma, J.C. (2019). Os objetivos de desenvolvimento do milênio e sua transição para os objetivos de desenvolvimento sustentável. *Ciência e Cultura, 71*(1), 33-39. https://doi.org/10.21800/2317-66602019000100011.

Saito, C.H. (2017). Concept Map to Visualize Opposite Perspectives of Rapa Nui History as a Whole. *Journal of Historical Archaeology & Anthropological Sciences*, 1(5): article 00029. https://doi.org/10.15406/jhaas.2017.01.00029

Schmink, M., & Gómez-García, M.A. (2015). *Under the canopy: Gender and forests in Amazonia. Occasional Paper 121*. Bogor, Indonesia: CIFOR. https://www.cifor.org/publications/pdf_files/OccPapers/OP-121.pdf. Accessed 03 August 2020.

Silva, R.J., Garavello, M.E.P.E., Nardoto, G.B., Mazzi, E.A., Martinelli, L.A. (2017). Factors influencing the food transition in riverine communities in the Brazilian Amazon. *Environment, Development and Sustainability, 19*(3), 1087-1102. https://doi.org/10.1007/s10668-016-9783-x

Silva, S.C.P., Pereira, H.S., Silva, R.G.P., Cabral, A.I.R., Saito, C.H., Laques, A.E. (2019). Efeitos do pagamento por serviços ambientais sobre a dinâmica da paisagem em uma unidade de conservação de uso sustentável no Amazonas. In Anppas. Anais do IX ENANPPAS-Encontro da Pós-Graduação e Pesquisa em Ambiente e Sociedade (pp. 748-759). Brasília: Anppas.

Tchansia, K. (2012). L'agriculture à Saint-Georges-de-l'Oyapock: bilan et perspectives. *Confins, 16*. https://doi.org/10.4000/confins.8045

UN (2014). The road to dignity by 2030: Ending poverty, transforming all lives and protecting the planet: Synthesis report of the Secretary-General on the post-2015 sustainable development agenda. *United Nations General Assembly A/69/700*. https://www.un.org/disabilities/documents/reports/SG_Synthesis_Report_Road_to_Dignity_by_2030.pdf. Accessed 03 August 2020.

UN (2015). Transforming our world: the 2030 Agenda for Sustainable Development. *United Nations General Assembly A/RES/70/1*. https://documents--dds-ny.un.org/doc/UNDOC/GEN/N15/291/89/PDF/N1529189.pdf?OpenElement. Accessed 03 August 2020.

van Vliet, N., Adams, C., Ima Célia Guimarães Vieira, I.C.G., Mertz, O. (2013). "Slash and Burn" and "Shifting" Cultivation Systems in Forest Agriculture Frontiers from the Brazilian Amazon. *Society & Natural Resources, 26*(12), 1454-1467. https://doi.org/10.1080/08941920.2013.820813

Viana, V.M. (2008). Bolsa Floresta (Forest Conservation allowance): an innovative mechanism to promote health in traditional communities in the Amazon. *Estudos Avançados 22*(64), 143-153. https://doi.org/10.1590/S0103-40142008000300009.

Viana, V., Tezza, J., Solidade, V., Marostica, S., Salviati, V., Soares, A. (2012). Impactos do Programa Bolsa Floresta: uma avaliação preliminar. *Inclusão Social, Brasília, 6*(1), 201-218. http://revista.ibict.br/inclusao/article/view/1703/1909. Accessed 03 August 2020.

Victora, C.G., Aquino, E.M., do Carmo Leal, M., Monteiro, C.A., Barros, F.C., Szwarcwald, C.L. (2011). Maternal and child health in Brazil: progress and challenges. *The Lancet, 377*(9780), 1863-1876.

Vogelmann, J.E., & Rock, B.N. (1988). Assessing forest damage in high-elevation coniferous forests in Vermont and New Hampshire using Thematic Mapper data. *Remote Sensing of Environment, 24*(2), 227-246. https://doi.org/10.1016/0034-4257(88)90027-2

Implementation of the SDGs Through Greening Household Responses to Water, Energy and Food Shortages in Newlands West, Durban

18

Muchaiteyi Togo and Hirshwyn B. Arulappan

Abstract

Newlands West, a suburb in Durban has been experiencing water and energy shortages, as well as scarcity in fruit and vegetable food items. Recurrent water and power cuts reflect the challenge of meeting household demand by local authorities. To meet their energy and water demand during water and power cuts, households devised their own coping strategies. More often, there is disregard of the environment as households are more worried about meeting their immediate needs rather than the long-term environmental effects of their coping strategies. This paper establishes the coping strategies employed by households and their willingness to adopt "green" options/ coping strategies in Newlands West, Durban. Data was gathered by means of a questionnaire. Most of the respondents confirmed that there were water and energy challenges in the area with many attesting to the fact that water cuts and power outages are becoming more frequent. Coping strategies employed by households include rainwater harvesting and water storage in large containers to meet their water needs, use of generators, gas and portable lighting to mitigate against energy challenges and buying in bulk and growing the food items that are in shortage. In addition, most households expressed interest in adopting green coping strategies. Current strategies are a mix of those strategies with potential to harm the environment and those which build towards sustainability and reduce impacts on the environment. Overall, results show that households can play a crucial role in implementing the Sustainable Development Goals (SDGs) through greening their coping strategies in the face of shortages of basic resources. The study recommends further research to establish acceptable and affordable greening options for household coping strategies.

Keywords

Water · Food · SDGs · Nexus · Coping strategies · Climate change · Alternative energy

1 Introduction

Global warming and cooling is a natural process that has occurred over millennia in the earth's natural history, with ice ages and warmer interglacial periods. In recent years however, with regard to fluctuations in global climate in the

M. Togo (✉) · H. B. Arulappan
Department of Environmental Sciences, University of South Africa, Florida, South Africa
e-mail: togom@unisa.ac.za

G. Nhamo et al. (eds.), *Sustainable Development Goals for Society Vol. 2*, Sustainable Development Goals Series, https://doi.org/10.1007/978-3-030-70952-5_18

latter half of the twentieth century, there has been a noticeable abnormality in the historical climate record. According to Matthews, Weaver, Meissner, Gillett and Eby (2004), the Intergovernmental Panel on Climate Change (IPCC) who represent the global scientific community concluded that human activities are largely accountable for the climate changes that are being experienced at present. Watson (2003) states that the change in the earth's climate is expected to negatively affect the socio-economic systems, which include water, agriculture, forestry and fisheries, land and water ecological systems and human health. Furthermore, it is likely that developing countries, such as South Africa, will be the worst affected countries, especially with regard to poorer communities.

Residents within Durban, specifically in Newlands West, have been experiencing numerous water and electricity cuts as well as a scarcity of natural food items like certain fruit and vegetables which are normally available in local stores. Such shortages have huge implications on the socio-economic well-being of households within Newlands West. According to Adger et al. (2003), social vulnerability to the risks related to climate change may intensify ongoing social and economic challenges, especially in sectors of societies that are dependent on resources that are sensitive to changes in climate. This paper explores the implementation of the SDGs in Newlands West through greening their responses to climate-related shortages of water energy and food. The research directly addresses four main SDGs, namely: SDG 2 (zero hunger), SDG 6 (clean water and sanitation), SDG 7 (affordable and clean energy) and SDG 13 (climate action).

The paper is divided into four major sections – literature review, methods, results and discussion. The last section contains a conclusion and recommendations.

2 Literature Review

2.1 Water Issues (SDG 6)

Access to water affects human survival and dignity. Water is a resource that cannot be substituted and should be available in sufficient quantities and quality for overall human well-being (Hildering 2004). According to Hildering (2004) the international water law stipulates that 40–50 litres of water are needed per day per person for household use, with the exact amount depending on climatic conditions. In contrast, the World Health Organization (WHO) and the United Nations International Children's Emergency Fund (UNICEF) estimate that 20 litres a day per person are adequate (Lankford et al. 2013). Academics set the figure at 85 litres per day per person for domestic use at household level for activities such as cooking, drinking and washing (Lankford et al. 2013). In South Africa, the majority of households have between eight and ten members (Earle et al. 2005), and therefore, the amount of water required to satisfy basic household needs is extensive. Occurrences of water shortages therefore have considerable implications on households due to this high number of household occupants.

According to Davison (2008) everyday water use within the more private area of the household is supported by strong values and social standards. Water use within households is associated with cultural principles of ethics, hygiene and public health. The changed habits around the flushing of toilets, wearing clean clothes and bathing on a regular basis have brought about a remarkable increase in water consumption per person over the twentieth century (Lindsay and Supski 2017). Lindsay and Supski's study on the change of household consumption practices after drought showed that household residents changed their gardening habits by using alternate water sources such as boreholes, and in some incidences, residents removed all their lawn and replaced it with paving. Other water practices include households reusing bath water to water plants, having shorter showers and utilising shower timers to notify users of how long they have been showering for (Lindsay and Supski 2017).

In the midst of household resource shortages, there are measures that may be put in place to lessen the impact of these shortages. Coping with water shortages typically implies devising temporary procedures intended to maintain or return to the current household situation that will allow society to function during changing conditions (Kelly and Adger 2000). According to Adeniji-Oloukoiet al. (2013), there is a link between population socio-economic features and the coping capability specifically for livelihood supports. This may allow for more efficient coping strategies being employed in higher-income areas. Adeniji-Oloukoi et al. (2013), who conducted a study on household coping strategies for water shortages in Nigeria, confirmed that households employed strategies such as rainwater harvesting, the purchasing of bottled water from shops and storing water in large containers. Households from poorer backgrounds chose to use alternate solutions such as digging boreholes to obtain water.

2.2 Energy Issues (SDGs 7 and 13)

Like water, energy is recognised as a basic human right in many countries. According to Tully (2006), an individual's right to electricity is already accepted under the international human rights law. By satisfying this right, individuals are able to adopt a suitable standard of living and improve living conditions as recognised by the International Covenant on Economic, Social and Cultural Rights (ICESCR) (Tully 2006). This means when electricity supplies diminish in households, the standard or living and living conditions within these dwellings decline as well. Extended periods of electricity shortages result in water receding, as water pumps require electricity to function (Kesselring 2017). This shows the nexus between water and energy resources and how one resource influences the other. Kesselring's study on electricity crisis in Zambia noted that many residents felt that the shortage of water is worse than the shortage of electricity. This is however based on each household's views and will differ depending household activities which rely on such resources. The South African Constitutional Court in the Grootboom judgment of 2001 determined that the right to suitable housing includes access to basic services which includes electricity (Tully 2006). Section 10(1) of the South African Electricity Act 41 of 1987 enforces responsibilities on electricity service providers to supply electricity to all applicants who are capable of paying for such services (Tully 2006). However, households that contribute to tax and electricity tariffs still experience shortage of electricity supplies.

According to Yue, Long and Chen (2013), encouraging the use of energy-efficient equipment and new energy development is not adequate to resolve problems of high-energy consumption and environmental pollution. However, altering households' energy consumption activities can possibly play a great role in resolving the problems. Households' energy consumption levels differ and are related to factors such as the type of houses, income levels, household configurations, locations, etc. (Druckman and Jackson 2008; O'Doherty et al. 2008). Kesselring's (2017) study on electricity crisis in Zambia mentions that households had to alter their daily practices and work sharing to align with scheduled power cuts. The study also adds that households which experience electricity shortages alter cooking practices by utilising wood or charcoal as energy sources and alter cleaning practices by hand washing clothes instead of using washing machines (Kesselring 2017). These practices may result in high levels of inconveniences especially in households with large family groups and may result in increased household gender roles.

In Zambia, households which experience electricity shortages develop coping plans such as equipping their dwellings with solar panels, making use of car batteries for power and the use of gas stoves (Kesselring 2017). Individuals charge their mobile phones at stalls, and with the help of cheap solar panels purchased from Chinese shops, they are able to listen to the radio. Households generally utilise various sources of energy (Kesselring 2017).

2.3 Food Shortage (SDG 2)

As with the other resources, the right to adequate food is a basic human right (Künnemann and Ratjen 2004). In a study conducted in Tanzania on climate variability and eating habits among households, climate variability was reported to influence food prices by lowering the levels of food production, leading to reduced food accessibility and household food insecurity (Saronga et al. 2016). Because of this, households are inclined to change their eating habits as a coping mechanism to food shortages (Saronga et al. 2016).

With concerns of shortage of natural food items, Walsh and van Rooyen (2015) conducted a study in the Free State Province of South Africa to determine food security in rural and urban communities. The study confirmed that households experienced food shortages, experienced changes in eating habits and relied more heavily on canned food items (Walsh and van Rooyen 2015). Urban households in the study also reported running out of money to buy food, which ultimately led to changes in feeding practices. This included relying on inadequate food items to nourish themselves, reducing the size of meals or skipping meals and providing children with less food than they needed (Walsh and van Rooyen 2015).

With regard to food shortages, Farzana et al. (2017) studied coping strategies related to food insecurity in Bangladesh. Households in Bangladesh adopted strategies such as borrowing food, consuming lower-quality food and consuming smaller quantities of food. Similar coping strategies were mentioned by Walsh and van Rooyen (2015) following their study in the Free State Province of South Africa which was aimed at determining food security in rural and urban communities. The study revealed coping strategies including assistance from family members or neighbours, borrowing food or money during food shortages and growing own food, and, in some instances, household members went to live elsewhere during periods of food shortages.

2.4 Aim of the Chapter

Recent times have seen a noticeable increase in the occurrences of drought in parts of South Africa, and this has had devastating impacts on the supply of water due to reduced dam levels. Climatic models foresee a likely decrease in the annual runoff into catchments from 2010 to 2040 by as much as 25% as a result of climate change (Shrestha et al. 2014). The city of Durban has experienced many adverse climate change impacts over recent years, namely, storm surges resulting from the rise in sea level, heat waves and hotter temperatures, change in rainfall and storm patterns, flooding and reduced drinking water supplies (Buechler and Hanson 2015). Because of the recurrence of these climatic events in Durban, it can be foreseen that imminent water shortages will be felt among urban households. These water shortages have also ultimately led to impacts in the natural food industry, which requires large volumes of water supply, in addition to changes in climatic conditions which is a major determinant of food production (Cobon et al. 2016). With regard to electricity, power stations use copious amounts of water in the production process, and water is the main source of power generation in hydropower stations across the country. Apart from the need for water, high electricity demands for cooling in a warmer climate has also affected supplies. These factors associated with climate change have greatly influenced resource shortages at household levels.

In Newlands West, households are suffering business downtime, household maintenance restrictions as well as negative effects on household sanitary systems. Lower-income households struggle with food insecurity worsened by the increasing cost of fruits and vegetables. There is a high likelihood that social vulnerability to risks related to climate change may intensify, especially in sectors of society that are dependent on climate-sensitive resources (Adger et al. 2003). Residents, as taxpayers, have a right to basic amenities such as a steady supply of water and electricity, and hence compensation/coping measures by service providers have to be adequate to meet their demand. In Newlands West, the measures employed by the municipality, for example, deployment of water tanks to affected areas (durban.gov 2017), have not been adequate and cannot be sustained over long periods of time. Even notifications of planned power cuts are not very effective. This research was therefore undertaken to establish the extent of the challenges faced, coping strategies and the understanding and employment of green coping measures. The contextual setting of the research in Durban is well-aligned to an existing gap in literature where there is not much research on Durban household coping strategies to water, energy and food shortages.

3 Methodological Orientation

3.1 Description of Study Area

This study was conducted in the suburb of Newlands West, in the areas of Hillgrove, Castle Hill and Parlock. These areas were chosen as they cover a large proportion of Newlands West with each displaying different socio-economic characteristics. Newlands West is located in the western suburbs of KwaZulu-Natal and falls under the municipality of eThekwini in the city of Durban. Durban is the largest city in the province and the third largest in the country (About eThekwini Municipality 2015). Its land area of 2297 square km is relatively larger than most South African cities. It is a sophisticated cosmopolitan city of over 3,442,398 people according to the 2011 Census (About eThekwini Municipality 2015).

According to Statistics South Africa (2017), Newlands West has a population of 50,627 with a population density of 3799 persons/km^2. The area has 12,222 households, of which 94.4% are formal housing, 94.6% have piped water, 96.8% have flush toilets connecting to sewage and 96.4% have electricity for lighting (Statistics South Africa 2017). The source of energy in Newlands West is Eskom with eThekwini Metropolitan Municipality as the distributor. The areas that form part of this study fall within eThekwini ward 37, and according to Wazimap (2016), household socio-economic statuses show that 47% of individuals are employed and earn an average annual income of R57 500. From observations however, there are noticeable differences between the area of Parlock and the other two. Parlock displays households from a higher-income bracket, while the areas of Hillgrove and Castle Hill display households from mainly the middle-income bracket. Food security is not stable especially during drought years. Generally, the food challenges/shortages are not as grave as water and energy challenges.

3.2 Research Design

This research is hinged on the theory of change (TOC) (Oberlack et al. 2019). Oberlack et al. (2019) cited Mason and Barnes who argue that TOC is basically constituted by mental representations and theoretical assumptions which explain how and why changes are generated from certain activities or initiatives. It has several steps from problem identification, goal setting, to measurement of outcomes and explaining the logic of the initiative. This paper however draws only on the first two steps of this theory, that is, identifying the problem to be addressed and the setting of goals.

The study was informed by a quantitative approach which falls under the positivist paradigm. It is based on descriptive research design. "Descriptive designs are complex designs that describe a phenomenon in order to answer research questions" (Taylor 2006: 173). The research is a case study of the Newlands West area. It focuses on real-life situations that cannot be manipulated by the researcher. A case study design was chosen as it allowed the researchers to preserve the complete features of real-life events while investigating possible scenarios. According to Shaban, Considine, Fry and Curtis (2017), a case study is concerned with particularisation, not generalisation, and hence, it is more concerned with specificity rather than how it contrasts from other cases.

3.3 Data Collection Method

A questionnaire was drawn up in order to obtain quality data from households. The structure of the questionnaire was in the form of closed-ended questions which were formulated based on what is already known in terms of the literature review and questions in line with the objectives of the study. Creswell (2014) notes that questionnaires generate a quantitative or numeric description of trends, attitudes or opinions of a population from a sample of the population.

Primary data collected included resource shortages that households face, household responses to resource shortages as well as the possibility of adopting green adaptive strategies among households. Identification of resource shortages aligns with TOC problem identification, while determining the possibility of adopting green adaptive strategies marks the beginning of goal setting (Oberlack et al. 2019). Selection of respondents was based on random sampling. Not every household could be included in the study and random sampling ensured that each household had an equal chance of being selected to be part of the research. This reduces bias.

Households in each of the three areas were identified on a map and assigned a number. The total number of houses was fed into a random number generator software and each number generated was selected for the study. If households that were selected were not willing to participate, a new number was generated on the software. The procedure for handing out the questionnaires involved approaching households that were chosen by random number generator software and finding out from the homeowners whether they were willing to participate in the study.

The sampling frame for this research was 3349 households from the 3 research areas. A sample size of 93 was determined with confidence level of 95% and 10% margin of error using the following Raosoft sample size calculation formula: (https://www.surveysystem.com/sscalc.htm):

$$\mathrm{ss} = \frac{Z^2 * (p) * (1-p)}{C^2}$$

where:

Z = Z value (e.g. 1.96 for 95% confidence level)

p = percentage picking a choice, expressed as decimal (.5 used for sample size needed)

c = confidence interval

Although considered as a weakness, the 10% margin of error allowed for a smaller number of households to be selected for this study due to time and budget constraints; however, a high confidence level of 95% remained. Apart from the calculated sample size of 93, a total of 99 households were selected from the 3 residential areas of Newlands West, a little more than the calculated sample size to cater for the possibility of spoiled questionnaires. Of the selected 99 households, 95 confirmed to have experienced resource shortages and therefore the required sample size was met.

3.4 Data Analysis

The data analysis techniques of this research involved use of descriptive statistics. The variables forming the basis of the analyses included resource shortages, perceived causes of shortages, coping strategies and willingness to adopt green coping mechanisms. The use of descriptive statistics helped to summarise gathered data to show prevalent challenges, strategies and opinions which then helped in reaching conclusions as far as the objectives are concerned. The data was summarised in graphs and tables. The data obtained was graphically illustrated using Microsoft Excel where the information was imported onto a spread sheet and graphs were constructed.

For contributors to be fairly selected, there was no discrimination based on age, sex, religion, race or disability. Time for data collection was scheduled with households to ensure data was collected when it was most convenient for the respondents. A consent form outlining what the research entails was provided to contributors so that they were aware of what they are contributing to. All of the information provided by respondents were not altered in any way to ensure data integrity. All personal information of participants remained confidential and was not disclosed through any form of media, thereby allowing for complete anonymity of participants.

4 Presentation of Results

4.1 The Nature of Experienced Resource Shortages

The research focused on three basic resources – water, energy and food. This section is meant to illustrate the extent of the shortages as well as give an indication of the gravity of these shortages among the surveyed households (Oberlack et al. 2019). About 96%of the households indicated that they have experienced basic resource shortages. This study is, therefore, premised on the views and experiences of 95 households (n = 95) who indicated that they experienced shortages in at least 1 of the 3 resources.

The 95 households which said they experienced resource shortages were asked to indicate the shortages they experienced from among the three (multiple answers were possible). Figure 18.1 shows that the majority of the 95 households experience water shortages (86%), followed closely by electricity shortages (74%), with some households experiencing food shortages (24%).

In terms of frequency of occurrence of the shortages per month, it turns out that water is the resource which is in shortage most often. Figure 18.2 summarises the average frequency of shortages among households per month.

Generally, most of the households experience shortages in each of the three resources once or twice a month, and the number of households facing shortages decreases as frequency increases. However, more households experience water shortages more frequently than the other two resources with 16% (n = 82) experiencing these shortages 5 or more times in a single month – compared to 12% (n = 70) for energy and 0% for food. The majority of households which experience energy shortages (64%; n = 70) experience these shortages only 1–2 times per month. Water is the resource that is in shortage less frequently among the three with 83% (n = 23) of the households experiencing it only 1–2 times in a month. None of the respondent

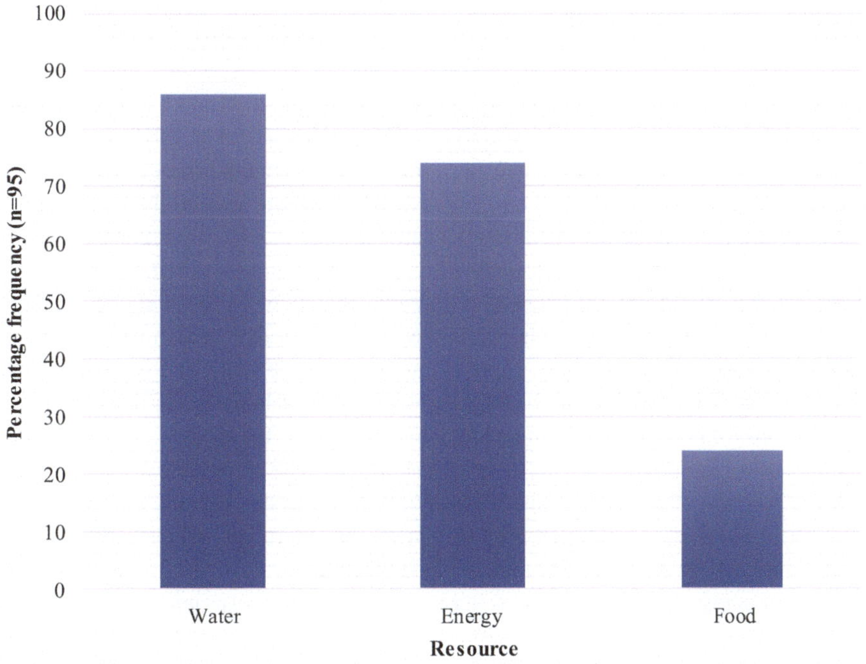

Fig. 18.1 Frequency of households experiencing water, energy and food shortages. (Source: Authors)

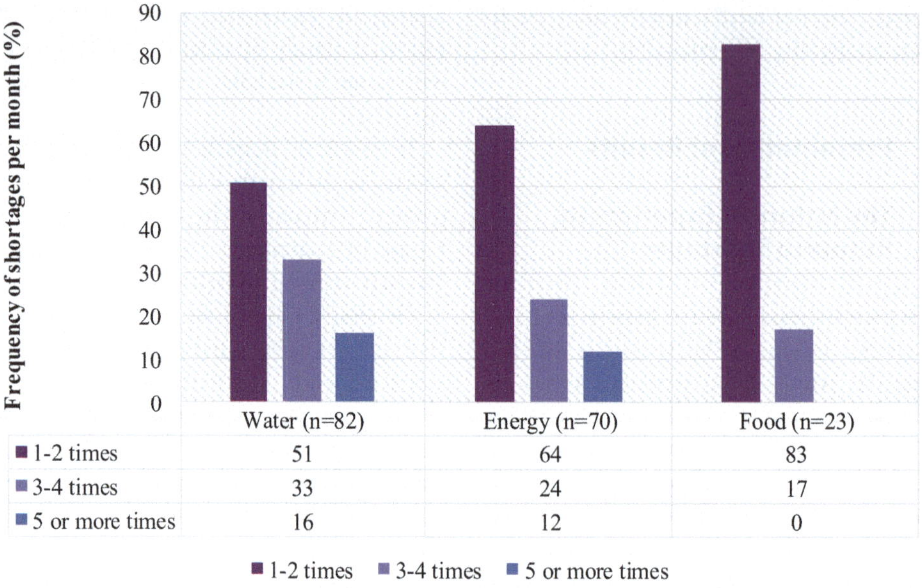

Fig. 18.2 Frequency of resource shortages per month. (Source: Authors)

Table 18.1 Duration of water and power outages

Resource	Duration of shortages (hours)		
	1–2 hours	3–4 hours	5 or more hours
Water ($n = 82$)	12%	39%	50%
Energy ($n = 70$)	9%	50%	41%

Source: Authors

households experience food shortages more than four times a month. Of note is the fact that only 23 households indicated that they experience any food challenges.

The research went on to establish the duration of water and electricity shortages. These two resources are the ones which are experienced by the majority of the respondent households (NB only 23 households indicated that they experience food shortages). The duration of shortages was captured in hours of power outages and water cuts as summarised in Table 18.1.

Water is the resource that most households indicated to be in shortage for longer periods compared to electricity. In half ($n = 82$) of the households experiencing water shortages, water cuts last for 5 or more hours. Very few households (12%) experience water cuts which last for 1–2 hours. Where energy is concerned, power outages in 50% of the households ($n = 70$) last for 3–4 hours.

4.2 Households' Understanding of Resource Shortages and Coping Strategies

Data in this section marks the beginning of goal setting (Oberlack et al. 2019) as the research was meant to understand the current practices with a view to explore potential for greening adaptive strategies as part of implementing specific SDGs. All the respondents in this research (95) indicated that they were aware of climate change. Most of them (77%) also believed that resource shortages have increased in recent years (77%), while 23% of households believe there has been no change in the degree of household resource shortages. The respondents were also asked their opinions on the causes of water and energy resource shortages. The results are shown in Table 18.2.

Most of the respondents attributed the shortage of both water and energy resources to climate change (56%). This is followed by wasteful usage and poor supply. Other mentioned causes are

Table 18.2 Perceptions on the causes of resource shortages

Causes of water shortages	Frequency	Cause of energy shortages	Frequency
Climate change	56%	Climate change	56%
Wasteful water usage	48%	Wasteful usage of energy	42%
Poor water supply	49%	Poor electricity supply	31%
Water pollution	15%	Power cable theft	32%
Unsure	8%	Unsure	8%

Source: Authors

water pollution (for water) and power cable theft (for energy resources).

Basic resources are pertinent to human survival, and without them, life is not possible. As a result, when they are in short supply, people come up with alternative ways to cope especially if the shortages are for prolonged periods. This research established that most households (86%) developed a number of strategies to cope with the shortages. Only 14% said they do not have any coping strategies in place.

Table 18.2 shows the coping strategies being employed by households. Out of the 95 households, in terms of water coping mechanisms, the majority store water in large containers (72%) during water shortages, followed by those who reduce their water consumption (26%) and lastly by those who practice rainwater harvesting (17%). With regard to coping mechanisms for electricity shortages, the majority use of gas stoves (68%), followed by the use of portable lighting (51%), use of power generators (24%) and last, use of solar energy (11%). Coping mechanisms for food shortages include bulk buying of scarce food items (17%) and reducing the consumption of food items that are in shortage (11%), while 7% of households grow their own food items. Households without any coping strategies were 14%.

While households have developed the coping mechanisms in Table 18.3, some of these are difficult to implement since the municipality rarely notifies them prior to water and power cuts. Because of this, households expressed dissatisfaction with municipal provisioning of water and energy. Households were therefore open to other ways of coping with the shortages with 90% ($n = 95$) of them believing that a more sustainable use of resources in households would help alleviate this problem. Only 10% felt that such methods will not make any difference to the current shortage of resources.

Some of the households are already utilising devices and gadgets which are water and energy efficient. The research did not establish the nature of the devices being used. It just established the proportion of respondent households which are using these devices just to get a sense of how open minded the residents are and to establish potential for them to accept innovative ideas. Figure 18.3 which captures the results of this investigation shows that 77% of the households use at least some kind of an energy-saving device and 23% do not. Households with water-saving mechanisms are 51%. One would expect more households to have water-saving strategies compared to those with energy-saving mechanisms since water outages are experienced by more households in comparison.

Most of the households (85%) indicate their willingness to green their homes by implementing coping strategies that do not impact negatively on the environment or at least that reduce their impacts on the environment. Some of the households were however unsure (15%), while a few (4%) said they were not willing to change their current strategies.

5 Discussion

In hierarchical order, most of the 95 households experience water shortages (86%), followed closely by electricity shortages (74%) and last by food shortages (24%). In terms of frequency of shortages per month, water shortages were experienced more frequently by more people compared to energy and water. In terms of dura-

Table 18.3 Coping mechanisms employed by households

Coping strategies for water shortages	% Frequency (n = 95)	Coping strategies for energy shortages	% Frequency (n = 95)	Coping strategies for food shortages	% Frequency (n = 95)
Storing water in large containers	72%	Utilising gas stoves	68%	Bulk buying of scarce food items	17%
Reducing quantities used in bathing and laundry	26	Portable lighting	51%	Reducing consumption of foods in shortage	11%
Rainwater harvesting	17%	Utilising generators	24%	Growing own food	7%
None	14%	Use of solar energy	11%	None	14%
		None	14%		

Source: Authors

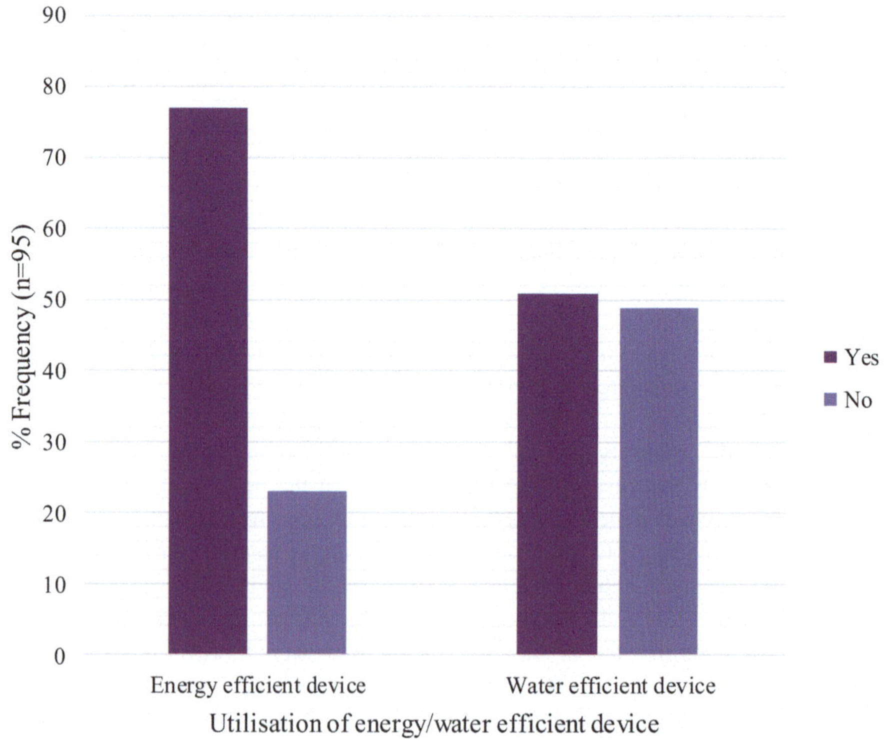

Fig. 18.3 Utilisation of water-/energy-saving devices among households. (Source: Authors)

tion of shortages, water is also the resource that is in shortage for longer periods of time compared to electricity.

These responses relate to the adverse effects of climate change experienced in recent years within the city of Durban especially in terms of heat waves and hotter temperatures, changes in rainfall and storm patterns, flooding and reduced drinking water supplies (Buechler and Hanson 2015). Erratic rainfall has been experienced in eThekwini in recent years, and, at one time, dam levels were critically low (Shrestha et al. 2014). South Africa is a water-scarce country; hence, it follows that water will be the resource in shortage in most of the households. Contributory factors to water shortage, particularly drinking water, include over-exploitation of natural water sources like rivers, declining dam water levels and a bad state of repair of municipal drinking water facilities (Donnenfeld et al. 2018). This shows that SDG 6 (clean water and sanitation) is far from being met among communities in the area of study. One third of countries have medium or high levels of water stress, and 60% out of 172 countries are unlikely to meet the targets of SDG 6 by 2030 (United Nations Economic and Social Council (UNESC) 2019). Water affects individuals' very survival and their inherent dignity and is a resource that cannot be substituted and should be available in sufficient quantities and quality to support the well-being of people (Hildering 2004).

Where energy is concerned, these results are in line with the blackouts that are being experienced by residents and businesses throughout the country. These planned blackouts (or load shedding) were introduced by Eskom in 2008 and have since then been scheduled on an on-and-off basis depending on the grid supply (Coetzee and Els 2016). SDG 7 (affordable and clean energy) remains unmet with some 800 million people remaining without access to electricity as of 2019 (UNESC 2019). In addition, 59% of energy supply in South Africa is from coal (Coetzee and Els 2016).

Hunger is actually said to be on the rise globally (UNESC 2019). In the studied area, fewer households (24%) reported food shortages. This could be because the research was done in an urban area where most households have a working adult who earns income to support the everyday needs of the household. The bulk of the food consumed in urban areas is bought and not directly produced from the farms. There is much less direct reliance on farming activities which are mainly dependent on natural weather elements. However, access to food is generally reduced even in urban areas when rainfed agriculture is affected in areas that supply the urban areas. Just as Saronga et al. (2016) established in Tanzania, this results in reduced food accessibility and household food security.

All the respondents said they are aware of climate change. Where the causes of the shortages are concerned, most respondents attributed them to climate change (56%) as stated in the Results section. It is striking that most respondents made a connection between climate change and resource shortages as in most cases, authorities are blamed for not being able to provide adequate resources to the communities they serve. The connection between water and energy shortages and climate change is in line with climatic models of South Africa that foresee a likely decrease in the annual runoff into catchments in the period from 2010–2040 by as much as 25% because of climate change (Shrestha et al. 2014).

Although data shows that the majority of households believe that climate change is the cause of resource shortages, there are other leading factors as well. According to the information gathered, these factors relate to supply issues and wasteful usage of resources, pollution and theft. Many of these factors are conditions that the eThekwini municipality has control of, and therefore, there is a clear challenge of service delivery in terms of controlling some of these factors in ways that enable it to deliver to its community.

Newlands West residents have been affected drastically considering that these resources are basic human rights that all households are entitled to (Tully 2006). One can conclude that the basic needs of households within the Newlands West area are not being met due to the shortages experienced. Water and food are among the basic needs which people cannot survive without. Shortage of these items will result in people finding ways to cope. In Newlands West, household coping strategies can be divided into those that are green and those that have a negative impact on the environment. Options like storing water in large containers, bulk buying of scarce food items and utilising generators may impact negatively on the environment, especially the use of generators which depend on fossil fuels that release carbon dioxide into the atmosphere. Options that are more friendly to the environment and reduce negative impacts include:

- Utilising gas stoves
- Use of solar energy
- Reducing water quantities used in bathing and laundry
- Rainwater harvesting
- Growing own food

In addition to this, most households claim to use water- and energy-saving devices (77% and 51%, respectively), while 85% indicated their willingness to green their coping strategies. These results have an inclination on awareness levels among respondents – all of them indicated that they are aware of climate change. Other studies have also established high climate change awareness levels in specific South African communities, an example being the research done by Oduniyi and Nkonki-Mandleni (2018). In this research however, residents were only asked to respond to whether they are aware of climate change or not, nothing more was established.

Marx-Pienaar and Erasmus (2014) argue that people need explicit knowledge or factual information about a problem for them to change. They explain explicit knowledge as factual information which includes a good understanding of the relevant concepts, problems, their causes and consequences. According to the authors, that is what triggers a willingness to change. The authors actually argue that people's explicit knowledge should be above average. While explicit knowledge has not been established in the research, many Newlands West residents may be possessing enough insight about climate change to prompt a willingness to change unsustainable practices.

Overall, some basic needs are not being met among Newlands West residents, especially water and energy. This has triggered a number of coping strategies. Not all the coping strategies are not green, but they are all meant to meet the resources that are in shortage. As such, the coping strategies are reflective of the contribution that the community is making in implementing some of the SDGs. Most importantly, some of their strategies have no impacts or have less impacts on the environment compared to a business-as-usual scenario. The willingness of the community to embrace green coping options shows their readiness to address the challenges in an environmentally friendly way, which translates into added impetus to implement SDGs.

6 Conclusion and Recommendations

The overall results of the study confirm that households within the Newlands West area are in shortage of some basic resources, particularly water and energy. All the households in this study were aware of climate change with more than half attributing the experienced resource shortages to climate change, among other factors. Households had to develop an array of coping strategies such as storing water in large containers, utilising gas stoves, utilising power generators, rainwater harvesting, growing own food, bulk buying of scarce food items and portable lighting. All the households are aware of climate change and many use some water- and energy-saving devices. Most of the households are also willing to adopt green alternative ways of coping with the shortages. These results show that the households are contributing to the implementation of the following SDGs through coping strategies: goal 7 (affordable and clean energy) and goal 13 (climate action). The other goals being implemented to a lesser extent are goal 2 (zero hunger) and goal 6 (clean water and sanitation). Most important is the fact that many households are open to greening their coping strategies. Thus, further research is recommended in order to establish other green options which are acceptable and affordable to households which they can implement to meet the basic resources which are in shortage.

References

About eThekwini Municipality (2015) Durban.gov.za. http://www.durban.gov.za/Discover_Durban/Pages/AboutEM.aspx. Accessed 14 Apr 2017

Adeniji-Oloukoi G, Urmilla B, Vadi M (2013) Households' coping strategies for climate variability related water shortages in Oke-Ogun region, Nigeria. Environmental Development 5:23-38

Adger WN, Huq S, Brown K, Conway D, Hulme M (2003) Adaptation to climate change in the developing world. Progress in development studies 3(3):179-195

Buechler S, Hanson AMS (Eds.) (2015) A political ecology of women, water and global environmental change (Vol. 15). Routledge, London

Cobon DH, Ewai M, Inape K, Bourke RM (2016) Food shortages are associated with droughts, floods, frosts and ENSO in Papua New Guinea. Agricultural Systems 145:150-164

Coetzee D, Els M (2016) The impact of load shedding on the construction industry in South Africa. 9th CIDB Postgraduate Conference: "Emerging trends in construction organisational practices and project management knowledge area". Pages 268-276. February 2-4, 2016, Cape Town, South Africa

Creswell JW (2014) A concise introduction to mixed methods research. Sage Publications, Nebraska-Lincoln

Davison G (2008) Down the gurgler: historical influences on Australian domestic water consumption. Troubled Waters: Confronting the Water Crisis in Australia's Cities. ANU Press, Canberra

Donnenfeld Z, Crookes C, Hedden S (2018) A delicate balance: water scarcity in South Africa. Southern Africa Report, 13 March 2018.: Institute for Security Studies, Pretoria

Druckman A, Jackson T (2008) Household energy consumption in the UK: A highly geographically and socio-economically disaggregated model. Energy Policy, 36(8): 3177-3192

Earle A, Goldin J, Kgomotso P (2005) Domestic Water Provision in the Democratic South Africa changes and challenges. The Nordic Africa Institute, Uppsala

Farzana FD, Rahman AS, Sultana S, Raihan MJ, Haque MA, Waid JL, Ahmed T (2017) Coping strategies related to food insecurity at the household level in Bangladesh. PloS one, 12(4):1-17

Hildering A, (2004) International law, sustainable development and water management. Netherlands, Eburon Publishers https://www.surveysystem.com/sscalc.htm

Kelly PM, Adger WN (2000) Theory and practice in assessing vulnerability to climate change and Facilitating adaptation. Climatic change, 47(4):325-352

Kesselring R (2017) The electricity crisis in Zambia: Blackouts and social stratification in new mining towns. Energy Research and Social Science, 30:94-102

Künnemann R, Epal-Ratjen S (2004) The right to food: A resource manual for NGOs. American Association for the Advancement of Science, Science and Human Rights Programme. New York, HURIDOCS (Human Rights Information and Documentation Systems, International)

Lankford B, Bakker K, Zeitoun M, Conway D (Eds.) (2013) Water security: Principles, perspectives and practices. Routledge, Canada

Lindsay J, Supski S (2017) Changing household water consumption practices after drought in three Australian cities. Geoforum, 84:51-58

Marx-Pienaar NJMM, Erasmus AC (2014) Status consciousness and knowledge as potential impediments of households' sustainable consumption practices of fresh produce amidst times of climate change. International Journal of Consumer Sciences; Special Issue: Food Safety, 38(4):419-426

Matthews HD, Weaver AJ, Meissner KJ, Gillett NP, Eby M (2004) Natural and anthropogenic climate change: incorporating historical land cover change, vegetation dynamics and the global carbon cycle. Climate Dynamics, 22(5):461-479

Oberlack, Christoph; Breu, Thomas Michael; Giger, Markus; Harari, Nicole; Herweg, Karl Günter; Mathez-Stiefel, Sarah-Lan; Messerli, Peter; Moser, Stephanie; Ott, Cordula; Providoli, Isabelle; Tribaldos, Theresa Margarete; Zimmermann, Anne; Schneider, Flurina (2019). *Theories of change in sustainability science: Understanding how change happens.* GAIA - Ökologische Perspektiven für Wissenschaft und Gesellschaft / ecological perspectives for science and society, 28(2), pp. 106-111. Oekom Verlag https://doi.org/10.14512/gaia.28.2.8

O'Doherty J, Lyons S, Tol RS (2008) Energy-using appliances and energy-saving features: Determinants of ownership in Ireland. Applied Energy, 85(7):650-662

Oduniyi OS, Antwi M, Nkonki-Mandleni B (2018) Determinants of Climate Change Awareness among Rural Farming Households in South Africa. Journal of Economics and Behavioral Studies, 10(5):116-124

Saronga NJ, Mosha IH, Kessy AT, Ezekiel MJ, Zizinga A, Kweka O, Kovats S (2016) "I eat two meals per day" impact of climate variability on eating habits among households in Rufiji district, Tanzania: a qualitative study. Agric and Food Secur 5(14):1-7

Shaban RZ, Considine J, Fry M, Curtis K (2017) Case study and case-based research in emergency nursing and care: Theoretical foundations and practical application in paramedic pre-hospital clinical judgment and decision-making of patients with mental illness. Australasian Emergency Nursing Journal, 20(1):17-24

Shrestha S, Babel MS, Pandey VP, (2014) Climate Change and Water Resources. CRC Press, Thailand

Statistics South Africa (2017) Statistics South Africa. Statssa.gov.za. http://www.statssa.gov.za/?page_id=4286&id=10408. Accessed 24 May 2017

Taylor BJ (2006). Research in nursing and health care: Evidence for practice. Cengage Learning Australia, Australia

Tully S (2006) The human right to access electricity. The Electricity Journal, 19(3):30-39

United Nations Economic and Social Council (2019) High-level political forum on sustainable development. Special edition: progress towards the Sustainable Development Goals. Report of the Secretary-General. 2019 session Agenda items 5 (a) and 6. United Nations

Walsh CM, Van Rooyen FC (2015) Household food security and hunger in rural and urban communities in the Free State Province, South Africa. Ecology of food and nutrition, 54(2):118-137

Watson RT (2003) Climate change: the political situation. Science, 302(5652):1925-1926

Wazimap (2016) Wazimap profile: Ward 37 (59500037), eThekwini, KwaZulu-Natal. Wazimap https://wazimap.co.za/profiles/ward-59500037-ethekwini-ward-37-59500037/#households. Accesses 12 Jul 2017

Yue T, Long, R, Chen H (2013) Factors influencing energy-saving behavior of urban households in Jiangsu Province. Energy Policy, 62:665-675

Part V

Conclusion and policy recommendations

19 Summary of Findings, Conclusions and Policy Recommendations

David Chikodzi, Kaitano Dube, and Godwell Nhamo

Abstract

The world has gone past 5 years of the Sustainable Development Goal (SDG) implementation, and their 2030 deadline is fast approaching. It is therefore imperative to make an inventory and a report of what has been done on the ground in some locations to inform further implementation strategies. This chapter summarises the key findings from the book, which has a special focus on food security, energy, climate action and biodiversity SDGs. In the study, the nexus approach in analysing the synergies and trade-offs in the implementation of the SDGs is recommended. The approach helps to provide balanced insights into the SDGs as a system rather than looking at them as individual elements. This entails systematic and integrated multidisciplinary approach to the SDGs to generate new data and updated scientific tools.

D. Chikodzi (✉) · G. Nhamo
Exxaro Chair in Business and Climate Change, Institute for Corporate Citizenship, UNISA, Johannesburg, South Africa
e-mail: chikod@unisa.ac.za; nhamog@unisa.ac.za

K. Dube
Ecotourism Management, Vaal University of Technology, Vanderbijlpark, South Africa

Keywords

Energy and society · SDG policy · SDG for society · Biodiversity engagements

1 Introduction

In September 2015, during the United Nations Sustainable Development Summit, global heads of states and leaders embraced the Sustainable Development Goals (SDGs). There are 17 SDGs aiming to provide a template through which the vast but intertwined problems facing humanity can be solved by 2030. Blesh et al. (2020) among other writers distinguish climate change, degradation of the environment, inequality and poverty as being key challenges currently facing mankind. SDG implementation endeavours to bring equilibrium to all the facets of sustainable development which include the economy, society, biophysical/ecosystems, well-engineered institutions and upright leadership (Gomez-Echeverri 2018). This is to recognise the understanding that it is impractical to achieve development which caters for all when the biophysical environment and extreme weather events are overlooked. By nature, the 17 SDGs together with their 169 targets are glued to each other and inseparable. Hence, an action aimed at implementing the SDGs needs to reflect on their synergies (Yillia 2016). The implementation of the SDGs has

G. Nhamo et al. (eds.), *Sustainable Development Goals for Society Vol. 2*, Sustainable Development Goals Series, https://doi.org/10.1007/978-3-030-70952-5_19

become pertinent as a result of the increasing over-exploitation of natural resources around the world. This is a direct outcome of the rising demand that has been fuelled by increasing economic prosperity as well as steep population growth in an era of climate variability and change (Ibid.).

Horan (2019) highlights the importance of formulating innovative and ground-breaking methods in dealing with challenges given in the SDG agenda. These methods must harness the synergies that occur among several goals in order to attain numerous SDGs at once through actions that produce enhanced co-benefits among the goals. In this respect, multi-stakeholder partnerships can be envisaged as the launchpad through which these synergies can be realised. The SDGs are a high-level global agenda, yet executable only at national and subnational levels by integrating them into local and national development planning. The implementation of the SDGs at the local level can be effected through a process termed "localisation" (United Cities and Local Governments 2019).

Localisation is the process of considering subnational contexts in the execution of the SDGs. This includes the processes of setting out goals and targets, to the determination of the means of implementation, and using indicators to measure and monitor progress (The Global Taskforce of Local and Regional Governments 2016). Localisation also relates to the manner in which regional/provincial and municipal/local governments enhance the SDGs through bottom-up approaches and to how SDGs can offer a template for local development strategies (United Cities and Local Governments 2019). This approach brings societies into the equation.

While the SDGs are global in nature, their achievement will depend to a large extent on the capability of countries to make them a reality at a local level. Most of the SDGs have targets that are directly related to the responsibilities of local and regional governments, particularly their role in delivering basic services and protection of the environment. This brings about the need to put local communities and their governments at the core of the 2030 Agenda. The Global Taskforce of Local and Regional Governments (2016) developed a roadmap for the localisation of the SDGs. They recommend stages that can be followed in the process which include awareness raising at subnational level, advocacy to include subnational perspectives in national SDG approaches and implementation, and monitoring and evaluation which are critical components of the continuous improvement cycle.

As countries are working towards attaining the SDGs, the challenge of domesticating them in relation to countrywide and local development priorities entails the convergence of technical, managerial, scientific and political viewpoints. A collaborative research approach must be taken in order to stay true to the SDGs' inclusive and bottom-up approach. This provides researchers across and within disciplines with boundless and novel opportunities to engage with the SDGs, especially at societal levels. SDGs, therefore, remain an agenda for society. Hence, there is a need to document their implementation across societies in this book. The world has gone past 5 years of SDG implementation, and the 2030 deadline is fast approaching. It is therefore imperative to make an inventory and a report of what has been done on the ground in some localities so as to inform further implementation strategies going forward as advocated by the plan-do-check concept of continuous improvement and also to facilitate peer-to-peer learning. The book documents SDG research focusing on how key stakeholders are engaging, localising and implementing them with a special focus on ending hunger, energy, climate action and biodiversity SDGs. This comes at a time when there is limited documented research on SDG implementation from the global south, especially Africa. Given the centrality of people and society in the 2030 Agenda for Sustainable Development, the book showcases the movement towards creating a better world for people and our planet now and into the future.

The next section summarises the main results from the book chapters and discusses the key issues addressed in the book.

2 Emerging Key Findings, Conclusions and Discussions

This section presents the key findings and conclusions from the book, as well as discussions around the key issues addressed. It will also be divided into sub-sections that coincide with the parts in this book.

2.1 Food Security and Sustainable Energy

This section had chapters that explored ways in which different methods were used to contribute to the attainment of SDG 2 (ending hunger) and SDG 7 (sustainable energy). One of the chapters analysed the contribution of responsible leadership in the South African Land Bank's ability to raise funds in support of its mandate. The mandate of the Bank is to support commercial and development agriculture as well as create a more inclusive and transformed sector. The chapter showed that responsive and responsible leadership was key in solving challenges raised by investors. This led to the strengthening of the bank's funding and diversity of investor base. The leadership identified legitimate concerns and expectations of investors and developed specific responses which exceeded regulatory requirements to address these issues.

Another chapter confronted issues of poverty, hunger and food insecurity while integrating lessons from local communities in Malawi and Zimbabwe. In the chapter, it was observed that public-private partnerships and contract farming arrangements yielded positive results for local farmers. However, historical problems were manifesting through poor access to agricultural inputs and markets and lack of active farmer participation in experts' interventions. Unless addressed, these problems were likely to compromise the fight against poverty and hunger in these communities.

Another chapter looked at the strategies aimed at minimising the fall armyworm (FAW) damage to the maize crop by altering planting time and using varied genotypes. It was then observed that the time of planting expressively influenced crop damage by the FAW. Early planting was shown to be effective in protecting the maize crop from FAW-induced damage, whereas late planting resulted in extensive crop damage and subsequent lower yields. Early planting mostly resulted in maize passing the most susceptible stages before the pest incidence increase.

Other chapters explored urban farming as a strategy for urban households to achieve food security as well as assessing the access to improved water and sanitation facilities in the smallholder farming sector. These concluded that urban farming had the potential to eradicate poverty and hunger, and that it could contribute to good health and wellbeing of households. However, it was observed that urban agriculture encountered challenges such as lack of inputs and appropriate policies, hence the practice needed to be supported by all stakeholders, including local authorities. It was shown that the marital status of household head, household size and farm equipment significantly influenced access to an improved water source. Land tenure, level of education of household head, livestock ownership and farm equipment significantly affected the access to improved toilet facilities. The observed improved access to water and sanitation facilities was attributed to partnerships between government, non-governmental and local communities in drilling and maintaining community boreholes.

Notwithstanding the progress witnessed in the fight against hunger since the Millennium Development Goals in 2014, there have been notable increases in populations suffering from hunger. The United Nations (2020) highlights that in 2019, an estimated 25.9% of the world population faced moderate to severe food insecurity up from 22.5% in 2014. Extensive famine and poverty remain some of the greatest challenges confronting many nations (UNstat 2019). Almost a fifth of the global population survive on less than US$1 per day, and a seventh of the world population suffer from the omnipresence of hunger. Mugambiwa and Tirivangasi (2017) observe the global south as being the most adversely affected by poverty and hunger. The

United Nations' Food and Agricultural Organization (FAO) (1996:1) defines food security "as a condition that exists when all people, at all times have physical and economic access to sufficient, safe and nutritious food that meets their dietary needs and food preferences for an active and healthy life". Perez-Escamilla (2017) and FAO (1996) highlight food security at local household level as being determined by factors such as the physical and economic obtainability of food and the capacity of the body to absorb and utilise vital nutrients. A balance in all these factors needs to be achieved if food security is to be reached at the household level. Food security also includes having a balanced diet measured through the safety and nutritional content of consumed food (Gil et al. 2019).

SDG 2 aims at ending hunger and malnutrition as well as doubling agronomic efficiency and earnings. Most of the global food production occurs on smallholder farms. A study by Rapsomanikis (2015) shows that close to 90% of the world's 570 million farmlands are smallholder farms (below 2 hectares). These farms are worked on by close to 1.5 billion people, of which the majority are from poor households. These smallholder farmers produce up to 80% of food in sub-Saharan Africa, Asia and parts of Latin America. These are also the areas hardest hit by the two problems of poverty and hunger, hence guaranteeing constant or improved production by smallholders goes a long way in fulfilling SDG as well as others that it has synergies with (Abraham and Pingali 2020). However, these farmers face a cocktail of challenges the world over. These challenges relate to the limited availability extension services, hybrid and high-yielding seeds, herbicides and fertilisers. These problems are then compounded by limited access to production machinery and skills as well as poor linkages with markets to sell their outputs (Pingali 2012).

Byerlee et al. (2009) and then Pingali (2010) view the characteristics of smallholder farmers as differing, given the context in which they occur. This also gives rise to numerous and diverse challenges. Historical, economic, institutional, geographical, political and social factors combine to influence the challenges and opportunities faced by these farmers. Efforts to improve the production and success of these farmers also need to consider these factors if they are going to be effective. There is also a need to note that some of these factors vary within and across countries and communities, hence the need for local community involvement in the design and implementation of any intervention strategies. Increased output from smallholder farmers will likely translate to increased income at the household level for the affected farmers. This is essential if the significant reduction of both poverty and hunger as well as gender equality is to be achieved given that about 1.5 billion individuals use small-scale farming as the dominant source of their livelihood (Gil et al. 2019).

Blesh et al. (2020) see an increase in studies that seek to have a deeper understanding of the interlinkages existing between agriculture and health as well as the environment. They attribute this scholar shift to changes in public understanding of these issues. For example, through the targets that aim to achieve sustainable agricultural production in an endeavour to end hunger, the target emphasises the well-known fact that large-scale monoculture plantation agriculture poses serious threats to important ecosystem services and functions that support sustainable food production (IPCC 2013). Some of the well-studied impacts of industrial agriculture include loss of both plant and animal species diversity, washing away of the topsoil, increase in greenhouse gas secretions, pollution and eutrophication of both ground and surface water bodies (Foley et al. 2011). FAO has recently adjusted the terminology from "food security" to "food and nutrition security", in order to highlight and bring awareness to the worldwide upsurge in micro-nutrient insufficiencies. In 2015, about 815 million persons around the world remained malnourished, with a further estimated 2 billion people hit by various micro-nutrient deficits (FAO, IFAD & WFP 2015).

Nilsson et al. (2016) highlight that the objectives of SDG 2 bridge disciplinary realms. Thus, they imply that transforming present political and economic models is necessary in simultaneously dealing with the issues of food security and ecological sustainability challenges. SDG 2's detailed targets address issues to do with the environment like conservation agriculture, biological diversity and adaptation to climate change and variability. This SDG also spans issues that address public health, for instance, combating malnutrition, hunger and dietary requirements for women. Further, the economic and social factors affecting food production and farmers are also considered. These include the revenue of farmers, access to markets, access to extension services, possibilities for value addition and unfair trade (Ibid.)

In order to address the sustainability issues that are fundamental for the achievement of SDG 2, ecological values have been progressively adapted to evaluate the resilience as well as the sustainability of agroecological systems (Schipanski et al. 2016). Agroecology is a broad area of study which covers issues of soil fertility, rangeland management, livestock management, pest and disease management and weed control. Since the main objective of managing agroecological systems is to produce enough food for people, the management of soil fertility needs to be scientifically done so as to replenish nutrients it recurrently loses as they are translocated from the topsoil to the crops. Management of soil fertility is, therefore, an important aspect of sustaining and improving crop production with time and therefore a foremost limitation to the productivity of agroecological systems. Blesh et al. (2020) observe agroecosystem diversification as contributing significantly to the maintenance of natural biological diversity, which also has the potential to contribute to more efficient management of pests and pollination. Conservation agriculture, crop rotation, intercropping and hedgerows have been proven to be agricultural best practices that enhance the presence of "natural enemies, increase herbivore suppression and significantly decrease damages to crops" (Letourneau et al. 2011). Agroecosystems that have higher measures of biological diversity have the capability to support complex webs of interrelating species. These complex webs give rise to natural and independent pest control which is robust and can decrease the requirements of applying expensive insecticides, which are damaging to both the health of the farmer and the ecosystems (Perfecto et al. 2014).

The commercialisation of agrarian production in the global south is an indispensable step towards the fulfilment of the poverty, social, food security and ecological SDGs. Increased access to factors of production is an important component in raising the yields of low productive agrarian systems in order to fulfil the food security and nutritional goals. Access to credit, insurance and research and development as well as extension services will help farmers adopt a hybrid, high-yielding as well as drought-resistant crop varieties. These factors have the potential to raise productivity levels, create climate-smart agriculture and, at the same time, reduce ecological damage. Equally important is the ability of the farmers to have access to key value chains so that they can add value to their products. This can include things such as having agreements with retailers or manufactures as well as other end users which will assist them to realise marketplace prospects and most likely realise higher returns from their produce (Blesh et al. 2020).

Investing in the agrarian sector is also significant in the endeavour to eliminate starvation and poverty as well as increasing food security. Figure 19.1 shows government expenditure on agriculture as a function of its contribution to the GDP (agriculture orientation index (AOI)). The figure shows that except for Central and Southern Asia and Oceania, the investment in agriculture by governments around the world has significantly declined relative to its contribution to the national economies from 2001 to 2017. UNstats (2019) also observes aid allocated to improve agricultural reduction around the world has also drastically plunged from about 25% of allocated aid in the mid-1980s to about 7% in 2017 which equates to about to US$12.6 billion. This notable de-investment in agriculture may derail the achievement of Goal 2 by 2030.

There were also chapters in the book that focused on sustainable energy. One of the chapters focused on identifying the leadership capacities demonstrated in the development and

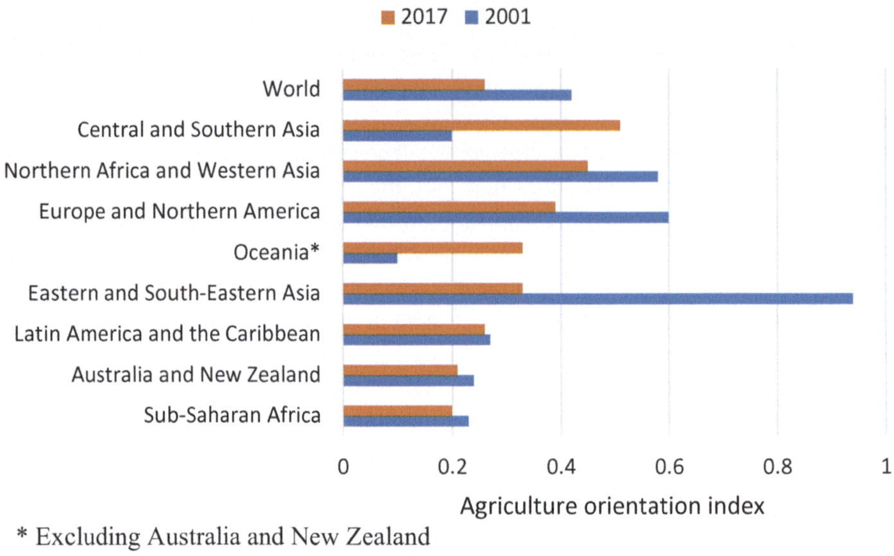

Fig. 19.1 Agriculture orientation index 2001–2017. * Excluding Australia and New Zealand. (Source: Authors, data from UNstats (2019))

implementation of the Botswana Biogas Project of the United Nations Development Programme (UNDP). The chapter had the view of promoting the production and utilisation of biogas as an alternative, clean energy source contributing to the achievement of SDG 7. Findings from the chapter point to the fact that for inclusive growth to be achieved in the African development context, governance and leadership need to be transformative and that shared leadership and collaborative problem-solving need to be key components of development practice.

Other chapters also explored the importance of responsible leadership in realising the energy SDG as well as other SDGs intertwined with it. This included a chapter on the private sector-led initiatives in renewable energy production. Findings showed that the profiled energy companies successfully realised several targets under SDGs 2 to 4 (ending hunger, health and quality education), SDGs 7 (sustainable energy) and 8 (sustainable jobs), as well as SDGs 13 (climate action) and 17 (partnerships) through renewable energy projects. This demonstrated the inter-reliant nature of the 17 SDGs. Findings from this chapter also show that creating shared value and transformational and bottom-up collaborations were the central leadership capabilities that enabled successful SDG-oriented project delivery.

Availability of energy is an important catalyst to the attainment of SDGs. Energy is a key enabler in the fight against hunger poverty, provision of quality education, healthcare, improved sanitation, support of economic progress as well as protection of the ecosystems (Santika et al. 2019). Since most SDGs have synergies with energy provision, the drive towards implementing them at all levels of society will likely see an increased strain in the energy sector of different countries, especially those in the global south. Figure 19.2 shows global access to electricity by region. It shows that in every region of the world, there have been notable increases in access to electricity since 2000 although some regions like sub-Saharan Africa and Oceania still seriously lag. Given that energy is a key enabler of other SDGs, these regions are unlikely to meet other SDGs if they do not significantly invest in the provision of electricity and other energy sources.

Razmjoo et al. (2020) observe the need for the world to shift from heavy dependence on fossil fuels to renewable sources. They maintain that this shift is slowly being accepted and adopted by

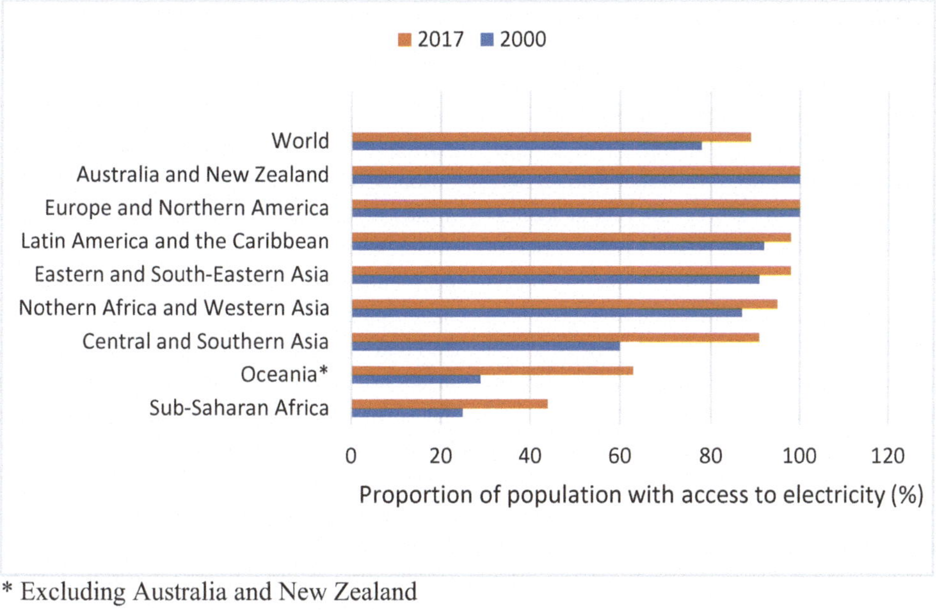

Fig. 19.2 Global access to electricity by region. * Excluding Australia and New Zealand. (Source: Authors, data from UNstats (2019))

progressive countries around the world. Among the most exploited renewable energies are wind, hydro and solar energy (Razmjoo et al. 2017). The distribution and accessibility of cheap renewable forms of energy has emerged as one of the key discussion points on sustainable development (Gugul et al. 2018). Deployment and availability of renewable energy sources has been shown to be positively correlated with an increase in human development (Roy et al. 2015). There have been observed modest to sharp growths in societal wellbeing in relation to energy consumption in the global south nations compared to the saturation levels reached in the global north. An increase in power utilised does not lead to noteworthy changes in human development and quality of life.

Santika et al. (2019) concede that the biggest challenges modern-day society faces are in formulating sustainable methods by which increases in the demand for energy brought about by SDG implementation can be accommodated without creating trade-offs with other SDGs to do with the environment. Three main targets for SDG 7 give the foundation for sustainable energy adoption. These include improving availability and obtainability of clean and contemporary energy, improving the national component of energy being harnessed from renewable energy sources as well as doubling energy efficiency at all levels of society. The United Nations (2019) posits the percentage of the populace to which electrical energy is accessible and the proportion of the populace with access to clean fuels and technology as being key pointers determining access to clean and modern energy. Other indicators of SDG 7 targets (2nd and 3rd) include the share of renewables in the entire energy utilised and power intensity as determined by prime power supply per unit of GDP. However, economic growth is the prime driver of increased utilisation of electricity; hence any attempt to reduce its consumption may produce knock-on effects on the economy except if it is achieved through energy efficiency. In an environment of energy efficiency, there is production of the same goods and services utilising less energy. On the other hand, raising the cost of energy to try and restrict demand may hurt economies as well as create trade-offs with other SDGs that deal with economic growth (Santika et al. 2019).

2.2 Climate Action for the SDGs

This section of the book focused on studies looking at how different societies, industries and sectors were responding to the threats of climate change around the world. Given that limited economic assessments on carbon supply and sequestration have been done on Africa's forests, one of the chapters assessed the possibilities of forests serving as the nexus for climate action and clean production and consumption. This was in an endeavour to help support policies that address broad-ranged socioeconomic and political impediments for the promotion of carbon supply and sequestration under SDG 15. The study observed that timber harvest in previous periods contributes to a decline in the per unit area of carbon supply; however, carbon supply was found to come from the management of new forest areas. Economic performance was found to have a negative effect on long-run carbon supply. Affordable, scalable solutions were shown to be key in enabling countries to leapfrog to cleaner and more resilient economies.

Two chapters in this section of the book looked at the aviation industry and how it was domesticating the climate change action SDG. One such chapter sort to document and provide a policy roadmap for the localisation of the SDGs within the aviation industry using examples from South Africa. It also highlighted some of the challenges that airports are going through in an endeavour to achieve sustainability. The chapter showed that some airports were making strides in localising the SDGs, thereby responding to their customer and community needs for sustainability. The other chapter examined the handling of the climate change action SDG by airports around the world under the voluntary mechanisms championed by Airport Carbon Accreditation (ACA). The study showed the ACA as one of the best coordinated programmes that have seen several airports across the world, combining efforts for the greater good of battling climate change. The programme started in Europe but has spread to all regions of the world and has allowed airports across the world to learn from each other in building a sustainable airport future.

An investigation was also done to show how universities were interacting with climate change SDG. This was reflected in a chapter that investigated ways in which the University of South Africa (UNISA) had localised the water and sanitation SDG as it was preparing itself for eventualities regarding climate change, water shortage and general groundwater conservation and environmental stewardship. It emerged that UNISA developed and implemented a Water Master Plan and had plans for rainwater harvesting, water use efficiency and wetland rehabilitation projects. These investments and engagements resulted in huge water and cost savings by minimising irrigation from boreholes and drawing water from the municipal system and building climate resilience.

Climate change has been argued to be the greatest challenge that modern civilisation is facing; thus, Goal 13 is concerned with "urgent action to combat climate change and its impacts" (Campbell et al. 2018). The impacts of climate change on our ability to fulfil the SDGs have been on the spotlight for some time (Gomez-Echeverri 2018). There seems to be a consensus that it will hamper the ability to achieve the set goals. The projected 2 °C heating of the Earth in the next 50 years potentially threatens the sensitive ecosystems and the ecosystem services that they provide as well as many livelihoods in an unparalleled manner (Wright et al. 2015). The anticipated negative effects of climate change are increases in extreme weather events, impaired economic growth, reduced agricultural production, loss of biological diversity, involuntary migration, civil unrest and increased incidences and areas vulnerable to tropical diseases (Ibid.). Regrettably, the burden of these impacts will disproportionately fall upon the global south nations who have limited capacity to adapt and are the least responsible for carbon emissions (Chong 2018). It has been argued by many that a 2 °C increase in mean global temperatures will have serious consequences for successfully implementing the SDGs. These are unlikely to be realised under such circumstances (Figueres 2015).

Climate change also impacts most of the SDGs. It is a threat multiplier, with the potential to worsen some of humanity's greatest challenges, including health, poverty, hunger, inequality and ecosystem preservation, among others. Conversely, addressing climate change also offers humanity the greatest chance to positively impact these goals (Zhenmin and Espinosa 2019). Climate change is not a distant threat, but a clear and present danger to humanity. This is proven by extreme weather events that were often described as once-in-a hundred years events but now occurring more frequently.

In implementing Goal 13, action and strategy must be informed by how it relates to other goals and the organisations working to implement them (Gomez-Echeverri 2018). SDG 13 specifically requires all countries to act on climate change together with its effects (United Nations 2015). It specifically deals with matters of execution like financing, policy and planning, capacity building at all levels of society and institutional setups. The issues it prioritises for action such as resilience, adaptation and mitigation are closely intertwined with other SDGs that influence the pace of climate change or are negatively affected by it like health, poverty and hunger.

Many studies have analysed the business case for addressing the impacts of climate change. Most of the findings provide a compelling case for assessing the synergies between sustainable development and climate in order to enhance them and circumvent or lessen the adverse trade-offs. A study by the Organisation for Economic Co-operation and Development -OECD (2015) assessed the ramifications of climate change on certain drivers of economic development such as productivity, availability of investment and labour. The quantitative analysis was modelled up to the year 2060. The study findings showed that the highest damaging impacts were envisioned for the agrarian and health sectors, with the largest burden of the damage being experienced in the global south, mostly in Africa and parts of Asia. However, there is still a chance for the world to avoid the disastrous impacts of climate change. This will only be possible if the world "acts now rather than later", with the costs of inaction argued to be greater in the long term than the costs of action (Gomez-Echeverri 2018).

2.3 Health, Water and Biodiversity Engagements

This part of the book looked at how societies around the world were localising SDG 15 through health, water and biodiversity engagements. It is from here that we learn of a project to protect forests in Brazil. A network was created which expanded the protected areas but allowed traditional livelihoods to operate within sustainable management rules. This led to grant programmes such as the Bolsa Floresta Programme (BFP), which aimed to protect the forest by offering financial support for family agricultural operations that avoided deforestation. The results showed complex interdependences between BFP and the SDGs as well as existing gaps and trade-offs during the implementation of the programme. Some recommendations were made to enhance the synergies between the BFP and the various SDGs with which it directly aligns with leading to the effective realisation of the goals. Scaling-up of the programme to other regions was seen to potentially bring positive benefits to local populations and a higher commitment to the SDGs at the local level.

In another chapter with a case study from Durban, South Africa, it was observed that households devised their coping strategies to meet their energy and water demand during water and power cuts. This was sometimes done with disregard of the environment as households were more worried about meeting their immediate needs rather than the long-term environmental effects of their coping strategies. The research done in Newlands West sought to establish the coping strategies employed by households and their willingness to adopt green coping strategies. The coping strategies employed by households included rainwater harvesting, water storage in large containers, use of generators, gas and portable lighting, buying food in bulk and growing own food. Most households expressed interest in adopting green coping strategies. Hence, they could play a crucial role in implementing the SDGs through greening their coping strategies in the face of shortages of basic resources. Junker et al. (2015) argue that landscapes have become intricate socio-ecological systems that human action and biophysical factors interact across several scales. They further highlight that the amalgamation of social and economic growth objectives with conservation stratagems as an approach for sustained resource conservation entails an in-depth understanding of the coupling that exists between human and bio-physical processes. The Global Biodiversity Outlook highlights that generally vicissitudes in biological diversity being witnessed on Earth are increasingly being observed to be mostly negative (SCBD 2014), hence the formulation Goal 15 which seeks to "protect, restore and promote sustainable use of terrestrial ecosystems, sustainably manage forests, combat desertification, and halt and reverse land degradation and halt biodiversity loss" (United Nations 2015:14). Bridgewater et al. (2015) argue that strategies which address biological diversity protection, rehabilitation and management mutually with societal development will be vital to the achievement of SDG 15. Figure 19.3 shows that close to 20% of the world's land surface area was degraded from 2000 to 2015, leading to substantial losses in ecosystem functions vital to societal wellbeing. Most of this degradation occurred in Oceania and Central and Southern Asia.

In terms of water resources, Garrick et al. (2017) suggest stages that need to be followed in designing societies that sustainably interact with the resource. These include the measurement that raises the need for data on the state of the catchment or basin, quantifying the demand for water use and projections for future use. This then can be followed by identifying the different conflicting uses of water and giving the many dimensions of its value as it applies to different users. This then leads to the incorporation of multiple values of water and the trade-offs among them to create a balanced decision support system. Finally, there is a need to create viable and responsive institutions with the capability to manage the resource. Lorenz and Kunstmann (2012) highlight that measuring and monitoring water use within catchments is critical and

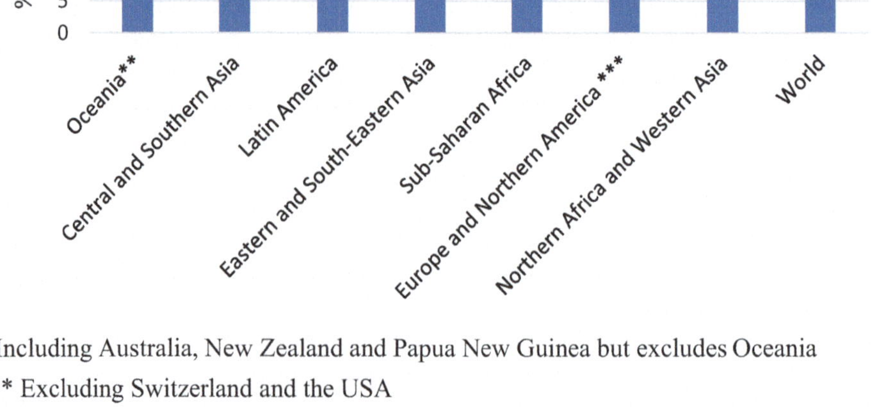

Fig. 19.3 Land degraded between 2000 and 2015. * Including Australia, New Zealand and Papua New Guinea but excludes Oceania. *** Excluding Switzerland and the USA. (Source: Authors, Data from UNstats (2019))

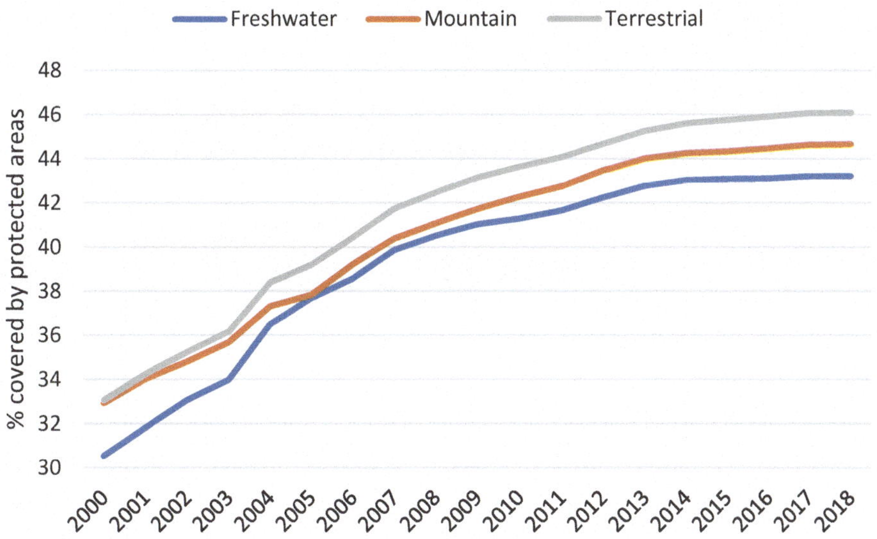

Fig. 19.4 Freshwater, terrestrial and mountain ecosystems under protected areas. (Source: Authors, Data from UNstats (2019))

unavoidable since this will inform management interventions necessary to improve management of the resource.

Preserving areas that are vital for terrestrial, freshwater and mountain biodiversity is important in safeguarding the longstanding and viable utilisation of natural resources. Since 2000, there has been a steep rise in the protection of these ecological systems; however, the pace of progress has notably slowed down since the year 2010. Figure 19.4 shows the surface area of freshwater, terrestrial and mountain ecosystems that occur in legally protected areas from 2000 to 2018. The area under each of these ecosystems in protected places rose by over 10% from 2000 to 2010. On the other hand, from 2010 to 2018, the coverage managed to rise by two percentage points. At this rate, less than 50% of these ecosystems will be covered by protected areas by 2030.

From 1990 to 2015, the Earth's surface area under forests plunged from about 31.7% to around 30% of the Earth's surface. Conversion of forest land to other land uses especially the development of infrastructure and agricultural expansion was noted to be the prime mover of this change (UNstats 2019). However, during the same period some places, especially those in the global north, increased the areas under forests. This was due to the deliberate efforts in tree planting, landscape rehabilitation as well as natural extension and encroachment. The drive towards reducing deforestation around the world has had mixed results with dichotomies between the global south and the global north. Figure 19.5 shows that the biggest decline in forest cover occurred largely in the global south countries, although there were notable increases in Eastern Asia. The global north (developed world) registered positive gains during the same time periods. The year on year pace of decline in forest cover around the world was 25% less between 2010 and 2015 when compared to the 2000 to 2005 era (Fig. 19.5). Helped by the significant decline in the pace of forest loss, the percentage area under protected forests and those under long-term conservation plans rose for all global regions.

3 Policy Recommendations

Through the adoption of the 2030 Agenda for Sustainable Development, the concept of integrated and all-inclusive thinking entered the global policy discourse in a highly visible way. Policymakers and different sectors of society can

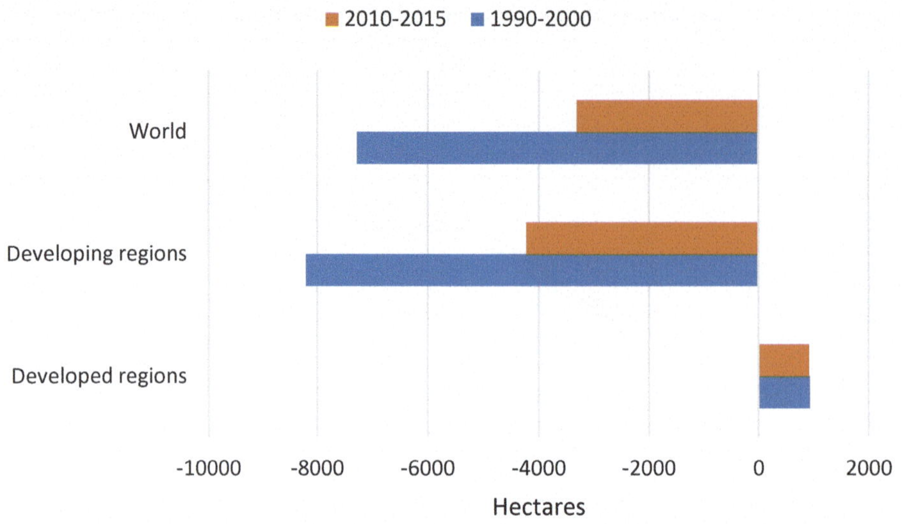

Fig. 19.5 Loss/gain in forest cover (hectares) from 2000 to 2015. (Source: Authors, Data from UNstats (2019))

no longer work in silos and develop policies based on assumptions from the sector but need to incorporate ideas from several sectors. In order to fill knowledge gaps in critical areas, there is need for a paradigm shift from scattered and separate research to systematic and integrated research and for different disciplines to share knowledge and collaborate at levels higher than what is at present. This will lead to new data, updated scientific tools and fresh, interdisciplinary perspectives that support the implementation of the SDGs.

Related to the above is the need to consider the nexus approach in analysing the synergies and trade-offs in the implementation of the SDGs. This will help to provide balanced insights into the SDGs as a system rather than looking at them as individual elements. Implementing individual SDGs separately is likely to provide outcomes that omit critical connections and feedback, hence producing sub-optimal solutions. Awareness of the linkages that exist among different SDGs can stimulate and direct global efforts on cohesive planning and execution of several SDGs. The nexus approach helps produce decision-making methods that are cross-cutting. This will also prevent a situation in which strategies, plans and partnerships to implement one set of SDGs undermines the other rather than advancing all.

Societal leaders need to go a step further than just the recognition of the synergies and interactions among the SDGs. There is an urgent necessity to mobilise more funds and enact new legislation, planning and assessment tool compatible with such thinking. In implementing the SDGs, robust policies with the necessary institutions and feedback mechanisms are key to achieving high efficiency. Implementing policies needs public support; hence, participatory processes become key to achieving this. Integrated stakeholder participation in decision-making improves policy design and implementation and prevents public opposition.

Success in achieving the 2030 Agenda will require more ambition and more political will. It will take unprecedented levels of collaborative, multilateral action in the form of increased efforts by nations and all segments of society to remain true to the agenda.

References

Abraham, M., & Pingali, P. (2020). Transforming Smallholder Agriculture to Achieve the SDGs. In S. Gomez y Paloma et al. (eds.), The Role of Smallholder Farms in Food and Nutrition Security, https://doi.org/10.1007/978-3-030-42148-9_9.

Blesh, J., Hoey, L., Jones, A.D., Friedmann, H., & Perfecto, I. (2020). Development pathways toward

"zero hunger". *World Development*, 118, 1–14. https://doi.org/10.1016/j.worlddev.2019.02.004.
Bridgewater, P., Régnier, M., & García, R.C. (2015). Implementing SDG 15: Can large-scale public programs help deliver biodiversity conservation, restoration and management, while assisting human development? *Natural Resources Forum*, 39, 214–223 DOI: https://doi.org/10.1111/1477-8947.12084.
Byerlee, D., de Janvry, A., & Sadoulet, E. (2009). Agriculture for development: Toward a new paradigm. *Annual Review of Resource Economics*, 1, 15–31.
Campbell, B.M., Hansen, J., Rioux, J., Stirling, C.M., Twomlow, S., & Wollenberg, E. (2018). Urgent action to combat climate change and its impacts (SDG 13): transforming agriculture and food systems. *Current Opinion in Environmental Sustainability*, 34,13–20. https://doi.org/10.1016/j.cosust.2018.06.005.
Chong, D. (2018). The Sustainable Development Goals and Climate Change. *Social Alternatives* Vol. 37(1):43-48.
FAO, IFAD & WFP. (2015). The state of food insecurity in the world 2015. Meeting the 2015 international hunger targets: Taking stock of uneven progress. Rome, Italy.
Figueres, C. (2015). Goal 13: Taking urgent action to combat climate change – SDGs and the Paris Climate Agreement, UN Chronicle, Vol. LI, No. 4. Retrieved from: https://unchronicle.un.org/article/goal-13-taking-urgentaction-combat-climate-change-sdgs-and-parisclimate-agreement (accessed 10/11/2020).
Foley, J.A., Ramankutty, N., Brauman, K.A., Cassidy, E.S., Gerber, J.S., Johnston, M., Mueller, N.D., O'Connell, C., Ray, D.K., & West, P.C. (2011). Solutions for a cultivated planet. Nature 2011, 478:337-342.
Food and Agriculture Organisation. (1996). World Food Summit. Rome (Italy). Retrieved from: http://www.fao.org/wfs/index_en.htm. (Accessed 20 August 2020).
Garrick, D.E., Hall, J.W., Dobson, A., Damania, R., Grafton, R.R., Hope, R., Hepburn, C., Bark, R., Boltz, F., De Stefano, L., O'Donnell, E., Matthews, N. and Money, A. (2017). Valuing water for sustainable development. Measurement and governance must advance together, Science 358 (6366), 1003-1005. DOI: https://doi.org/10.1126/science.aao4942.
Gil, J.D.B., Reidsma,P., Giller, K., Todman, L., Whitmore, A., Van Ittersum, M. (2019). Sustainable development goal 2: Improved targets and indicators for agriculture and food security. Ambio 2019, 48:685–698. https://doi.org/10.1007/s13280-018-1101-4.
Gomez-Echeverri, L. (2018). Climate and development: enhancing impact through stronger linkages in the implementation of the Paris Agreement and the Sustainable Development Goals (SDGs).Phil. Trans. R. Soc. A 376: 20160444. https://doi.org/10.1098/rsta.2016.0444.
Gugul, G. N., Koksal, M. A., & Ugursal, V.I. (2018). Techno-economical analysis of building envelope and renewable energy technology retrofits to single family homes. *Energy for Sustainable Development*, 45,159–70. doi:https://doi.org/10.1016/j.esd.2018.06.006.
Horan, D.A. (2019). New Approach to Partnerships for SDG Transformations. *Sustainability*, 11, 4947.
IPCC. (2013). Climate change 2013. The physical science basis. Summary for policymakers. United Nations. Retrieved from http://www.ipcc.ch/.
Junker, J.B, Mundry, C., Stephens, R., Lormie, C., Tweh, M., & Kühl, C.H.S. (2015). Education and access to fish but not economic development predict chimpanzee and mammal occurrence in West Africa. *Biological Conservation*, 182, 27-35.
Letourneau, D.K., Armbrecht, I., Rivera, B.S., Lerma, J.M., Carmona, E.J., Daza, M.C., & López, S.D. (2011). Does plant diversity benefit agroecosystems? A synthetic review. *Ecological Applications*, 21, 9–21.
Lorenz, C., & Kunstmann, H. (2012). *Journal of Hydrometeorology*, 13, 1397
Mugambiwa, S.S., & Tirivangasi, H.M. (2017). Climate change: A threat towards achieving 'Sustainable Development Goal number two'(end hunger, achieve food security and improved nutrition and promote sustainable agriculture) in South Africa, Jàmbá. *Journal of Disaster Risk Studies*, 9, 1–6.
Nilsson, M., Griggs, D., & Visback, M. (2016). Map the interactions between sustainable development goals. *Nature*, 534, 320-322.
OECD. (2015). The economic consequences of climate change. OECD Publishing. Retrieved from: http:// www.oecd-ilibrary.org/environment/the-economic-consequences-of-climate-change_9789264235410-en. (Accessed 20 August 2020).
Perez-Escamilla, R. (2017). Food Security and the 2015–2030 Sustainable Development Goals: From Human to Planetary Health. Current Developments in Nutrition.
Perfecto, I., Vandermeer, J., and Philpott, S. M. (2014). Complex ecological interactions in the coffee agroecosystem. *Annual Review of Ecology, Evolution, and Systematics*, 45, 137–158.
Pingali, P. (2010). Agriculture renaissance: Making "agriculture for development" work in the 21st century (Chap. 74). In P. Pingali&R. Evenson (Eds.), Handbook of agricultural economics, (Vol.4, pp. 3867–3894). Elsevier. https://doi.org/10.1016/S1574-0072(09)04074-2.
Pingali, P. (2015). Agricultural policy and nutrition outcomes—Getting beyond the preoccupation with staple grains. *Food Security*, 7(3), 583–591. https://doi.org/10.1007/s12571-015-0461-x.
Rapsomanikis, G. (2015). The economic lives of smallholder farmers: An analysis based on household data from nine countries. Food and Agriculture Organization of the United Nations, Rome.
Razmjoo, A., M., Qolipour, R., Shirmohammadi, S., Mohammadreza, H., & Faraji, I. (2017). Techno-economic evaluation of standalone hybrid solar-wind systems for small residential districts in the central desert of Iran. *Environmental Progress & Sustainable Energy*, 36 (4):1194–207. doi:https://doi.org/10.1002/ep.12554.

Razmjoo, A.A., Sumper, A., & Davarpanah, A. (2020). Energy sustainability analysis based on SDGs for developing countries, Energy Sources, Part A: Recovery, Utilization, and Environmental Effects, 42:9, 1041-1056, DOI:https://doi.org/10.1080/15567036.2019.1602215.

Roy, H., Jayaraj, R., & Gupta, A. (2015). Energy consumption and human development: a global perspective. *Econ. Policy Energy Environ*, 1, 111–131.

Santika, W.G., Anisuzzaman, M., Bahri, P.A., Shafiullah, G.M., Rupf, G.V., & Urmee, T. (2019). From goals to joules: a quantitative approach of interlinkages between energy and the Sustainable Development Goals. *Energy Res. Soc. Sci*, 50, 201–214. https://doi.org/10.1016/j.erss.2018.11.016.

Schipanski, M.E., MacDonald, G.K., Rosenzweig, S., Chappell, M.J., Bennett, E.M., Kerr, R.B., Blesh, J., Crews, T., Drinkwater, L., & Lundgren, J.G. (2016). Realising resilient food systems. *Bioscience*, biw052.

Secretariat of the Convention on Biological Diversity-SCBD. (2014). Global Biodiversity Outlook 4. SCBD, Montréal, Canada.

The Global Taskforce of Local and Regional Governments. (2016). Roadmap for localising the SDGs: implementation and monitoring at subnational level. Retrieved from: https://www.global-taskforce.org/. (Accessed 20 August 2019).

United Cities and Local Governments. (2019). Towards the localisation of the SDGs. Retrieved from: https://www.uclg.org/sites/default/files/towards_the_localization_of_the_sdgs_0.pdf. (Accessed 15 August 2020).

United Nations Statistics. (2019). Sustainable Development Goal indicators website. Retrieved from: https://unstats.un.org/sdgs/report/2019/Goal-/. (Accessed 10 August 2020)

United Nations. (2015). Transforming our World: The 2030 Agenda for Sustainable Development. New York: United Nations Secretariat.

United Nations. (2020). The Sustainable Development Goals Report 2020. Retrieved from: https://unstats.un.org/sdgs/report/2020/ (Accessed 10 August 2020).

Wright, H., Huq, S., & Reeves, J. (2015). Impact of Climate Change on Least Developed Countries: Are the SDGs possible? International Institute for Environment and Development Briefing Paper. Retrieved from: http://pubs.iied.org/17298IIED (Accessed 20 August 2020).

Yillia, P.T. (2016). Water-Energy-Food nexus: framing the opportunities, challenges and synergies for implementing the SDGs. *Osterreichische Wasser- und Abfallwirtschaft*, 68, 86–98. DOI https://doi.org/10.1007/s00506-016-0297-4.

Zhenmin, L. and Espinosa, P. (2019). Tackling climate change to accelerate sustainable development. *Nature Climate Change*, 9, 493–496. https://doi.org/10.1038/s41558-019-0519-4.

Correction to: Preventing Fall Armyworm (Spodoptera Frugiperda JE Smith) Damage in Maize by Altering Planting Time and Using Varied Genotypes

Leonard Nyabanga, Ronald Mandumbu, Joyful T. Rugare, Never Mafuse, Emmanuel Zivenge, Handsen Tibugari, George Nyamadzawo, and Christopher T. Gadzirayi

Correction to:
Chapter 4 in G. Nhamo et al. (eds.), *Sustainable Development Goals for Society Vol. 2*, Sustainable Development Goals Series, https://doi.org/10.1007/978-3-030-70952-5_4

The original version of this chapter was inadvertently published without updating the correct spelling of the chapter author name "Handsen Tibugari". Now, this has been corrected in the chapter proof and front matter.

The updated online version of this chapter can be found at
https://doi.org/10.1007/978-3-030-70952-5_4.

G. Nhamo et al. (eds.), *Sustainable Development Goals for Society Vol. 2*, Sustainable Development Goals Series, https://doi.org/10.1007/978-3-030-70952-5_20

Index

A
Abandoned land category, 169
Africa
  biogas technology, 96, 97
  leadership, 94, 95
  sustainable development, 92, 93
Africa Rising narrative
  development process, 35
  economic environment, 34
  economic growth, 34
  Malawi and Zimbabwe, 39
  SDGs, 34, 35
  tobacco marketing system, 42
African agriculture, 35
African Development Bank (AfdB), 22
African land tenure, 169
Agenda for Sustainable Development (AfSD), 191
Agrarian production, 291
Agrarian sector, 291
Agribusiness, 73
Agricultural Development and Marketing Corporation (ADMARC), 38
Agricultural inputs, 289
Agricultural sector, 24, 39
Agriculture, 12, 19, 20
Agriculture orientation index (AOI), 291, 292
Agriculture sector, 12
Agriculture *vs.* health, 290
Agro-ecological regions, 226
Agroecology, 291
Agroecosystem diversification, 291
Agroecosystems, 291
Agroforestry systems (AFS), 264
Agronomic techniques, 60
Airline industry, 178, 179
Air manufacturers, 178
Air manufacturing industry, 179
Air pollution, 238
Airport Carbon Accreditation (ACA) programme, 180
  accreditation levels, 240–242
  ACI, 239
  Africa, 245
  airports, 240–243, 294
  annual reports, 239, 240
  Australia, 246
  aviation sustainability, 245
  carbon audit, 240
  carbon emissions, 241
  carbon neutrality, airports, 238
  carbon optimisation, 241
  carbon reduction, 241
  carbon reduction measures, selected airports, 246–248
  content and thematic analysis, 239
  critical document analysis, 239, 240
  European region, 242–245, 248
  Inasmuch, 243, 248
  North America, 244, 245
  policy, 246
  SDG 13, climate action, 248
  voluntary project programme, 240
  WBCSD, 239
  WRI, 239
Airport industry, 239
Airports
  ACA programme, 238, 240
  air pollution, 238
  aviation sector, 177
  bird populations, 238
  bird species, 238
  CAEP, 179
  carbon audit, 240
  carbon emission reduction, 243
  carbon emissions, 239, 240
  carbon-neutral programme, 243
  coastal areas, 239
  electricity, 185
  environmental challenges, 178, 179, 187
  extreme weather events, 239
  ground vehicles, 185
  host communities
    ecological demands, 179
    economic wellbeing, 179
    social demands, 179
    South Africa, 180
  land and atmospheric pollution, 238

G. Nhamo et al. (eds.), *Sustainable Development Goals for Society Vol. 2*, Sustainable Development Goals Series, https://doi.org/10.1007/978-3-030-70952-5

Airports (*cont.*)
management and reduction, 242
smuggling illicit drugs, 238
South Africa, 180
sustainability, 179, 180, 237, 238
temperature, 239
transport hubs, 186
voluntary project programme, 240
Airports Council International (ACI), 239
Alternative energy, 282
Amazon biome, 256
Amazon forest, 262
Amazon River, 257
Analytical model, 169
Anthropocene, 124
Anthropogenic air pollution, 4
Anti-politics machine, 40
Aridity, 234
Artificial wetlands, 208, 209
Assigned amount units (AAUs), 162
Automatic meter reading (AMR), 196
Aviation industry, 294
Aviation sector
accredited airports, 243
airports, 177
biodiversity, 186
carbon footprint, 238
climate change, 238
consumers, 179
contrail effect, 245
environmental challenges, 178
green initiatives, 238
ICAO, 178
inasmuch, 248
KMI Airport, 186
roadmap lays out plans, 178
SDGs implementation, 187
sustainability, 178, 179
Aviation sustainability, 245

B
Biodiversity, 11, 186, 262
Biodiversity engagements, 295
Biodiversity loss, 10, 11
Biogas, 96, 97
Biogas technology, 94
Africa, 96, 97
logic framework model, 97
Biomass, 96
Biophysical environment, 3, 4
Biotechnological approach, 234
Bolsa Floresta Programme (BFP), 295
Bolsa Floresta Programme (BFP), Brazilian Amazon
AFS, 264
Amazon forest, 262
association component, 256, 266
capoeira areas, 264
cash transfer programme, 263
cassava, 257, 264, 266
community, 257
concept map, 259, 262
Conservation and Sustainability Centres, 265
deforestation, 259, 263
eating habits, 264
electricity, 264
family component, 256
family income, 267
fieldwork results, 258
financial compensation, 258
food insecurity, 264
food security, 264
Forest Conservation Grant Fund, 256
gender, 259, 265
income component, 256, 265
interviewees, 263
Luz para Todos programme, 264
MDGs, 267
nutritional security, 264
Payment programme, 266
payments, 265
political maturation, 266
poverty, 259
primary forest deforestation, 262
production cost, 263
production diversification, 266
quality education, 259
RDS-Uatumã, 257, 258, 264–266
responsible consumption and production, 259
SDG and targets, 259–261
SDG-4, 265
SDGs, 256
SDGs and targets, 259
social component, 257, 265
soil fertility, 257
Sustainable Development Reserve, 256
sustainable production development, 256
Swidden areas, primary forest, 261, 263
tourism, 265
Uatumã Sustainable Development Reserve, 256
water supply, plants, 264
zero deforestation, primary forest, 259
zero hunger, 259
Botswana
biogas technology installations, African region, 96
economic development, 91
energy, 92
firewood, 92
poverty, 92
rural households, 92
rural population, 92
sustainable development, 92, 93
Botswana Biogas Project, 292
Bottom-up approach, 132
Bottom-up collaboration, 131, 132
Brazil, Russia, India, China, and South Africa (the BRICS), 34
Broad-based black economic empowerment (BBBEE), 132

Business Council for Sustainable Development (WBCSD), 239
By-products, 96

C
Cameroon's carbon supply, 169
Cameroon's forests, 163, 164
Carbon audit, 240
Carbon capture
  long-run, 167, 168
  short-run, 168–170
Carbon dioxide (CO_2), 158, 159
Carbon emissions, 162, 163, 238, 294
Carbon footprint, 248
Carbon markets, 162
Carbon-neutral airport, 243, 247
Carbon Offsetting and Reduction Scheme for International Aviation (CORSIA), 246
Carbon policy, 169
Carbon sequestration
  Cameroon's forests, 163, 164
  economic-ecological modelling, 164–166
  nature and source of data, 166
Carbon supply, 294
Cash transfer programme, 141, 263
Cassava, 264
Certified emission reductions (CERs), 159, 246
Chief Financial Officer (CFO), 22
Child-headed households, 218
Civil Aviation Organization (ICAO), 178
Civil Protection Organisation in Zimbabwe, 232
Clean Development Mechanism (CDM), 158, 159, 169
Climate action, 117, 156
Climate change, 42, 98, 116, 118, 124, 178, 183
  Africa, 212, 226
  Africa's growth momentum, 158
  agriculture production, 213
  carbon capture (*see* Carbon capture)
  carbon dioxide (CO_2), 159
  carbon-neutral/carbon sinks, 159
  carbon sequestration (*see* Carbon sequestration)
  climate change-induced extreme weather events, 8
  COP, 159
  COP 21, 160
  COP 22, 160
  cost-effective carbon sequestration, 160
  cost-effective mitigation, 160
  developing countries, 118
  double agricultural productivity, 156
  drivers, economic development, 295
  drylands leads desertification, 158
  energy production, 8
  energy-related emissions, 8
  extreme weather events, 294
  food insecurity, 7
  forest management, 156
  GHG emission target setting, 4, 118, 158–160
  global environmental challenges, 5
  global forests, 11
  and global warming, 158
  green energy, 8–10
  human behaviours, 212
  human progress, 157
  human survival dilemma, 211
  humanity, 295
  hunger, 7
  hydroelectricity, 8
  industries and sectors, 294
  integrated approaches, 156
  Kyoto Protocol's market mechanisms, 158, 159
  land use changes, 158
  leaders, 108
  livelihood strategies, dry regions, 234
  LULUCF, 159
  MDGs, 226
  mitigation and adaptation, 160
  modern civilisation, 294
  NDCs, 160
  poverty eradication, 158
  rainfall patterns, Zimbabwe, 226
  renewable energy investment, 8
  RL elements, 108
  RL theory, 117
  SCP, 158
  SDG targets, 225
  SDGs, 212, 294, 295
  sequestered carbon, 159
  SFM, 157
  socioeconomic and biophysical environment, 225
  stakeholders, 109
  sub-Saharan region, 212
  super wicked problem, 111, 112, 115
  sustainable development, 155
  sustainable forests, 161–163
  temperatures, 213
  TPES, 9
  UNFCCC, 158–160
  weather patterns, 157
  WEF, 108
  wicked problem, 117, 118
  Zimbabwe's vulnerable communities (*see* Zimbabwe's vulnerable communities, climate change)
Climate change-induced extreme weather events, 10
Climate justice, 118
Climate-sensitive tasks, 214
Climate-smart agriculture (CSA), 214, 217–219
Climate-smart development, 219
Collaborative leaders, 128
Collaborative leadership
  Energy Company X, 133
  SDGs, 127, 128
Collaborative research approach, 288
Colonialism, 36
Committee on Aviation Environmental Protection (CAEP), 179
Communication, 116
Community-Based Disaster Risk Reduction (CBDRR), 218

Competency requirement, 102
Conference of the Parties (COP), 159
Congo Basin, 163, 164
Conservation and Sustainability Centres, 265
Contract farming
  Chiredzi, 41, 42
  Rumphi, 41, 42
Coping strategies, 295
Coping strategies, Newlands West, Durban
  basic human rights, 281
  causes of resource shortages, 278, 279
  clean water and sanitation, 281
  climate change, 274, 280–282
  climate change awareness levels, 282
  data analysis, 276
  data collection method, 275, 276
  duration, water and electricity shortages, 278
  energy, 273, 281
  factual information, 282
  food shortage, 273, 274
  households, 274, 277, 279, 280, 282
  hunger, 281
  options, 281
  research design, 275
  residents, 274
  resource shortages experience, 277–279
  SDGs, 282
  study area, 275
  water, 272, 273
  water shortages, 274
Corporate social responsibility (CSR), 29, 178
Corporation for Public Deposits (CPD), 22
Cost-effective carbon sequestration, 160
Cost-effective mitigation, 160
Creating Shared Value (CSV)
  community, 131
  Energy Company X, 130, 131, 134
  financial success, 130
  leadership styles, 133
Crop losses, 49, 50
Cultural pest management strategy, 51
Customary Land Registration Act, 38

D
Daveyton Campus wetland, 205, 207
Deforestation, 157, 259, 297
Descriptive statistics, 166
Development Action for Marginalized Rural Area (DAMRA), 41, 43
Development Finance Institution (DFI), 24
Development partners, 143, 144
Development projects, 102
Director of Planning and Development (DPD), 40
Disaster risk management (DRM), 214
Disaster risk reduction, 218, 219
District Development Fund Coordinator (DDFC), 40, 43
Double agricultural productivity, 156
Drinking water, 80
Drive climate action
  change agent, 117
  communication, 116, 117
  leader, 116
  private sector, 118
Drivers, climate action
  climate change, 118
  climate justice, 118
  sense of care, 117, 119
  social justice, 117–119
Drought-resistant crops, 230, 234, 291
Droughts, 12
Drought-tolerant crops, 214, 220
Drylands leads desertification, 158

E
Early planted crop, 51
Early planting, 51, 54, 57, 58, 60, 289
Earth's climate system, 161
Earth's surface, 157
ECM parameter, 169
Ecological resource stock, 164
Economic and social development, 158
Economic-ecological modelling
  carbon sequestration, 164–166
Economic growth, 156, 161
Education, 87
Electricity, 92, 126, 273, 292, 293
Emission reduction units (ERUs), 159
Emissions trading, 159
Employment and Training Program, 37
Energy, 92, 273, 281, 292, 293
Energy Company X
  communities, 132
  CSV, 130, 134
  ED, 126, 132
  energy production, 126
  IPPs, 126
  leadership roles, 129, 131
  SDG targets, 128
  SED, 126, 132
  South Africa, 126
  transformational leadership, 133
Energy efficiencies, 126, 293
Energy-efficient equipment, 273
Energy sector, 9, 162
Energy security challenges, 126
Enterprise development (ED), 126
Environmental challenges, 3–5, 12, 13
Environmental degradation, 10, 264
Environmental institutions, 112
Environmental pollution, 186
Environmental poverty, 64
Equity, 26
Equity method, 107–109
  approach, 113
  Climate Action, 111
  climate action implementation, 114, 115
  climate change, 111, 112
  corporate sector, 112

drive climate action, 116, 117
drivers, climate action, 117, 118
environmental institutions, 112
GHG emission target setting, 110, 112, 115
long-term target setting *vs.* short-term decisions, 113, 114
objectives, 110
Promethium Carbon, 110, 112
respect, 114
RL, 113
SBTi, 112, 113
single case study, 110, 111
stakeholder engagement, 119
stakeholders, 113
trust, 114
vision, 115
Eradicating poverty, 125
Erratic rainfall, 280
European Investment Bank (EIB), 27

F

Fall armyworm (*Spodoptera frugiperda*)
AEH, 54
Africa's maize-producing countries, 48
African breeding community, 54
agronomic management practices, 60
analysis, 52
arthropod pest management, 51
average kernel score, 58, 59
biotic constraints, 48
crop losses, 49
cultural pest management strategy, 51
data collection, 52
early planting, 58, 60
Europe, 48
foliar damage ratings, 52, 53
food security, 48
host range, 50
household level, 49
land preparation, 52
larva feeding, developing maize cob, 50
leaf average score, 54
life cycle, 49
Maize, 48
maize genotypes, 51, 53, 60
maize varieties, 53
maize yield *vs.* time of planting, 56, 57
maize yields, 54, 59
monitoring, 52
observations, 52
percent maize discard *vs.* time of planting, 56, 57
planting date manipulation, 51
potential economic impacts, 48
SDGs, 48
sub-Saharan Africa, 48, 49
time of planting
late planted crop, 56
leaf damage, 54, 55
maize damage, 54, 55
maize plants damaged, reproductive stage, 54, 55
maize yield, 54
vs. leaf average score, 57, 58
trial location, 51
Zimbabwe, 50, 51
Farm workers
definition, 139
economic growth, 139
food aid/cash, 141
food and nutrition security (*see* Food and nutrition security, farm workers in Masvingo, Zimbabwe)
FTLRP, 140
hunger and malnutrition, 141
Masvingo, Zimbabwe, 140
social protection (*see* Social protection)
Farming communities, 40
Fast Track Land Reform Programme (FTLRP), 80, 140
FAW, *see* Fall armyworm (*Spodoptera frugiperda*)
FAW-induced damage, 289
Fertiliser Input Subsidy Programme (FISP), 38, 40
Field crops, 50
Fighting poverty and hunger, Malawi and Zimbabwe
African governments, 36
colonialism, 36
domestic autonomy, 38
FISP, 38
integrated rural agricultural projects, 38
PAAP, 37
PRSP, 38
rural societies, 37
SAPs, 37, 38
SDF, 37
Smallholder Authorities, 38
socialist policy, 36
state's plan, 37
World Bank (WB), 37
World Food Programme, 39
ZIMASSET, 37
Firewood, 92
Focus group discussions, 228
Food aid/cash, 141
Food and Agricultural Organization (FAO), 66–67, 143, 290
Food and Agriculture Organization (FAO) database (FAOSTAT), 166
Food and Nutrition Council (FNC), 143
Food and nutrition security, 290
Food and Nutrition Security Policy (FNSP), 144
Food and nutrition security, farm workers in Masvingo, Zimbabwe
FTLRP, 140
land reforms, 140
SDGs, 141
social protection (*see* Social protection)
vulnerable groups, 141
women, 142
Food insecurity, 12, 264
Food security, 19, 39, 65, 75, 217, 264, 290, 291
Food shortage, 273, 274

Forest Conservation Grant Fund, 256
Forest cover, 297, 298
Forest degradation, 157
Forest ecosystems, 156, 158
Forest losses, 11
Forest management activities, 162
Forests, 10, 297
Fossil fuels, 92, 292
Freshwater, 297
Fuel, 92

G
Gender, 142
Gender equality, 24, 183
Gender inequalities, 142
Gender-targeted social protection systems, 142
Geographic information system (GIS), 198
GHG emission target setting, 111–113, 115
GHG emissions
  GHG polluting countries, 4, 5
  global average temperature, 5, 6
  Paris Agreement, 5
  sectors, 4
Global agriculture industry, 12
Global food supply chain, 12
Global Goals, 156
Global indicators, 194
Global Partnership on Forest Landscape Restoration, 156
Global warming, 158, 164, 271
Globally Responsible Leadership Initiative (GRLI) model, 99
Gold panning, 232
Governance, 148
Government of Zimbabwe (GoZ), 213
Governments, 23
Green Airports Recognition, 180
Green aviation, 238, 246
Green energy project, 8
Green technology, 180
Greenhouse gas (GHG) emissions, 4, 158–160
  carbon fuels, 178
Gross Domestic Product (GDP), 34
Grounded theory, 111
Gweru urban community, 67

H
Harmonised Social Cash Transfer Program (HSCT), 146
Healthy ecosystems, 170
Higher education institutions and SDG 6
  climate change, 195
  global implementation, sustainability strategies, 195
  leakages, 197
  rainwater harvest, 195
  students awareness, 195
  universities' commitments, water SDG, 195, 196
  University of Bristol water consumption, 195, 197
  water and sanitation, 194
  water conservation methods, 194
  water efficiency, 195
    Edge Hill University, 196
    University of Plymouth, 195
    urine separation at source, 197
  Water Use Policy, 195
High-value vegetables, 183, 184
Host plant resistance, 51
Household resource shortages, 272
Households, 80, 276, 282, 295
Households' energy consumption levels, 273
Human development, 156
Human life, 158
Human security, 213
Hunger, 64, 65, 67, 68, 70–74, 76, 281, 289, 290
Hydro energy, 9
Hydroelectricity, 8, 248

I
IIASA, 162
Independent Power Producers (IPPs), 126, 134
Individual's right to electricity, 273
Industrial agriculture, 290
Influence pathway, 28
Infrastructure development, 182
Initiative for Climate Action Transparency (ICAT), 108
Input Subsidy Programme (ISP), 40
Integrated approaches, 156
Intended Nationally Determined Contributions (INDCs), 159
Intercrops, 230
Intergovernmental Panel on Climate Change (IPCC 2013), 5, 112, 158, 272
Intergovernmental Science-Policy Platform on Biodiversity and Ecosystem Services (IPBES), 10
Interim Poverty Reduction Strategy Paper (I-PRSP), Zimbabwe, 33, 37, 39, 41
Internal Saving and Lending Club (ISLC), 231
Internal Savings and Lending Schemes (ISALs), 232–234
International Civil Aviation Organization, 239
International Covenant on Economic, Social and Cultural Rights (ICESCR), 273
International Tropical Timber Organization (ITTO), 166
Irrigation, 200, 207, 217, 230
Irrigation agriculture, 229
Irrigation scheme, 219, 220

J
Joint Implementation (JI), 159

K
Key informant interviews, 228
Knowledge-based influence pathway, 28–29
Kreditanstalt für Wiederaufbau (KfW), 27
Kruger Mpumalanga International (KMI) Airport, South Africa

agriculture, 187
community projects, 187
environment, 187
field observations, 181
interviews, 181
location, 180, 181
Mpumalanga Province, 180
post-fieldwork data, 181
private aircrafts, 180
SDGs
environment, 183, 185, 186
society, 182, 183, 186
tourist iconic sites, 180
Kyoto Protocol, 156
Kyoto Protocol's market mechanisms, 158, 159

L
Land Bank
CFO, 22, 25
commercial banking sector, 26
CPD, 22
DFI, 24
EIB, 27
financial sustainability, 22
funding, 26
inductive analysis, 25
interview method, 25
KfW, 27
Land Bank Act, 1944, 21, 22
Land Settlement Act, 1912, 20
liquidity concerns, 26
loan recovery processes, 22
PIC, 22
poor credit decision-making, 22
public commitment, 27
responsible leadership capacities, 25, 26
responsible leadership theory, 22–25
SDG targets, 23, 24
short-term commercial funding, 20
South African market, 22
Strauss Commission, 21
Land Bank Act, 21, 22
Land degradation, 219, 296
Land Settlement Act, 1912, 20
Land use changes, 158
Land Use, Land Use Change and Forestry (LULUCF), 159
Leadership, 289
Africa, 94, 95
capacities, 94
definition, 124
factors, 95
local levels, society, 94
political, 94
responsibility, 95
SDGs, 94
sustainable development, 94
Leadership capabilities, 130, 133
Leadership capacities, 291
Leadership in Energy and Environmental Design (LEED), 247
Leadership qualities, 102
Leadership theory, 114
Legal frameworks, social protection
FNSP, 144
NAP for OVC, 145
NSPPF, 144
Zimbabwean constitution, 144
Livelihood development, Zimbabwe
biophysical environment, 226
climate change, 234
climate-induced vulnerability, 226
craftwork, 233
demographic survey, respondents, 228–229
drought-resistant crops, 230
face-to-face interviews, 228
focus group discussions, 228
gold panning, 232
Gwanda, 230
in-depth responses, 228
irrigation, 230
KII Nyanyadzi, 233
resilience strategies, 233
strategies, 233
study area, 226, 227
Livelihood enabler, 75
Livelihood resilience strategies, 234
Livelihood security, 226, 229
Livelihood sustainability, 229
Localisation, 288
Long-run carbon capture, 167, 168
Luz para Todos programme, 264

M
Maize (*Zea mays* L.)
FAW (*see* Fall armyworm (*Spodoptera frugiperda*))
foliar damage, 50
genotypes, 50, 51, 53
selected hybrids, 52
smallholder, 50
Spodoptera frugiperda, 49
sub-Saharan Africa, 48, 49
Maize yield, 56
Malawi and Zimbabwe
agricultural inputs, 40, 41
community participation, experts' interventions, 42–44
contract farming
Chiredzi, 42
Rumphi, 41, 42
domesticating SDGs, 39
farming communities, 36
markets, 40, 41
NGOs, 36
poverty and hunger, 35
Rumphi, 36, 44
Malawi Goals Development Strategy (MGDS) III, 33
Malnutrition, 35

Maternal health care, 256
Mbuyane Communal Property Association (MCPA), 182
Meter Data Management System (MDMS), 202
Millennium Development Goals (MDGs), 125, 226, 255
Ministry of Public Service, Labour and Social Welfare (MPSLSW), 142, 143
Monitoring and evaluation (M&E) framework, 215
Mountain biodiversity, 297
Mountain ecosystems, 186, 297
Municipal supply system, 207

N

National Action Plan for Orphans and Vulnerable Children (NAP for OVC), 145
National Development Plan (NDP), 93
National Smallholder Farmers Association of Malawi (NASFAM), 41, 42
National Social Protection Policy Framework (NSPPF), 144
Nationally Determined Contributions (NDCs), 160, 162
Nexus, 273
Noise pollution, 237
Non-governmental organisations (NGOs), 80, 127, 143, 220, 231
Non-timber forest products (NTFPs), 231, 232
Nutritional community gardens, 218
Nutritional security, 264
Nyanyadzi irrigation scheme, 231

O

Occupational schemes, 143
Off-farm resilience strategies, 231–234
On-farm resilience strategies, 229–231, 234
Oviposition, 49

P

Paris Agreement, 162
Paris Climate Agreement, 170
Payment programme, 266
Planting date manipulation, 51
Plastic substitution, 186
Policy promoting REDD+, 169
Policymakers, 297
Political leadership, 94, 95
Political systems, 101
Politicization, 40
Politics, 40
Poverty, 35, 64, 65, 67, 68, 70–76, 92, 182, 183, 289, 290
Poverty Alleviation Action Plan (PAAP), 37
Poverty eradication, 158
Poverty mitigation strategy, 44
Poverty Reduction and Strategy Papers (PRSP), 38, 39
Primary forest deforestation, 262
Private aircrafts, 180
Private voluntary organisation (PVO), 142
Productive Asset Creation, 143
Promethium Carbon, 111
Protect forest ecosystem, 257
Protected areas, 256, 297
Public health, 291
Public Investment Corporation (PIC), 22
Public work programmes (PWPs), 141

R

Rainwater, 195
Rainwater harvesting, 192, 195, 199–201, 203, 207, 208, 294
Randomised complete block design (RCBD), 52
Reducing Emissions from Deforestation and Forest Degradation (REDD+), 158–159
Remote sensing, 198
Renewable energy, 98, 126
Renewable Energy Independent Power Producer Procurement Programme (REIPPPP), 126, 129, 134
Renewable energy investment, 8
Renewable energy sources, 92, 96, 97, 126, 293
Renewable natural gas (RNG), 247
Resilience strategies, Zimbabwe
  definition, 226
  ethical considerations, 228
  face-to-face interviews, 228
  focus group discussions, 228
  key informant interviews, 228
  livelihood development, 233, 234
  off-farm, 231–233
  on-farm, 229–231
  population, 234
  sample size, 234
  sampling strategy, 227
  study area, 226, 227
  target population, 227
Resistance, 51
Resource shortages, 276, 281
Responsibility, 95
Responsibility leadership dimensions, Land Bank
  CFO, 27
  decision-making, 28
  forward-looking orientation, 28
  loan covenants, 28
  multiple stakeholders, 27
  shared responsibility, 28
Responsible climate leadership
  conceptual framework, 118, 119
Responsible consumption, 156
Responsible leader drivers, 110
Responsible leadership (RL)
  climate change, 108, 118
  definition, 108
  drive climate action, 119
  drivers, 110
  economic environment, 107
  elements, 108, 110
  energy SDG, 292
  Equity Method (*see* Equity Method)
  roles, 109, 110, 119

shareholder engagements, 109
stakeholder relations, 109
Responsible leadership theory (RLT)
economic perspective, 23
GRLI model, 99
influence pathway, 26, 28, 29
interview questions, 100
interviews, 99, 103
knowledge-based pathway, 23
leadership influence, development projects, 100, 101
leadership qualities, 102
leadership, 99
macro-level analysis, 25
meso-level analysis, 25
micro-level analysis, 25
multi-stakeholder environment, 22
organisational and individual level, 22
organisational level, 20
outcomes, 23
macro-level, 29
meso-level, 29, 30
micro-level, 30
political systems influence, 101
private sector, 20
psychological pathway, 23
social connection, 96
stakeholder perspective, 23
stakeholder relations, 101
systems thinking, 102
UNDP Botswana Biogas project, 95, 96
Rural District Councils (RDCs), 219
Rural Industries Innovation Centre (RIIC), 96
Rural livelihood diversification, 218

S
Sanitation, 80
Sanitation facility, 80
SBL wetland, 206, 207
SCP, 161
SDG 2 "Zero hunger", 23
SDG for society, 288
SDG localisation
global indicators, 194
higher education institutions (*see* Higher education institutions and SDG 6)
local government level, 193
process, 193
SDG 6, 194
UNISA, 194
universities, 193
water and sanitation, 194
SDG policy, 297, 298
SDG transformation, 162
SDG-12, 156
SDG-15, 156
SECA Project, Bulilima District
agricultural production, 218
benefits, 217
Bulilima District, 219
CBDRR, 218
challenges, 220
child-headed households, 218
climate change, 214
climate change adaptation strategies, 221
climate change-related mitigation/ adaptation interventions, 216
climate-smart development, 219
community gardens, 218
community livelihoods, 214
CSA, 214, 217, 219
description, 215, 216
disaster risk reduction, 218, 219
dissemination mechanisms, 217
DRM, 214, 218
FGD, 215, 217
food security, 217
food shortages, 214
households, 217
interviews, 217
irrigation, 217, 219
knowledge and information, 221
land degradation, 219
livelihood-centred approaches, 215
M&E framework, 215
natural resource management challenges, 219
nutritional community gardens, 218
participatory methods, 220
partnerships models, 214
recommendations, 221, 222
rural livelihood diversification, 218
rural women, 218
safe water supply, 215
SDGS, 215
secondary data, 216
soil management-based practices, 217
solar-powered irrigation systems, 219
water harvesting, 217
water supply, 219
Secondary plant metabolites, 51
Secondary plant substances, 51
Self-awareness, 102, 132
Sequestered carbon, 159
SGD-13, 156
Shared leadership, 128
Shared responsibility, 28
Short-run carbon capture, 168–170
Slash-and-burn agriculture, 257
Small, Medium and Micro Enterprises (SMMEs), 126
Smallholder Authorities, 38
Smallholder farmers, 290
Smallholder farming, 212
Smallholder farming communities, 86–87
Smallholder maize, 50
Smart water meter system, 203
Social capita, 102
Social Development Fund (SDF), 37
Social Economic Development (SED), 126
Social entrepreneurs, 30
Social inclusion, 265

Social justice, 117–119
Social networks, 101
Social policies, 81, 142
Social protection
administrative frameworks, 147, 148
categories, 141
definition, 140
deprivation, 141
development literature, 140
development partners, 143, 144
farm workers, 146
farm workers, Masvingo, 147, 148, 150
FGDs, 145, 146, 149
FNC, 143
food aid/cash, 141
food and nutrition insecurity, 140
food and nutrition security, 140, 145–148
framework, 149
gender, 142
governance, 148
HSCT, 146
hunger and malnutrition, 140, 148, 149
institutional frameworks, 146–148
KIIs, 145, 146, 149
lack of adequate budgetary support, 148
legal frameworks, 144, 145
macro-economic challenges, Zimbabwe, 146
Masvingo, Zimbabwe, 140
MPSLSW, 142, 143, 146
normative content, 140
NSSA, 143
policy, 146
poverty, 141
programmes, 140
public policies, 140
PWPs, 141
social assistance mechanisms, 141
stakeholders, 149
Zimbabwean government, 147
Social safety nets, 141
Social vulnerability, 272
Social Welfare Program, 37
Societal leaders, 298
Socio-ecological systems, 296
Soil fertility, 257, 291
Soil management-based practices, 217
Solar energy, 8
Solar-powered irrigation systems, 219
South Africa
food security, 19
Land Bank (*see* Land Bank)
the National Development Plan, 19–20
water stress, 208
Special Maize Production Programme, 39
Stakeholder inclusive approach, 28
Stakeholder relations, 109
Stakeholders, 128
Stormwater harvesting, 208
Strategic Plan for Biodiversity 2011–2020, 156
Structural Adjustment Programs (SAPs), 37
Sub-Saharan Africa, 157
Supporting Enhanced Climate Action Project (SECA Project), *see* SECA Projectm Bulilima District
Sustainable agricultural production, 290
Sustainable Consumption and Production (SCP), 158
Sustainable development, 64, 65, 70, 72, 75, 92–94, 96, 124, 156, 158, 159, 161, 162, 170, 171, 287
objectives, 158
Sustainable development goals (SDGs), 33, 155–156
biodiversity, 10, 11
Brazil, 255, 256
climate change, 8, 10
collaborative leadership, 127, 128
collaborative research approach, 288
Energy Company X, 128, 129
global food security, 12, 13
implementation strategies, 125, 288
inclusive and transformational vision, 125
Land Bank, 23, 24
leadership theories, 127
leadership, 94
localisation, 192, 288
marine degradation, 10, 11
MDGs, 125
multi-stakeholder partnerships, 288
societal levels, 288
sustainable development, 124, 192, 287
targets, 288
transferability, 130
transformational leadership, 127
UNGA, 124
urban farming, 65, 66
World Meteorological Organization (2019), 5
Sustainable energy, 291
Sustainable FAW management techniques, 50
Sustainable food security, 12
Sustainable Forest Management (SFM), 157
Sustainable forests
climate action, 161–163
life on land, 161–163
Sustainable leadership, 95
Sustainable livelihoods framework, 66–68
Sustainable management, 156
forests and halting deforestations, 157
implementation, 158
Systems thinking, 102

T

Terrestrial ecosystems, 297
Terrestrial forests, 10
Theory of change (TOC), 275
Total primary energy supply (TPES), 9
Tourism, 265
Trade-offs, 296
Transformational leadership
Energy Company X, 133
open communication, 128
organisational aspirations, 134
SDGs, 127, 133, 134

Transport sharing, 185
Tropical forests, 156
Trust, 114

U
Uatumã River, 257
Uatumã Sustainable Development Reserve (RDS-Uatumã), 256–259, 264–267
UN Convention on Biological Diversity, 156
UN Convention to Combat Desertification, 156
UNDP Botswana Biogas project
  analysis approach, 99
  Botswana National Development Plan 11, 97
  clean energy, 98
  data collection methods, 98, 99
  environmental project, 102
  implementation, 97, 98
  installation, biogas plants, 97
  leadership capacities, 98
  renewable energy, 97, 98
  RLT (*see* Responsible leadership theory (RLT))
  sample, 98, 99
  small-scale biogas plants, rural communities, 97
  sustainable leadership, 95
  sustainable management, agro-waste, 98
UNFCCC's Clean Development Mechanism, 163
UNISA, water and sanitation SDG
  climate change, 193, 198
  daily consumption of water, Muckleneuk Campus, 200, 202
  document analysis, 199
  GIS, 198
  in-depth interviews, 199
  irrigation, 200
  limitation, 200
  monitoring and evaluation, 203
  rainwater harvesting
    infrastructure, 203, 204
    master plans, 199
    opportunities, 201
  remote sensing, 198
  SDG 6, 198
  smart water meter system, 203
  Students' Representative Council, 200
  study area, 198
  sustainable development, 198
  time-series water consumption, 202, 203
  University Council, 200
  water consumption, 200, 202
  water efficiency, 199, 201, 204
  water-scarce country, 193
  Water Use Policy, 203
  water utilisation, 200
  weekly water consumption, 202
  wetlands, 205, 207, 208
United Kingdom's Department for International Development (DFID), 144
United Nations Children's Emergency Fund (UNICEF), 80
United Nations Development Programme (UNDP), 143, 193, 292
United Nations Framework Convention on Climate Change (UNFCCC), 158–160
United Nations General Assembly (UNGA), 124
United Nations International Children's Emergency Fund (UNICEF), 143–144
United States Agency for International Development (USAID), 144
University of South Africa (UNISA)
  Advanced Metering Infrastructure, 209
  climate change, 294
  open distance learning institution, 192
  SDG 6 localisation, 208, 209
  SDG 6 targets, 192, 193
  sustainability policy, 192
  water and sanitation SDG (*see* UNISA, water and sanitation SDG)
  Water Master Plan, 294
Unrestricted error correction model, 168
Urban authorities, 68
Urban development, 64
Urban farming, 289
  data analysis, 69
  environmental poverty, 64
  food insecurity, 64, 70
  food poverty, 71
  food security, 65
  Gweru urban, 65, 68, 69
  households interview, Gweru urban, 69
  hunger, 71
  hunger reduction, 71, 72
  livelihood enabler, 70
  populations, 64
  poverty, 64, 65
  poverty reduction, 71, 72
    policy implications, 75
    urban households' welfare, 73–75
  qualitative research approach, 69
  rain-fed crops, 71
  SDGs, 64–66
  sustainable development, 64, 65, 70
  sustainable livelihoods framework, 66–68
  urban development, 64
  urban households, 64
  urban populations, 64
Urban food security, 65
Urban households, 64, 74, 274, 289
Urban households' welfare
  agribusiness, 73
  entrepreneurship, 73
  SDGs domestication challenges, 74, 75
Urbanization, 10

V
Vector error correction (VECM) method, 166
Vegetable crops, 50
Vending, 231
Vulnerability, 226–228, 232

W

Water, 191, 272, 278, 281
Water and sanitation access, Zimbabwe
  agricultural productivity, 81
  bivariate analysis model, 82
  bivariate model estimates, 84, 86
  bivariate probit model, 82–84
  Blair toilets, 87
  contextual factors, 81
  descriptive statistics, socio-economic variables, 84, 85
  education, 87
  farm equipment, 86, 87
  farming households, water sources, 84, 85
  FTLRP, 80
  government resource allocation, 81
  households, 80, 81
  household size, 85
  household socioeconomic factors, 81
  land redistribution policies, 85
  land reform beneficiaries, 85
  land tenure, 81, 84, 86
  lower-income countries, 81
  marital status, household head, 85
  married household heads, 86
  Mashonaland central province map, 83
  NGOs, 80
  pit latrines, 87
  psychological factors, 81
  rural areas, 80
  sanitation facilities, smallholder farming households, 84, 85
  Shamva district, 82
  smallholder farmers, 83
  smallholder farming sector, 87
  social policy, 81
  sustainable development, 81
  urban areas, 82
  WASH programmes, 81
  water sources, 82
  water supply, 82
Water efficiency, 192, 195, 196, 199, 201, 203, 208
Water Efficiency and Alternative Supply Plan, 199
Water harvesting, 234
Water practices, 272
Water quality, 193
Water resources, 296
Water, sanitation and hygiene (WASH) programmes, 81
Water security, 13
Water shortage, 280
Water source supply, 207
Water sources, 80, 84, 85, 272
Water Use Policy, 195, 203
Weeds, 50
Wetland rehabilitation projects, 205, 294
Wetlands
  artificial, 208, 209
  Daveyton Campus, 205, 207
  irrigation, 207
  SBL, 205–207
  SDG 6 implementation, 208
  stormwater harvesting, 208
  UNISA, 207, 208
  water efficiency, 208
  wetland rehabilitation projects, 205
Wind energy, 8
World Bank (WB), 37
World Economic Forum's (WEF), 108
World Food Programme (WFP), 39, 142, 143
World Health Organization (WHO), 80
World Meteorological Organization (2019), 5
World Resources Institute (WRI), 239

Z

Zimbabwe Agenda for Sustainable Socio-Economic Transformation (ZIMASSET), 37
Zimbabwe's vulnerable communities, climate change
  agro-ecological zones, 214
  districts, 214
  drought-tolerant crops, 214
  economy, 213
  factors, 213
  food insecurity, 213
  GoZ, 213
  human security, 213
  malnutrition, 213
  poverty, 213
  risk, 213
  women, Bulilima District, 214